Biomolecular Stereodynamics

Biomolecular Stereodynamics, Volume III

Proceedings of the Fourth Conversation in the Discipline Biomolecular Stereodynamics held at the State University of New York at Albany, June 4-8, 1985

Edited by

R. H. Sarma & M. H. Sarma

Institute of Biomolecular Stereodynamics
State University of New York at Albany and
Naitonal Foundation for Cancer Research

Adenine Press, P.O. Box 355
Guilderland, New York 12084

Adenine Press
Post Office Box 355
Guilderland, New York 12084

Library of Congress Cataloging-in-Publication Data
(Revised for vol. 3 & 4)

Biomolecular stereodynamics.

    Vol. 3—   edited by R.H. Sarma & M.H. Sarma.
    Includes bibliographical references and indexes.
    Contents: v. 1-2. Proceedings of the Second SUNYA Conversation in the Discipline Biomolecular Stereodynamics held at the State University of New York at Albany April 26-29, 1981 under the auspices of the Department of Chemistry and organized by the University's Institute of Biomolecular Stereodynamics—v. 3-4. Proceedings of the Fourth Conversation in the Discipline Biomolecular Stereodynamics held at the State University of New York at Albany, June 4-8, 1985.
    1. Biomolecules—Congresses. 2. Stereology—Congresses. I. Sarma, Ramaswamy H., 1939-   .
II. Sarma, M.H. (Mukti H.), 1940-   .   III. SUNYA Conversation in the Discipline Biomolecular Stereodynamics (2nd : 1981 : State University of New York at Albany) IV. State University of New York at Albany. Dept. of Chemistry. V. State University of New York at Albany. Institute of Biomolecular Stereodynamics. VI. Conversation in Biomolecular Stereodynamics (4th : 1985 : State University of New York at Albany)
QH506.B554   574.8'8   81-14867
ISBN 0-940030-00-4 (v. 1)
ISBN 0-940030-01-2 (v. 2)
ISBN 0-940030-14-4 (v. 3)
ISBN 0-940030-18-7 (v. 4)

Made in New York, USA

Preface

These are the proceedings of the Fourth Conversation in Biomolecular Stereo-dynamics held at the State University of New York at Albany June 04-08, 1985 under the auspices of the Department of Chemistry and organized by the University's Center for Biological Macromolecules and the Institute of Biomolecular Stereodynamics. Over 500 scientists from 15 countries were gathered at Albany in early June 1985 for the Fourth Conversation. These volumes essentially constitute the invited presentations. The papers contributed at the poster sessions of the congress have already appeared in the various issues of the *Journal of Biomolecular Structure & Dynamics* released after the conference.

The conference and these volumes were made possible by generous support from public and private institutions. We are immensly grateful to the following institutions for supporting the congress:

State University of New York
State University of New York at Albany: Office of the President, Office of Vice-President for Research, Dean, College of Science and Mathematics, Department of Biological Sciences, Chemistry and Physics
Adenine Press
Domtar, Inc
General Electric Company
Hoffmann-La Roche Inc
IBM Instruments Inc
Lederle Laboratories
Merck & Co, Inc
Naitonal Foundation for Cancer Research
National Institutes of Health
Shell Development Company
Smith, Kline and French Laboratories
Upjohn Company
Wilmad Glass Company

We thank our friends and colleagues who have either served in the Organizing Committee or have provided valuable suggestions or have physically helped us in several ways. We must particularly mention: M. M. Dhingra, G. Gupta, C. W. Hilbers, N. R. Kallenbach, S. Manrao, W. D. Phillips, A. G. Redfield, A. Rich, N. C. Seeman and A. H.-J. Wang. We acknowledge with gratitude the help from Virginia Dollar and Charles Heller of Chemistry, John Elliot of Biology, Al Dasher of the College of Science and Mathematics and Don Bielecki of the conference department of SUNYA.

We thank David L. Beveridge for editing two of the articles in these volumes *viz.*, that by E. W. Prohofsky and the one by F. Vovelle and J. M. Goodfellow.

Most of all we thank, congratulate and applaud the participants of the Fourth Conversation for thier notable contributions and their continuing and unabated dedication to the discipline of biological structure, dynamics, interactions and expression.

We are looking forward to the pleasure of being your hosts for the Fifth Conversation June2-6, 1987.

Ramaswamy & Mukti H. Sarma
Albany, New York
March 20, 1986

CONTENTS

Volume III

CONTENTS

Volume IV

9

*Biomolecular Stereodynamics III, Proceedings of the Fourth Conversation in the
Discipline Biomolecular Stereodynamics, State University of New York,
Albany, NY, June 04-09, 1985, Eds., Ramaswamy H. Sarma & Mukti H. Sarma,
ISBN 0-940030-14-4, Adenine Press, ©Adenine Press 1986.*

The Structure of the Histone Octamer and its Dynamics in Chromatin Function

**Rufus W. Burlingame, Warner E. Love*, Thomas H. Eickbush[†],
and Evangelos N. Moudrianakis**
Department of Biology, Mudd Hall, and
*Department of Biophysics, Jenkins Hall
Johns Hopkins University
Baltimore, Md. 21218

Abstract

A two pronged approach was employed to elucidate the physiology of the eukariotic chromosome by studying the structures and physical chemical properties of its two major components, DNA and histones. We have demonstrated that when the electrostatic charge density and water activity in the immediate environment of the double helix are appropriate, DNA can fold into a chromatin-like structure in the absence of other macromolecules, mimicing the compaction it normally assumes *in vivo* where it is complexed with the $(H2A-H2B-H3-H4)_2$ histone octamer. Under certain *in vitro* conditions, the isolated histones are also organized into an octamer. This complex dissociates into its physiological subunits, the $(H3-H4)_2$ tetramer and the H2A-H2B dimer, when the hydrogen bonds linking the subunits are disrupted. We have determined the structure of the histone octamer by x-ray crystallographic techniques. The octamer has the overal shape of a rugby ball 110Å long and 65Å wide. Its suface is scoured by grooves and ridges. It exhibits a tripartite internal organization corresponding to two H2A-H2B dimers flanking the central $(H3-H4)_2$ tetramer. DNA was model-built around the octamer, and the resulting structure yields insight into the mechanism of compaction and decompaction of chromatin. We propose that the functional transitions of chromatin *in vivo* are regulated by subtle modulations in the degree of charge neutralization and hydration of the double helix, and in the state of association of the histone subunits in the histone-DNA complex.

Introduction

The genetic information of living systems is encoded in the structure of the linear polyelectrolyte DNA. Free DNA molecules are extremely asymmetrical and extended, while DNA molecules found inside eucaryotic nuclei are complexed with other molecular species and highly compacted. The major components complexed with the DNA in eucaryotic chromosomes are the histones, and the interaction of DNA with histones causes at least the first and second orders of DNA compaction in the chromosome.

[†]Present address: Department of Biology, University of Rochester, Rochester, N.Y. 14627.

11

Most of the current research in the field of chromatin structure is based on the discovery that the histone components of the chromatin are organized along the DNA backbone in repeated arrays. It is generally accepted that this periodic arrangement of histones along the DNA gives rise to the nucleosomes obtained after nuclease digestion (1,2) and the "beads" seen in electron micrographs of chromatin (3,4). Even though the histones are the major proteins interacting with the DNA in the chromosome, many other macromolecules, such as non-histone proteins, RNA, and perhaps complex carbohydrates, are an integral part of the chromosome. Furthermore, ions and small metabolites can enter the nucleus from the cytoplasm and affect the chromosome's architecture. As a first step to understanding the physiology of the eucaryotic chromosome, we have explored a simpler system consisting of the two major components in the chromosome, the DNA and the four core histones. This DNA-histone complex must undergo considerable conformational changes during the functional transitions of the chromosome in the course of the cell cycle. In order to understand the individual contributions of each of the two components in the DNA-histone complex, we have studied them separately. With these studies, we have begun to answer the following three questions. (1) What types of compaction are intrinsic properties of the DNA itself? (2) What types of interactions take place between the core histones in solution free of DNA? (3) What are the structures of the histones and the DNA-histone complex? Knowledge of the properties and structures of the two major individual components of chromatin will allow a better understanding of the structure and function of the chromosome.

The Self-directed Compaction of DNA Helices in the Absence of Any Other Macromolecules

Several lines of evidence suggest that the DNA helix has an active role in its own compaction. Changes both in the degree of charge neutralization of the DNA phosphates (5,6) and in the dielectric constant of the DNA solution (7,8) have been shown to elicit extensive changes in the hydrodynamic properties of the DNA helix. Following up on these observations, we have found in an electron microscopic study of DNA compaction that the double helix, in the absence of other macro-molecules, has the intrinsic potential to direct its own packaging into two distinct forms (9). Under conditions of a minimal charge shielding environment for the DNA (i.e. 1 mM Tris, pH 7.5, and 70% to 95% ethanol), both linear and covalently closed circular DNA molecules form left-hand supercoiled fibers with an outer diameter of 80-90 A. The lengths of the generated fibers are directly proportional to the DNA molecular weight and correspond to an 8.6-fold compaction, while the cross-sectional areas of the fibers remain invariant, that is, the fiber diameter is independent of the molecular weight of the DNA. In electron micrographs, the fibers formed from naked DNA (Fig. 1a) often show knobby or beaded regions remarkably similar in appearance to chromatin (Fig. 1b). Thus, DNA has the intrinsic ability to condense into a chromatin-like structure when it is in an environment of the appropriate ionic strength and dielectric constant (9). We have suggested that this type of compaction may be related to the compaction caused by histones in chromatin. Since chromatin is often treated with stains (salts) and then dried (alcohol),

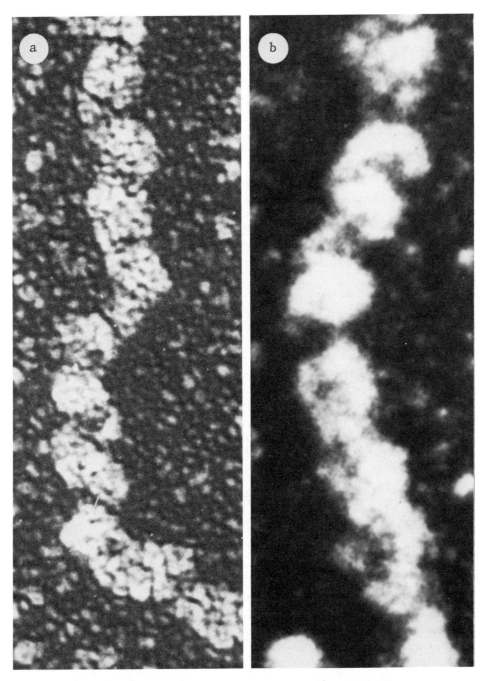

Figure 1. (a) Protein-free lambda DNA directly collapsed into a fresh, untreated carbon-coated grid from 1 mM Tris (pH 7.5) with 95% ethanol, air dried, and shadowed with platinum (9,64). The knobby DNA fiber has an outer diameter of approximately 90 A. (b) Formaldehyde-fixed chromatin deposited from approximately 20 mM cacodylate buffer, pH 7.5, onto fresh carbon-coated film, air dried, and shadowed with gold (64,74). The knobby chromatin fiber has an outer diameter of 100-110 A. No correction for thickness of the metal-shadowed layer was made in either figure.

one must be careful in attributing all bead-like structures in chromatin to DNA-histone complexes, since some may have been introduced into the DNA by the preparation procedure.

The second mode of condensation of the DNA takes place under conditions of full charge-neutralization (greater than 0.15 M ammonium acetate in 95% ethanol). Under such conditions, rods with a length of about 2000 A are formed from both linear and closed circular DNA. In contrast to the first mode of compaction, the length of the rod is independent of the molecular weight of the DNA, while the cross-sectional area of the rod increases linearly with the molecular weight of the DNA (9). Within these rods, the DNA is folded back and forth in a hairpin fashion. This type of compaction may be related to the packaging of DNA into viral capsids. Under appropriate conditions, the rods bend, their ends coalesce, and they appear like toroids. This study demonstrated that the electrostatic charge density and the water activity in the immediate microenvironment of the DNA helix determine the mode of compaction of the DNA double-helix.

Current models of chromatin structure suggest that the DNA is wrapped twice around the octameric histone core in a left-handed supercoil approximately 105 A in outer diameter to form the core particle, and that histone H1 interacts with this particle and with linker DNA to form the nucleosome (10). Naked DNA can form a left-handed supercoil of about the same diameter when water is removed from its molecular domain by alcohol and the charges of the phosphate backbone are partly neutralized by salt (9). We consider it probable that the combined hydrophobic and electrostatic characteristics of the histones cause a similar collapse of the DNA helix by expelling water from the DNA grooves. Thus, in chromatin, DNA-histone interactions modulate the supercoiling of the DNA. Furthermore, we suggest that histone-histone interactions provide the basis for a biologically meaningful control of the compaction of the supercoiled DNA in chromatin.

The Subunit Organization of the $(H2A-H2B-H3-H4)_2$ Histone Octamer as Revealed by Solution Studies

In order to understand better the histone-histone interactions, we have studied the physical-chemical properties of the four core histones in solution in the absence of DNA. In 2 M NaCl at neutral pH and 4° C, the core histones form an $(H2A-H2B-H3-H4)_2$ octamer (11,12). At either lower ionic strength, or non-neutral pH, or higher temperature, or in the presence of urea, the histone octamer dissociates into its subunits, i.e. two H2A-H2B dimers and one $(H3-H4)_2$ tetramer (12). Thus, the octamer is assembled by two sets of protein-protein interactions. The first set involves mostly hydrophobic interactions and is responsible for the stabilization of the H2A-H2B dimer and the $(H3-H4)_2$ tetramer subunits. The second set involves the weak association of one $(H3-H4)_2$ tetramer with two H2A-H2B dimers to form an octamer. These weak interactions are derived primarily from hydrogen bonds between the dimer and tetramer subunits (12,13). The range of the pH-dependent association implies that these hydrogen bonds are between histidine and lysine, or histidine and tyrosine (12). When the individual subunits associate, they first form an (H2A-H2B)(H3-H4)$_2$ hexamer intermediate (12,13,14). The binding of the first dimer to one side of the tetramer enhances by a factor of 4 the intrinsic binding affinity of the other dimer-binding site in the tetramer for the second dimer (13,14). The presence of an allosteric link between the two sites offers an attractive hypothesis for the basis of the observed positive cooperativity.

These studies have conclusively demonstrated that in 2 M NaCl at neutral pH, the four core histone polypeptides exist as an octamer in equilibrium with its physiological subunits, the H2A-H2B dimer and the (H3-H4)$_2$ tetramer. There is no evidence for the existence of an H2A-H2B-H3-H4 half octamer under any of the conditions used in the physical-chemical studies we have performed (12,13,14). We would like to suggest that the histones in chromatin have the same subunit organization as they have in 2 M NaCl, and that the same forces controlling the equilibrium between the octamer and its subunits in solution free of DNA control the equilibrium between the octamer and its subunits when they are bound to DNA in chromatin.

The Structure of the Histone Octamer at 3.3 A Resolution

Using the information derived from the previous studies concerning the physical-chemical properties of the histones in solution, we were able to crystallize the (H3-H4)$_2$ tetramer (15) and the (H2A-H2B-H3-H4)$_2$ octamer (16). The proteins in the washed octamer crystals were recovered intact and in equimolar amounts. The crystals diffract to 2.8 A resolution in an x-ray beam from a rotating anode generator. There is one half octamer per asymmetric unit, demonstrating that this protein complex has perfect twofold symmetry to at least this resolution.

We have recently determined the structure of the (H2A-H2B-H3-H4)$_2$ histone octamer to 3.3 A resolution using single crystal x-ray diffraction techniques (17). In order to phase the x-ray reflections, a heavy atom derivative was prepared in which there was a mercury atom bound to a single site in the crystallographic asymmetric unit. The data were collected at the Mark II area detector in the laboratory of N. H. Xuong at the University of California at San Diego (18). The initial phases used for the determination of this structure were calculated from the isomorphous and anomalous scattering differences between a native crystal and the heavy atom derivative crystal. These phases were then improved by using the iterated single isomorphous replacement (ISIR) method developed by B. C. Wang at the V.A. Medical Center in Pittsburgh (19). Figure 2a is an electron density map calculated with phases derived solely from the isomorphous and anomalous scattering differences from the single derivative. Figure 2b shows the identical sections as Figure 2a, and is contoured at the same starting level, but the phases used in the calculation have been improved by the ISIR procedure. The organization of the protein density in the electron density map as well as the shape and volume of the octamer in the map, are all nearly the same before and after the ISIR procedure. However, after the ISIR procedure, the solvent region has been flattened, and as a result the protein density is sharper and the well known artifactually high peak of density at

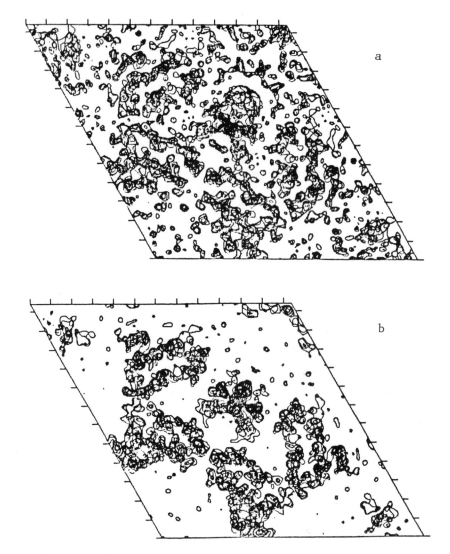

Figure 2. The electron density map viewed down the crystallographic c axis. The x and y axes from −0.25 to 0.75, while z runs from −0.054 to 0.054. Both maps are contoured starting at one sigma. Nine sections were superimposed to yield the section shown, so for clarity the contour lines were drawn at intervals of two sigma for Fig. 2a and three sigma for Fig. 2b. (a) The electron density map that was calculated with phases generated from the isomorphous and anomalous scattering differences between the native and the single mercury derivative crystals. (b) The same region of electron density as shown in Fig. 2a, but with the phases processed with the ISIR procedure. The boundaries of the protein are very similar in the two maps. In the map using the ISIR phases, the solvent has been flattened, and the well known artifactually high peak at the heavy atom position consequently has been greatly reduced.

the heavy atom position is much reduced. Because there is 65% solvent in the crystal, and because the positions of the two classes of crystallographic twofold axes put restrictions on the boundary of a single octamer, the exact boundary of more than 95% of the octamer can be unambiguously defined (17).

The octamer has the shape of a rugby ball 110 A long and 65-70 A in diameter. The most striking feature of the structure of the histone octamer determined by us is its tripartite organization. A central wedge about 30 A wide at its tip, 75 A wide at its base, and about 65 A high is flanked by two flattened balls each about 40 A in diameter (Figs. 2 and 4-7). Solvent channels 4-14 A wide and up to 40 A long separate each flattened ball from the central mass (Figs. 2 and 4). The mass in the central region is loosely packed, containing solvent channels 5-15 A wide and 20-40 A long. The central density contains four contiguous regions which correspond to four polypeptide chains. The mass of each of the flattened balls is tightly packed and contains long stretches of alpha helix. We have been able to trace two polypeptide chains, almost from one end to the other, within each flattened ball. Because it has been shown that the (H3-H4)$_2$ tetramer and the H2A-H2B dimer are stable entities in solution (12,14), the four polypeptide chains in the central wedge have been assigned to H3 and H4, and the two chains in each approximately 40 A diameter ball have been assigned to H2A and H2B. The size, shape, and internal organization of the histone octamer that we have found are radically different from the size, shape, and internal organization attributed to the histone octamer in earlier low resolution electron and x-ray diffraction studies (20,21,22).

The identity of the individual chains was determined by analysis of the biochemical properties of the crystal and information contained in the electron density map of the protein. An effective single-site mercurial derivative was obtained only after the protein in the crystal was reduced by dithiothreitol. Since the chicken erythrocyte histone octamer contains only two cysteines (amino acid at position 110 of each H3), the biochemical data strongly suggest that the cysteine of histone H3 was specifically labeled by the mercury. Thus, the density contiguous with each mercury binding site was assigned to each H3, both of which are located in the central wedge (Fig. 2). The other regions of density within the wedge were assigned to the two H4s. Within the approximately 40 A diameter balls, the H2B chain was identified by characteristic lumps of side-chain density that were interpreted as tyrosines and methionines, and corresponded to a pattern consistent with the sequence of H2B. The other chain was thus assigned to H2A. Figure 3a shows part of the 37 A long alpha helix from H2A, part of a long helix in H2B, and a short helix from H2B. Figure 3b shows the long helical arm of H3.

In order to understand more fully the structure of the histone octamer, we built a balsa wood model using the 3.3 A data scaled at 1 cm per 3.0 A. It was built in two halves, and virtually all of the electron density in an asymmetric unit was used for each half. The two halves were joined around the twofold axis of symmetry within the octamer. The contour level is one-eighth that of the highest peak on the map. At this contour level, there were only three minor connections between the dimer and the tetramer, all at the site of the H2B-H4 interface. The chains in the dimer are contiguous from one end to the other, as is all of the density within the (H3-H4)$_2$ tetramer.

We were able to measure directly the volume of the balsa wood model and thus

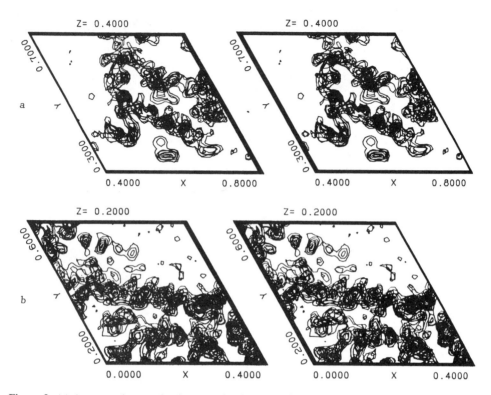

Figure 3. (a) A stereo electron density map showing part of the 37 A long helix in H2A (running diagonally), part of the long helix in H2B (running perpendicular to the plane of the paper), and a short helix in H2B (running perpendicular to the x axis in the plane of the paper). (b) A stereo electron density map showing the long helical arm of H3. The axes are labeled in Figs. 3a and 3b.

calculate the hydrated volume of the histone octamer. The model was placed inside a thin, unsealed, water-tight, plastic bag and was submerged in water which caused the bag to conform to the surface topography of the model. Extending the Archimedes principle, we determined that the volume of the octamer is approximately 184,000 A^3, which is equivalent to the volume of a sphere of ca. 35.3 A radius.

We have taken stereo photographs of the histone octamer from a variety of angles, and these are shown in Figures 4-7. Figure 4a shows the octamer directly down the molecular (and crystallographic) twofold axis from the "front", the face where the protein's twofold axis intersects the centrally located tetramer at its narrowest point. We consider this angle to be the most informative view of the octamer because the tripartite organization of the density is clearly visible. Figure 4b is a tracing of Figure 4a in which the individual polypeptide domains have been delimited. The leftmost and rightmost edges on the long axis of the ellipsoid are the "ends". In all of the photographs of the balsa wood model, the dimers are dark and the tetramer is light. Figure 4c, our previously published model of the histone octamer based on the results of solution studies (12), is remarkably similar to the actual structure.

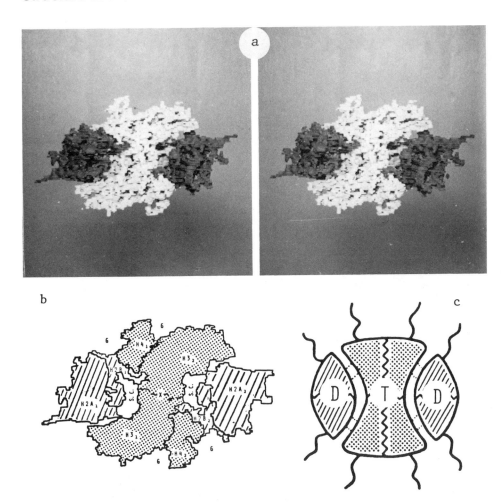

Figure 4. (a) The balsa wood model, viewed from the front, directly down the twofold axis. The tripartite organization is clearly visible since large solvent channels (S.C. in Fig. 4b) separate each dimer from the tetramer. The dimers have been painted dark and the tetramer light to emphasize the subunit organization of the octamer. There are grooves on the surface of the protein (G in Fig. 4b) at the bottom of the dimer-tetramer contact interface on the right side and at the top of the interface on the left side. Other grooves can be seen at the top and bottom of the tetramer. Along the 110 A axis, the histone tetramer has the characteristics of a left-handed propeller. At the front, the COOH-terminal arm of each H2B extends as a bridge to the tetramer. (b) A tracing of Fig. 4a in which the polypeptide domains are delineated. The numerical subscripts given to the histones identify their order along the DNA supercoil and do not correspond to any internal organization within the tetramer, since $H4_1$ binds most tightly with $H3_2$. The "X" marks the twofold axis, which is perpendicular to the plane of the page. (c) Our earlier model for the histone octamer based on the results of solution physical-chemical studies (12). In Figs. 4b and 4c, the striped regions are the dimers and the stipled region is the tetramer.

Figure 5 shows the octamer approximately 45 degrees from Figure 4a, while Figure 6 shows the octamer from one end, 90 degrees from Figure 4a. These views give a good idea of the shapes and relative orientation of the subunits. Furthermore, the grooves and ridges along the surface of the octamer can be discerned. A view down

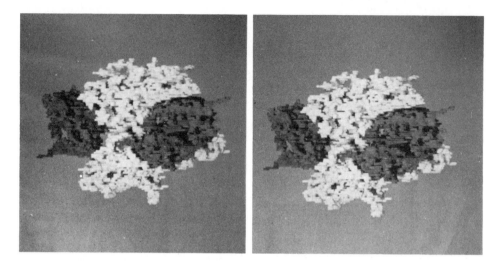

Figure 5. The balsa wood model, rotated 45 degrees about the c axis from Fig. 4. The horizontal metal rod through the middle of the octamer represents the twofold axis. The long helical arm of H3 runs horizontally in the plane of the photograph from the upper center to the upper right. From this angle, it is seen as the most solid wood density within the tetramer. One of the helices from the flat ledge of H2A can be seen in the upper right, just below the long helical arm of H3.

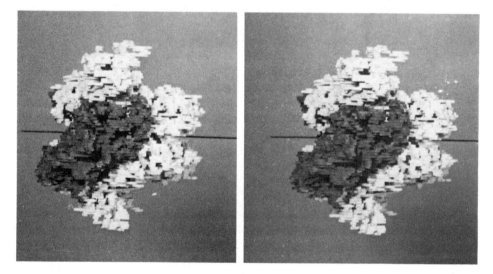

Figure 6. The balsa wood model, viewed from the end, 90 degrees from Fig. 4. The orientation of the dimer subunit with respect to the tetramer is clear from this view.

the molecular twofold axis from the back is shown in Figure 7. From this view, the left-handed pitch of the back lobes of the tetramer are clearly visible. A solvent channel about 15 Å wide and 15 Å deep can be seen separating the two back lobes. However, this channel does not cut through the octamer, and it does not appear that the tetramer could split into two halves along this or any other channel. These views of the octamer clearly show its tripartite organization, its rugby-ball shape, and a series of grooves and ridges on its surface.

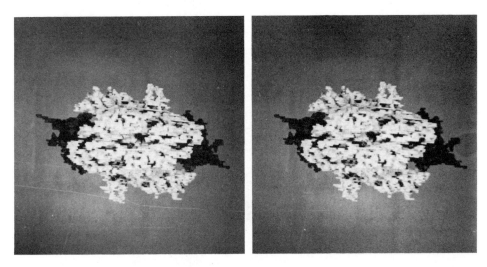

Figure 7. The balsa wood model viewed from the back, down the twofold axis. A solvent channel about 15 A wide, 15 A deep, and 40 A long runs diagonally across the tetramer, separating the two back lobes. It does not separate the entire tetramer into two halves since at the front of the molecule (Fig. 4) there is tight contact between H3 and H4, and between the two H3s. The general rugby ball shape is apparent from this view.

Model-building DNA around the Balsa Wood Model of the Histone Octamer

Since at least one function of the histone octamer is to compact the DNA in the chromosome, we considered it informative to model-build DNA around the balsa wood model of the histone octamer. Neutron diffraction studies established that in nucleosomes, the DNA is wrapped around the outside of the histone core (23,24). Several additional types of evidence have indicated that nucleosome core particles have overall twofold symmetry (25,26,27). Therefore, in nucleosome core particles the pseudo-twofold axis of symmetry of the DNA must be coincident with the twofold axis of symmetry of the octamer. This puts a severe constraint on the possible path of the DNA around the octamer. Guided by this constraint, we examined the surface of the balsa wood model for morphological clues to DNA binding sites. As can be seen by examining the stereo views (Figs. 4-7) of the balsa wood model, a series of grooves and ridges traverse the surface of the octamer in a discontinuous left-handed spiral path. Tubing, with a diameter correctly scaled to represent DNA, was placed around the model so that it passed through the twofold axis and followed the path dictated by these grooves and ridges (Figs. 8-12). In a few places along this path around the octamer, the DNA rests in grooves on the protein surface. At several other places it follows ridges that protrude from the surface.

Two complete turns of DNA saturate the probable DNA-binding surface of the protein and yield a structure with a length of 110 A and a diameter of 100-110 A, containing about 168 base pairs of DNA (16 turns with 10.4 base pairs per turn). The number of base pairs of DNA in the model is consistent with the number of base pairs determined by nuclease digestion studies (28). The order of binding of individual polypeptides along the strand of model-built DNA agrees well with the crosslinking results of Mirzabekov (26), even though those results were not used to position the DNA around the octamer. However, the shape and length of our model-built nucleosome core particle are radically different from the shape and length reported in some other studies (20,21,22,27,29). In addition, the internal organization of the histone octamer that we have directly determined and which is consistent with crosslinking results is very different from the organization that has been hypothesized elsewhere (20). An explanation for these large discrepancies will be the subject of a separate paper.

Figure 8. The balsa wood model with DNA, viewed from the front, directly down the molecular twofold axis. This is the same view as in Fig. 4, but a tube representing DNA has been placed around the model in a path suggested by features on the surface of the octamer. On a scale relative to the balsa wood model, the tube representing DNA has a diameter of 20 A, and a repeat distance for one turn of the double helix of 34 A, thus making the DNA correctly proportioned to the protein complex. The DNA looks like a spring wrapped around the tripartite core, with the ends of the DNA interacting with the H2A-H2B dimers, which protrude past the turn of the DNA at each end. The DNA might be able to roll from this position (see Figs. 13 and 14). In the arrangement shown, the center-to-center distance between the DNA ends is 75 A, which yields an average pitch for the DNA supercoil of 37 A. The identification of the polypeptide chains is the same as in Fig. 4b.

Figure 8 shows the model-built DNA-histone complex from the front. The DNA appears like a spring around the tripartite core, with the dimers interacting with the ends of the spring. Figures 9 and 10 show the model 45 degrees and 90 degrees away from Figure 8, respectively. It is clear that there are spaces between the DNA and the histone octamer at a number of points along the path of the DNA, particularly

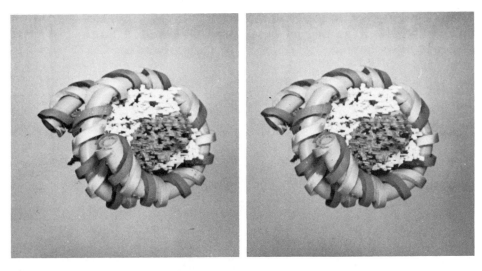

Figure 9. The octamer model with DNA, rotated 45 degrees about the c axis from Fig. 8. In the upper center of the photo, protein can be seen between the grooves of the model-built DNA helix.

Figure 10. The octamer model with DNA, viewed from the end. From this angle, it is clear that in a number of places, there is a separation between the protein and the DNA, particularly at the junction where the DNA passes from the tetramer binding region to the dimer binding region.

at the junction where the DNA passes between the dimer and tetramer. It also appears that the DNA makes more contact with the tetramer than with the dimer, which indicates that the tetramer directs most of the supercoiling of the DNA in the nucleosome. Figure 11, a view of the model from the back, demonstrates that the pitch of the DNA follows the left-handed pitch of the back lobes of the tetramer. Figure 12 shows the model from the top. The exposed ledge of H2A can be seen on the right side of the model.

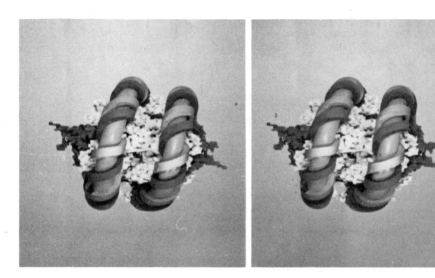

Figure 11. The octamer model with DNA, viewed down the twofold axis from the back. The pitch of the DNA follows the left-handed screw pitch of the back lobes on either end of the particle. Only from this angle does the particle give the illusion of being bipartite, because the tetramer and DNA obscure the dimers and because only parts of two turns of the DNA are visible.

Figure 12. The octamer model with DNA, viewed almost from the top. The model has been rotated approximately 45 degrees about the long axis of the octamer. If the DNA double helix continued in the path it is following, it would extend past the flat ledge of H2A on the right without directly binding to the ledge.

Figures 8-12 show the DNA in the path we feel it is most likely to follow when it associates with the octamer in chromatin. However, the results of a few well accepted low-resolution studies (20,21,22,29) have been interpreted to show that the DNA-histone complex has a diameter of about 110 A (as we have found), but a length of only 55 or 60 A instead of 110 A. The discrepancies between the structures are so great that they cannot be explained simply by the higher resolution of our own study. In addition, since the structure of the histone "octamer" deduced by the MRC group is the same in the absence of DNA (20) and in the presence of DNA (21,22), the differences between the structure we determined (17) and the earlier models (20,21,22) cannot be caused by the presence of DNA around the histones in some of those studies (43).

Because of questions raised concerning the length of the nucleosome particle that we have model-built, we tried to place DNA around the balsa wood model in other ways, the aim being to achieve shorter and more "flat" nucleosome models. We found that placing the DNA around the model 90 degrees away from the structure we have shown in Figures 8-12 yielded a complex with the shortest length, 70 A, but the diameter of this model-built particle is 140 A and there is no obvious path for the DNA to follow. It is also possible to wrap the DNA at an angle approximately 30-45 degrees to the left (Fig. 13) or approximately 45 degrees to the right (Fig. 14) from our preferred position (Figs. 8-12). In both of these orientations, the length of the model-built nucleosome is at least 100 A, but the ends of the particle are flatter. Based on the features of the histone octamer and the available information concerning the path of the DNA in the nucleosome (21,26,30,39), we believe that our preferred orientation is the one most likely to be found in vivo. Other orientations may also occur in chromatin, each with its own characteristic probability of existence.

Figure 13. The "left-tilt" model. In this orientation, the DNA helix does not bind along the center spine of the tetramer, but instead crosses it.

Figure 14. The "right-tilt" model. In this orientation, the DNA helix follows the "propeller" made up of the long helical arms of H3.

We are exploring two ways that could provide direct information concerning the path of the DNA around the octamer. First, we have soaked oligomers of defined sequence of DNA into the crystals and obtained changes in the reflections from screened precession photographs, but we have not yet visualized a DNA helix in the DNA-derivatized crystals. Second, we are fitting the amino acid sequences to our electron density map. Upon its completion, the fitted sequence will provide a number of useful insights. We anticipate that the path of the DNA around the octamer will be revealed by a left-handed spiral trail of positively charged side chains, bounded perhaps by patches of negative charge. In addition, the residues involved in the inter-subunit interfaces will also be placed, and thus yield more detailed information concerning the exact hydrogen bonds stabilizing the octamer. The shape and the surface contacts of the individual histone polypeptides may explain the extreme evolutionary stability of the amino acid sequences of H3 and H4, and of some of the domains of H2A and H2B. The positions of the amino acids will also give us insight concerning the functions of post-translational modifications in the octamer. Finally, the successful fitting of the amino acids into the electron density map will provide independent evidence for the correctness of our structure, though as documented in the following section, large amounts of data already are in agreement with our structure.

Interpreting Other Primary Data in the Literature in Light of the Structure We Have Determined

The structure of the histone octamer that we have determined by x-ray crystallographic techniques is significantly different from most (20,21,22), but not all (30), current models for histone structure. However, our structure is consistent with most of the primary data in the literature, but not with the current *interpretation* of these data. For example, in virtually all electron micrographs of "beads-on-a-string" chromatin, nucleosomes, or core particles, all the beads are at least 100 A in diameter (3,30,31,32,33,34,36,51,64). In order for the micrographs to be consistent with a model of a core particle 55 A long, it has been *assumed* that all nucleosomes sit on their "flat" face with the 55 A length perpendicular to the electron beam (32,33). However, this assumption is not necessarily correct, and in fact a number of studies show that nucleosomes and the "beads" in chromatin are oriented randomly on the electron microscope grid (30,34,35,36,37). If nucleosomes do sit randomly on the electron microscope grid, then our nearly equi-dimensional model-built nucleosome fits the observed data much better than the model for the nucleosome core particle with a 55 A length.

A recent study of nucleosome core particles by electron spectroscopic imaging and image reconstruction (30) yields a structure for the histone octamer that is very similar to the structure we have determined using x-ray crystallographic techniques (17). That electron microscopically determined structure has the same size, shape, central cavity, and protein protrusions that we have found. The novel technique of electron spectroscopic imaging has not been widely used for biological preparations, but the fact that the structure of the histone octamer deduced by that technique agrees well with the structure of the histone octamer determined by x-ray crystallography implies that under appropriate conditions, electron spectroscopic imaging yields high quality and valid information.

At times, the results of neutron diffraction studies of chromatin, nucleosomes, and core particles in solution have been interpreted in ways that are consistent with the structure we have determined (23,24,38), while at other times interpretations of this kind of data appear inconsistent with our structure (39,40). However, neutron diffraction studies yield low resolution data on the spherically averaged structure of the molecule in solution. The interpretation of that type of data is model-dependent, and it is well known that it is difficult to differentiate between oblate and prolate ellipsoids with axial ratios between 0.5 and 2 using neutron diffraction (38,41). The interpretation of the study of Braddock et al. (39) supports the 55 A long model for the core particle. However, the measured 34 A radius of gyration for the histone octamer was found to be "substantially greater than that expected of a core" consisting of a flat disk or wedge-shaped model. To resolve this discrepancy, the authors assigned 25% of the histone mass to flexible "tails", and the remaining 75% of the mass of the octamer to the volume of an equivalent cylinder 40 A long by 70 A in diameter, i.e. 153,860 A^3. If we assume that the remaining 25% of the protein occupies the volume expected for an "average" protein of that mass, i.e. 34,250 A^3 (20), then the calculated volume of the whole octamer would be 188,110 A^3. This compares well with the volume of 184,000 A^3 we have estimated by the extension of the Archimedes principle, but is substantially larger than the 137,000 A^3 that was *assumed* to be the volume of the histone octamer in the structures developed by the MRC group (20,21,22). Furthermore, the best fit for the pitch of the DNA in the core particle was found to be 37 A (Fig. 5 of ref. 39). This is the pitch that we found when we model-built DNA around the balsa wood model of the histone octamer (Fig. 7 of ref. 17), and should be compared with the pitch of 27 A in the other models (20,21,22,29). Thus, as with the electron microscopic data, even though current interpretation of the data is at odds with the structure we have determined, the primary data is consistent with our structure.

Hydrodynamic data such as ultracentrifugation and gel chromatography cannot accurately measure the size of a macromolecule because they measure values that are dependent on the spherically averaged shape of the molecule, and the calculated radius of the molecule is a function of the cube root of the measured volume. Nonetheless, the reported $s_{w,20}$-value for the histone octamer is 4.8 (42). Using the volume of the octamer we measured and the partial specific volume of 0.753 which was measured for the octamer in 2 M NaCl (12), the calculated s-value for our anhydrous structure in 2 M NaCl is 4.9 (43). We will more fully address the calculation of the s-value in another paper. In gel chromatography, the octamer (molecular weight of 108,000 daltons) migrates through a Sephadex G-100 column with an apparent molecular weight of 135,000 daltons, while it migrates through a sucrose gradient as if its molecular weight were only 55,000 daltons (12). The radically different behavior exhibited by the octamer in these two transport methods implies that the complex is either non-globular in shape and/or has a very high partial specific volume (44,45). The structure we have determined is in better agreement with these data than the model of a short dense, wedge-shape disk.

The x-ray (46,47) and neutron (24) diffraction patterns of chromatin consist of the first and higher orders from a Bragg spacing of 110 A, i.e. 1/110, 1/55, 1/37, 1/27. Our model-built nucleosome is roughly spherical, and 110 A in diameter. The pitch of the DNA is about 37 A, and the tripartite protein core is roughly divided into three 37 A long pieces (17). It has been shown that both the protein and DNA contribute to the 37 A reflection (24). Thus, the first order reflections from the pitch of the DNA superhelix and from the internal arrangement of the protein would superimpose on the third order from the diameter of the whole particle. This new interpretation of the x-ray and neutron scattering data accounts for all the observed reflections and does not give rise to reflections that are not observed. We conclude that the *primary data* from many different sources is *more consistent* with the structure of the histone octamer as we have determined it than with the model portraying the histone octamer and the core particle as flat disks.

What differences exist in the structure of the histone octamer when it is in solutions of high ionic strength free from DNA and when it is complexed with DNA in low ionic strength? Seven different probes have shown that the histone octamer in 2 M NaCl and near neutral pH has a very similar structure to the histone octamer in chromatin and nucleosomes at low ionic strength. These probes are circular dichroism (52,53), infrared (54) and Laser Raman spectroscopy (52), electron spin resonance (55), protein-protein crosslinking (11), accessibility of the cysteine at position 110 of H3 to sulfhydryl probes (56), susceptibility of the termini of the histones to trypsin (57,58), and the radius of gyration as measured by solvent contrast neutron diffraction

(23). An eigth probe, NMR, shows that the histones in the core particle have the same spectra in both 0.6 M NaCl when they are complexed with DNA and in 2 M NaCl when they have dissociated from the DNA (59). Also, trypsinized histones in core particles have the same NMR spectra in low ionic strength as in 2 M NaCl (60). It remains to be experimentally determined whether the histone octamer in 2.4 M ammonium sulfate, 100 mM sodium pyrophosphate pH 6.2-6.5 (our crystallization buffer), has approximately the same structure as the octamer in 2 M NaCl at near neutral pH and the octamer in chromatin. The information available to us at this time from preliminary physical-chemical solution studies indicates that the histones adopt approximately the same structure in an ammonium sulfate solution as in a NaCl solution. Furthermore, there is no reason to expect that there would be a major rearrangement in the histones due to ammonium sulfate since, for example, the structure of hemoglobin is virtually the same when it is crystallized from ammonium sulphate, sodium and/or potassium phosphate, or polyethylene glycol (61). Thus, a broad-based set of data supports our proposal that the structure of our model-built nucleosome is very similar to the most common state that the dynamic DNA-histone complex assumes in vivo.

A series of nucleosomes is present along the chromatin fiber. The mass per unit length along the fiber corresponds to one nucleosome per 110 A, as determined by the low angle X-ray (48) and neutron (40,41) scattering curves, by the relaxation time of dinucleosomes as measured with electric dichroism techniques (50), and by the sixfold compaction of DNA caused by the histones (31,51). These data are also consistent with the structure we have determined, and put constraints on possible ways that nucleosomes might be arranged in chromatin.

Synthesis of These Studies into a Model for Chromatin Structure and Function

We can now attempt to develop a plausible model for the first order of chromatin condensation, based on the properties of DNA in solution (9), the properties of the histones in solution (12,13,14), and the structure of the histone octamer (17). As seen in Figure 8, the DNA double helix can be wrapped with a pitch of 37 A around the histone octamer in approximately two left-handed superhelical turns, yielding a particle with an average outer diameter of 105 A and a length of 110 A. The last quarter of each DNA turn interacts with the dimer subunit. This type of condensation accommodates the inherent compaction property of the DNA double helix (9) that in this case is induced by the histones. The DNA-histone complex could start to decondense if the hydrogen bonds holding the dimer in contact with the tetramer were broken at the same time that the short region of DNA spanning the dimer-tetramer interface became more hydrated. The DNA would straighten at the point where it became more hydrated, and the dimer would dissociate from the tetramer while both protein subunits remained bound to the DNA. Once the two DNA-dimer complexes separated from the DNA-tetramer complex, conformational changes could occur within the porous DNA-tetramer complex (62,63), leading to a further decompaction. Thus, factors controlling the hydrogen bonds between the dimer and tetramer subunits, and the degree of hydration and charge-neutralization of the

DNA double helix at key points along its path around the histone octamer, could regulate the degree of compaction of chromatin. This model is in agreement with our earlier proposals concerning the architecture of the DNA-histone complex in chromatin (64), and is an extension of that model based on the evidence that has accumulated since that time.

Rather than review the data in the literature that supports the above model, as we have done previously (17,64), we would like to present some of the implications of this model. In solution free of DNA, the association and dissociation of the dimers from the tetramer were found to be cooperative, passing through an (H2A-H2B)(H3-H4)$_2$ hexamer intermediate (13,14). If a cooperative hexamer intermediate occurs when the octamer is associated with DNA during the compaction-decompaction cycle of nucleosomes in chromatin, then chromatin might have properties that have not previously been considered, due to the cooperativity and asymmetry inherent in the DNA-hexamer particle. There are indications that a DNA-hexamer complex can exist (65,66) and that an allosteric change can occur at the H2B-H4 contact interface. The allosterism might have been measured by changes in the species of crosslinked H2B-H4 dimers found under conditions where low ionic strength caused the dissociation of the dimers from the tetramer in chromatin (67), and under conditions where formaldehyde crosslinking might have induced such an allosteric change (68).

Up to this point, most models of chromatin have assumed that the nucleosome core particle has twofold symmetry, which is probably correct. However, the nucleosome contains only one copy of H1 (10), so at high resolution it must be asymmetric because the sequence of H1 is asymmetric. An asymmetric DNA-hexamer complex could direct the binding of H1 in a particular orientation. Furthermore, in an asymmetric particle, it would be possible to rearrange the DNA-hexamer interactions in such a way that all contacts between protein and DNA were with just the Watson strand or just the Crick strand. Such an arrangement would be impossible in a symmetric particle with two turns of DNA wrapped around it. If only one strand of the DNA formed major interactions with the hexamer, then it could be possible to separate the two strands of DNA so that one strand became free of protein, while the other single strand remained bound to the histone hexamer or the separated dimer and tetramer subunits. It has been shown that single-stranded DNA can form a nucleosome-like particle when complexed with histones (69), and that the histones elute from single stranded DNA at the same ionic strength and in the same pattern as from double stranded DNA (70). Thus, a single stranded DNA-histone complex could well be an intermediate in transcription and/or replication.

We would like to propose further that there is cooperativity among histone octamers along the chromatin fiber. Since the mass per unit length along the chromatin fiber corresponds to one nucleosome per 110 A (31,40,48,49,50,51) and the histone octamer is 110 A long, we believe that the histones are in contact along the length of the chromatin fiber. This contact is probably between the ledges of H2A in adjacent octamers, and is modulated by H1. It has been shown that very large oligomers of histones are formed when chromatin is crosslinked with dimethylsuberimidate (11), implying that the histones are in close proximity. Furthermore, H2A-H2A homodimers are formed when the histones in nuclei are crosslinked with formaldehyde (68). From the structure of the histones that we have determined, it would not be likely for one H2A to be crosslinked to another H2A in the same octamer, so this crosslink most likely occurs with H2A molecules in neighboring octamers. It has been shown that the distribution of histones reconstituted along DNA is non-random, implying that the histones bind cooperatively to regions that already contain bound histones (71). Furthermore, the extent of unfolding of the DNA as measured by melting profiles and circular dichroism depends on the number of nucleosomes in the oligomer being studied, implying that there are interactions between neighboring nucleosomes (72). One can envision the nucleosomes along a loop of DNA (73) unfolding one after the other in a cooperative fashion, changing the supercoil density of the loop of DNA as the supercoil within each nucleosome is reduced when the DNA-dimer complex dissociates from the DNA-tetramer complex.

Our data support a model in which the chromatin fiber is comprised of a series of contiguous and interacting DNA-histone and histone-histone domains. The DNA-tetramer domain is flanked by two DNA-dimer domains, while the DNA-dimer domain can also abut and influence an adjacent DNA-dimer complex. This model portrays the chromatin fiber as a dynamic continuum of DNA and protein rather than as a DNA helix wound around "spools" of discrete and passive histone octamers. Certainly the chromatin fiber appears continuous in electron micrographs that were prepared in 0.15 M NaCl (33,51,64), or "physiologic ionic strength". During the compaction-decompaction cycle in chromatin, the DNA-histone complex may pass through a stage where a string of independent and well defined nucleosomes exist along the fiber, but it is likely that at certain times in the cell cycle there is extensive contact among neighboring nucleosomal domains and the chromatin is compacted, while at other times the subnucleosomal subunits are well separated and the chromatin fiber is extended. Thus, even though the concept of the nucleosome and the core particle as fixed repeating units in the chromatin fiber has been useful in directing research over the last decade, it may be more profitable to modify and expand that model to include interactions and cooperativity both among neighboring nucleosomes and among the subnucleosomal domains.

Acknowledgements

We would like to thank Dr. B.C. Wang for providing the stereo electron density maps, and Dr. Mario Amzel for providing us access to his computing facilities. This work was supported by grants from the National Institutes of Health.

References and Footnotes

1. Hewish, D., and Burgoyne, L., *Biochem. Biophys. Res. Commun. 52,* 504 (1973).
2. Rill, R. and Van Holde, K.E., *J. Biol. Chem. 248,* 1080 (1973).
3. Woodcock, C.I.F., *J. Cell Biol. 59,* 368a (1973).

4. Olins, A.L. and Olins, D.E., *Science 183,* 330 (1974).
5. Wang, J.C., *J. Mol. Biol. 43,* 25 (1969).
6. Rinehart, F.P., and Hearst, J.E., *Arch. Biochem. Biophys. 152,* 723 (1972).
7. Herskovits, T.T., Singer, S.J. and Geiduschek, E.P., *Arch. Biochem. Biophys. 94,* 99 (1961).
8. Lerman, L.S., *Proc. Nat. Acad. Sci. U.S.A. 68,* 1886 (1971).
9. Eickbush, T.H., and Moudrianakis, E.N., *Cell 13,* 295 (1978).
10. Kornberg, R.D., *Annu. Rev. Biochem. 46,* 931 (1977); McGhee, J.D. and Felsenfeld, G., *Annu. Rev. Biochem. 49,* 1115 (1980).
11. Kornberg, R.D. and Thomas, J.O., *Science 184,* 865 (1974); Thomas, J.O. and Kornberg, R.D., *Proc. Nat. Acad. Sci. U.S.A. 72,* 2626 (1975).
12. Eickbush, T.H. and Moudrianakis, E.N., *Biochemistry 17,* 4955 (1978).
13. Benedict, R.C., Moudrianakis, E.N. and Ackers, G.K., *Biochemistry 23,* 1214 (1984).
14. Godfrey, J.E., Eickbush, T.H. and Moudrianakis, E.N., *Biochemistry 19,* 1339 (1980).
15. Lattman, E., Burlingame, R., Hatch, C. and Moudrianakis, E.N., *Science 216,* 1016 (1982).
16. Burlingame, R.W., Love, W.E. and Moudrianakis, E.N., *Science 223,* 423 (1984).
17. Burlingame, R.W., Love, W.E., Wang, B.C., Hamlin, R., Xuong, N.H. and Moudrianakis, E.N., *Science 228,* 546 (1985).
18. Xuong, N.H., Freer, S., Hamlin, R., Nielson, C. and Vernon, W., *Acta Cryst. A34,* 289 (1978); Hamlin, R., *Trans. Amer. Cryst. Assoc. 18,* 95 (1982).
19. Wang, B.C., *Acta Cryst. A40,* C12 (1984); Wang, B.C. in *Diffraction Methods for Biological Molecules,* volume of *Methods of Enzymology,* Ed. H. Wycoff, Academic Press, New York, in press.
20. Klug, A., Rhodes, D., Smith, J., Finch, J.T. and Thomas, J.O., *Nature (London) 287,* 509 (1980).
21. Bentley, G.A., Lewit-Bentley, A., Finch, J.T., Podjarny, A.D. and Roth, M., *J. Mol. Biol. 176,* 55 (1984).
22. Richmond, T.J., Finch, J.T., Rhodes, D. and Klug, A., *Nature (London) 311,* 532 (1984).
23. Pardon, J.F., Worcester, D.L., Wooley, J.C., Tatchell, K., Van Holde, K.E. and Richards, B.M., *Nucleic Acids Res. 2,* 2164 (1975).
24. Bradbury, E.M., Hjelm, R.P., Carpenter, B.G., Baldwin, J.P., Kneale, G.G. and Hancock, R., in *The Molecular Biology of the Mammalian Genetic Apparatus 1,* Ed. P.O.P. T'so, North Holland Publishing Co., Amsderdam p. 53 (1977).
25. Camerini-Otero, R.D., Sollner-Webb, B. and Felsenfeld, G., *Cell 8,* 333 (1976).
26. Mirzabekov, A.D., Shick, V.V., Belyavsky, A.V. and Bavykin, S.G., *Proc. Nat. Acad. Sci. U.S.A. 75,* 4184 (1978); Belyavsky, A.V., Bavykin, S.G., Goguadze, E.G. and Mirzabekov, A.D., *J. Mol. Biol. 139,* 519 (1980).
27. Uberbacher, E.C. and Bunnick, G.J., *J. Biomol. Str. Dyn. 2,* 1033 (1985).
28. Simpson, R.T., *Biochemistry 17,* 5524 (1978).
29. Finch, J.T., Lutter, L.C., Rhodes, D., Brown, R.S., Rushton, B., Levitt, M. and Klug, A., *Nature (London) 269,* 29 (1977).
30. Harauz, G., and Ottensmeyer, F.P., *Science 226,* 936 (1984).
31. Oudet, P., Gross-Bellard, M. and Chambon, P., *Cell 4,* 281 (1975).
32. Langmore, J.P. and Wooley, J.C., *Proc. Nat. Acad. Sci. U.S.A. 72,* 2691 (1975).
33. Thoma, F., Koller, T. and Klug, A., *J. Cell Biol. 83,* 403 (1979).
34. Olins, A.L., Breillatt, J.P., Carlson, R.D., Senior, M.B., Wright, E.B. and Olins, D.E., in *The Molecular Biology of the Mammalian Genetic Apparatus 1,* Ed. P.O.P. Ts'o, North Holland Publishing Co., Amsterdam, p. 211 (1977).
35. Olins, A.L., *Cold Spring Harb. Symp. Quant. Biol. 42,* 325 (1977).
36. Bazett-Jones, D.P. and Ottensmeyer, F.P., *Can. J. Biochem. 60,* 364 (1981).
37. Bazett-Jones, D.P. and Ottensmeyer, F.P., *Science 211,* 169 (1981).
38. Hjelm, R.P., Kneale, G.G., Suau, F., Baldwin, J.P., Bradbury, E.M. and Ibel, K., *Cell 10,* 139 (1977).
39. Braddock, G.W., Baldwin, J.P. and Bradbury, E.M., *Biopolymers 20,* 327 (1981).
40. Pardon, J.F., Cotter, R.I., Lilley, D.M.J., Worcester, D.L., Campbell, A.M., Wooley, J.C. and Richards, B.M., *Cold Spring Harbor Symp. Quant. Biol. 42,* 11 (1977).
41. Yang, J.T. and Wu, C-S.C., *Biochemistry 16,* 5785 (1977).
42. Thomas, J.O. and Butler, P.J.G., *J. Mol. Biol. 116,* 769 (1977).
43. Moudrianakis, E.N., Love, W.E., Wang, B.C., Xuong, N.H. and Burlingame, R.W., *Science 229,* 1110 (1985) Klug, A., Finch, J.T., and Richmond, T.J., *Science 229,* 1109 (1985).

44. Martin, R.G. and Ames, B.N., *J. Biol. Chem. 236,* 1373 (1961).
45. Siegel, L.M. and Monty, K.J., *Biochim. Biophys. Acta 112,* 346 (1966).
46. Pardon, J.F., Wilkins, M.H.F. and Richards, B.M., *Nature (London) 215,* 508 (1967).
47. Richards, B.M. and Pardon, J.F., *Exp. Cell Res. 62,* 184 (1970).
48. Sperling, L., and Tardieu, A., *FEBS Lett. 64,* 89 (1976).
49. Suau, P., Bradbury, E.M. and Baldwin, J.P., *Eur. J. Biochem. 97,* 593 (1979).
50. Crothers, D.M., Dattagupta, N., Hogan, M., Klevan, L. and Lee, K.S., *Biochemistry 17,* 4525 (1978).
51. Griffith, J.D., *Science 187,* 1202 (1975).
52. Thomas, G.J., Prescott, B. and Olins, D.E., *Science 197,* 385 (1977).
53. Bidney, D.L. and Reeck, G.R., *Biochemistry 16,* 1844 (1977).
54. Cotter, R.I. and Lilley, D.M.J., *FEBS Lett. 82,* 63 (1977).
55. Hyde, J.E. and Walker, I.O., *Biochim. Biophys. Acta 490,* 261 (1977).
56. Wong, N.T.N. and Candido, E.P.M., *J. Biol. Chem. 253,* 8263 (1978).
57. Weintraub, H., Palter, K. and Van Lente, F., *Cell 6,* 85 (1975).
58. Hatch, C.L., Bonner, W.M. and Moudrianakis, E.N., *Biochemistry 22,* 3016 (1983).
59. Lilley, D.M.J., Pardon, J.F. and Richards, B.M., *Biochemistry 16,* 2853 (1977).
60. Diaz, B.M. and Walker, I.O., *Bioscience Rep. 3,* 283 (1983).
61. Ward, K.B., Wishner, B.C., Lattman, E.E. and Love, W.E., *J. Mol. Biol. 98,* 161 (1975); Ladner, R.C., Heidner, E.J., and Perutz, M.F., *J. Mol. Biol. 114,* 385 (1977); Love, W.E., Fitzgerald, P.M.D., Hanson, J.C., Royer, W.E. and Ringle, W.M. in *Biochemical and Clinical Aspects of Hemoglobin Abnormalities,* Academic Press (1978).
62. Burch, J.B.E. and Martinson, H.G., *Nucleic Acids Res. 9,* 4367 (1981).
63. Dietrich, A.E., Axel, R. and Cantor, C.R., *J. Mol. Biol. 129,* 589 (1979).
64. Moudrianakis, E.N., Anderson, P.L., Eickbush, T.H., Longfellow, D.E. and Rubin, R.L., in *The Molecular Biology of the Mammalian Genetic Apparatus 1,* Ed. P.O.P. T'so, North Holland Publishing Co., Amsterdam, p. 301 (1977).
65. Rill, R.L. and Nelson, D.A., *Cold Spring Harbor Symp. Quant. Biol. 42,* 475 (1977); Nelson, D.A., Mencke, A.J., Chambers, S.A., Oosterhof, D.K., and Rill, R.L., *Biochemistry 21,* 4350 (1982); Bair, B.W. and Rhodes, D., *Nature (London) 301,* 482 (1983).
66. Jordano, J., Nieto, M.A. and Palacian, E., *J. Biol. Chem. 260,* 9382 (1985).
67. Martinson, H.G., True, R.J. and Burch, J.B.E., *Biochemistry 18,* 1082 (1979).
68. Jackson, V., *Cell 15,* 945 (1978).
69. Palter, K.B., Foe, V.E. and Alberts, B.M., *Cell 18,* 451 (1979).
70. Palter, K.B. and Alberts, B.M., *J. Biol. Chem. 254,* 11160 (1979).
71. Rubin, R.L. and Moudrianakis, E.N., *J. Mol. Biol. 67,* 361 (1972).
72. Mandel, R. and Fasman, G.D., *Nucleic Acids Res. 3,* 1839 (1976).
73. Poulson, J.R. and Laemmli, U.K., *Cell 12,* 817 (1977); Igo-Kimenes, T. and Zachau, H.G., *Cold Spring Harbor Symp. Quant. Biol. 42,* 109 (1977).
74. Anderson, P.L., *PhD Thesis,* Johns Hopkins University, Baltimore (1971).

Biomolecular Stereodynamics III, Proceedings of the Fourth Conversation in the
Discipline Biomolecular Stereodynamics, State University of New York,
Albany, NY, June 04-09, 1985, Eds., Ramaswamy H. Sarma & Mukti H. Sarma,
ISBN 0-940030-14-4, Adenine Press, ©Adenine Press 1986.

A Different View Point on the Chromatin Higher Order Structure: Steric Exclusion Effects

L.E. Ulanovsky and E.N. Trifonov

The Department of Polymer Research,
The Weizmann Institute of Science, Rehovot, Israel

Abstract

The dependence of mutual orientation of adjacent nucleosomes on the length of the DNA linker between them and on the DNA helical repeat length is analyzed. A nontrivial distribution of linker lengths, modulated by sterical exclusion effects has been obtained. As the linker is progressively elongated, the mutual screw-like rotation of the neighboring nucleosomes leads to collision of the two nucleosomes at certain values of the linker length. The lengths in the intervals 1 to 6, 12 to 16 and 22 to 25 base-pairs have been found to be sterically excluded. On the basis of the calculations of the DNA path in alpha-satellite cromatins specific limitations on the number of turns of the DNA superhelix in the nucleosome are imposed.

Introduction

The DNA path in the nucleosome appears to be a smooth superhelix. This view has recently been supported by a variety of theoretical and experimental evidence (1-4). DNA component of a chromatin higher order structure can be viewed as an assembly of the nucleosomal DNA superhelices in certain orientations relative to each other, connected by DNA linkers of various lengths.

DNase digestion experiments strongly indicate that the DNA minor groove attacked by the DNase is oriented outwards with respect to the histone core at regular distances from either of the ends of the nucleosomal DNA, namely those multiple of about 10.4 base-pairs (5-7). Thus, the histone core is apparently spatially fixed with respect to the minor and major grooves of the nucleosomal DNA. This implies that the mutual orientation of two adjacent nucleosomes in space is determined by the length of the DNA linker between them as long as the linker is of regular double stranded DNA structure. In particular, each increase in the linker length by one base-pair should result not only in about 3.4 Å shift, but also in a rotation of the two nucleosomes with respect to each other by about 34° around the axis of their linker. Figure 1 illustrates several such screw-like translations along the double helix of the linker DNA.

35

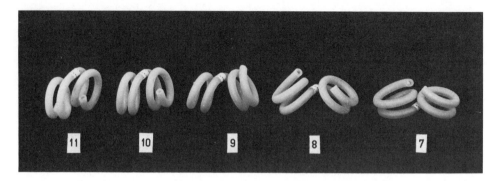

Figure 1. Wire models of various orientations of adjacent nucleosomes. In this example the linker between the nucleosomes is shown to have five length values L from 7 to 11 base-pairs. For simplicity, the DNA paths are shown without histones. It can be seen that each increase in the linker length L by one base-pair causes an approximately 34° screwlike turn of one nucleosome with respect to the other following the double helix of the linker DNA, which would lead to a steric hindrance for linkers immediately longer than 11 and shorter than 7 base-pairs. The contact between the two nucleosomes is clearly seen in the extreme left and right pictures of Figure 1 (Body-Body I interaction). The linkers are shown white with dots indicating phosphates.

Noteworthy, that in vivo the linker lengths can vary widely within one chromatin sample. For example in SV40 minichromosome the electron microscopy data (Saragosti et al., 1980) reveal that the linkers differ in length on the average by 40-50 base-pairs in both directions from the average value of about 55 base-pairs. Thus, the frequently used concept of "repeat length" of chromatin should not be understood literally, but rather as an average value.

Only two basic relative orientations of adjacent nucleosomes have been discussed so far (9-11): "parallel" and "antiparallel", depending on the linker length between them, about either 10n or 10n+5 base-pairs (see, however (12)). These two orientations correspond to the two extremes of Figure 1. There are no reasons, however, why intermediate orientations (intermediate linker lengths), as shown in Figure 1, should not be considered as well.

One might suppose, therefore, that the knowledge of the linker length in any nucleosome dimer should suffice, in principle, to predict the mutual orientation of the adjacent nucleosomes, assuming, as in the simplest cases above, straight double-helical linkers and certain standard geometry of the nucleosome. However, as our calculations show, it is also necessary to know the length of the double-helical repeat of the DNA in the nucleosome with a very high accuracy.

The dependence of mutual orientation of nucleosomes on the linker length and on the double-helical repeat of DNA is analyzed in the present paper. Certain specific lengths of linkers are found to result in steric hindrance of two adjacent nucleosomes, like in Figure 1, thereby imposing restrictions on the range of permitted linker lengths. This and some other types of steric hindrance are studied here using computer simulated paths of the DNA in chromatin.

Results and Discussion

Basic Assumptions:

In our calculations, the DNA axis path in each nucleosome was assumed to be an ideal left-handed superhelix having the pitch of 25.8 Å (3). It was checked that deviations from this value within ± 2 Å do not significantly affect our results.

The linker DNA which connects the nucleosomes was considered to be straight and tangential to the nucleosomal DNA at the point of conjunction between the linker and the superhelix of the nucleosomal DNA (no kink at the conjunction point). This assumption of straight linkers is only appropriate for small linker lengths. For lengths larger than or comparable to the DNA persistence length, about 150 base-pairs in natural conditions (13), the linker DNA has to be considered flexible rather than rigid.

The nucleosomal DNA was assumed, for the purpose of calculation, to contain an integral number of the helical repeats of the DNA. Being separated by the integral number of DNA helical repeats, the beginning and the end of the nucleosomal DNA have identical phases of the DNA double helix. Therefore, two adjacent nucleosomes with a hypothetical zero length linker between them form one continuous DNA superhelix of twice the number of turns as compared to one nucleosome. These two nucleosomes are geometrically parallel, that is, turned with respect to each other by the angle $\varphi = 0$. The superhelices of two adjacent nucleosomes connected by a straight linker of a nonzero length L are turned one with respect to the other (see Figure 1) by the angle φ proportional to the length L of the linker DNA (in base-pairs):

$$\varphi = L/P \times 360° \tag{1}$$

where P is the helical repeat of linker DNA in base-pairs. The repeat P of the linker DNA double helix is assumed to be equal to that of free DNA in solution: P = 10.55 base-pairs per repeat (14,15). The possibility of deviations of the pitch P from the value P = 10.55 will be considered below.

For the purpose of the calculations the length of the nucleosomal DNA was taken to be 145 b.p. × 3.33 Å = 483 Å. The length of 166 base-pairs, corresponding to the "chromatosome" 16 could also be taken for the calculations instead of 145 base-pairs, without resulting in any significant difference in the exclusion effects discussed in this paper (see also below).

Under these assumptions, the 3-dimensional path of the DNA axis comprising two or more adjacent nucleosomes has been calculated by computer for different values of the two following variable parameters: a) the number of turns N in the nucleosomal superhelix, scanned in the interval 1.35 to 2.15 turns; b) the linker lengths L (integer), scanned in the interval 0 to 52 base-pairs. As a result, certain combinations of values of these two parameters, N and L, which lead to steric hindrance, form areas of exclusion in the 2-dimensional diagram with N and L as axes (see below). A pair of values of N and L was considered to be a hindrance case if, in the calculated DNA path, the minimal distance between DNA axes of two different elements (two different nucleosomes, nucleosome and linker or two different linkers) was less than 20 Å. The final results are insensitive to variations of this figure.

Should we know how much the histones protrude out of the nucleosome, we could take into account the protein-protein and protein-DNA contacts. The steric exclusion areas presented in this paper refer just to the DNA-DNA contacts, and therefore can only be somewhat enlarged by considering the protein component as well.

Steric Exclusion Diagrams

Four types of hindrance were investigated:

1) Body-Body I: interaction between two neighbouring nucleosomes (see Figure 1).

2) Linker-Body: interaction between a nucleosome and the second linker of the adjacent nucleosome (see Figure 2A).

3) Linker-Linker: interaction between a linker and the next but one linker (see Figure 2B).

4) Body-Body II: interaction between a nucleosome and its second neighbour with the two linkers between them being of equal length (see Figure 3).

The results of the calculations are presented in diagrams in Figure 4 and Figure 5 which show the hindrance areas for certain values of the linker length L and the number of superhelical turns in the nucleosome N.

The most extensive areas of exclusion are formed by the Body-Body I type of hindrance between a nucleosome and its closest neighbour (see Figure 1). These areas are shown by wide shading in Figures 4 and 5, and form three vertical zones of

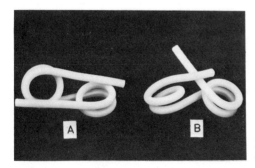

Figure 2. Linker-Body (A) and Linker-Linker (B) interactions (see text).

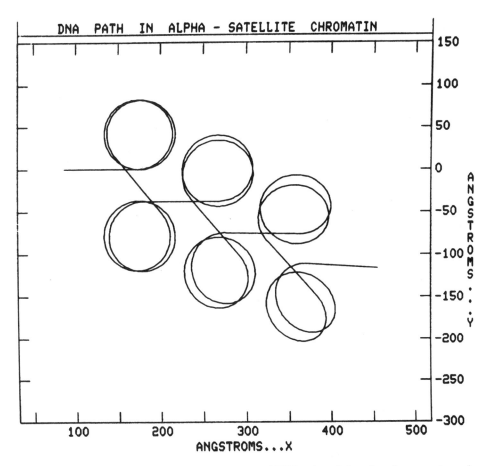

Figure 3. A computer drawn projection of the computed DNA axis path in a 6 nucleosome piece of tandem alpha-satellite chromatin having the DNA sequence repeat of 171 base-pairs (the linker length $L = 26$) and the number of superhelical turns in the nucleosome $N = 1.86$. Only the DNA axis is shown. It can be seen that each "nucleosome" is in close contact with its two immediate neighbours (Body-Body I interaction) as well as with its two second neighbours (Body-Body II interaction) ensuring a very compact chromatin structure (about 20 Å distance between DNA axes of contacting nucleosomes).

exclusion comprising the forbidden linker lengths L in the intervals 1 to 6, 12 to 16 and 22 to 25 base-pairs. One can also see a fourth exclusion zone of L equal to 33 and 34 base-pairs for small values of N.

A similar distribution of avoided linker lengths was found recently by computerized sequence-directed mapping of nucleosomes (17,18). A discrete periodical distribution of linker lengths was also found experimentally (9-11). Unfortunately, the accuracy of these experimental data does not allow a more detailed quantitative comparison with the theoretical predictions here.

Additional restricted areas are provided by two other types of hindrance shown in Figure 4 by dense inclined shadings. They result from Linker-Body and Linker-

Linker interactions (see Figure 2) and form several small exclusion areas, all below N = 1.84.

Finally, the Body-Body II type of hindrance between a nucleosome and its second neighbour with the two linkers between them being of equal length is shown by vertical shading in Figure 5. This type of hindrance (see Figure 3) might be of a special interest in connection with the regular structure of satellite chromatins as discussed below.

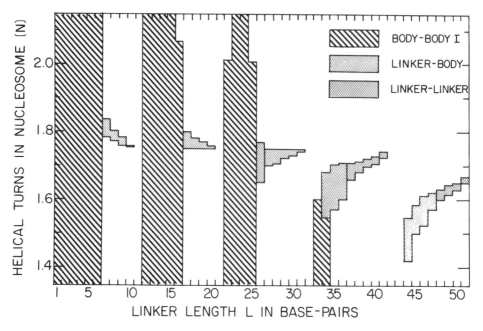

Figure 4. Diagram of steric exclusion in chromatin. The horizontal axis is the linker length L in base-pairs. The vertical axis is the number of turns N in DNA superhelix in the nucleosome. The wide inclined shading shows the Body-Body I type exclusion areas. The dense inclined shadings show the Linker-Body and the Linker-Linker type exclusion areas.

From the computer simulations it follows that in some cases different types of exclusion areas overlap. This is not shown in Figures 4 and 5 for the sake of simplicity.

Variations of Basic Assumptions and Parameters

The question arises: how much change will occur in the resulting distribution of the steric exclusion areas shown in Figures 4 and 5, if our idealized assumptions about the structure of the nucleosome and linkers are changed?

The helical repeat of the linker DNA was assumed to be P = 10.55 base-pairs. P affects the diagrams of Figures 4 and 5 through equation (1). Therefore, the scale of the horizontal axes of the diagrams is inversely proportional to the value of P. Any decrease (increase) of P would lead to a contraction (expansion) of the diagrams in Figures 4 and 5 along the horizontal axis in the same proportion (up to a few percent for reasonable values of P). Strictly speaking, the helical repeat value P can characterize the pitch of the natural linker DNA only on the average. Individual DNA sequences might have slightly different values of P due to differences in helical twist angles between different combinations of adjacent base-pairs (19).

In general, the linkers connecting the nucleosomes in chromatin are not exactly straight. The deformation might be caused by thermal motion, by proteins bound to the linkers, or both. The curvature of linkers can lead to smearing of the contours of the exclusion areas, especially for long linkers which can be bent at a significant overall bending angle. The diagrams shown in Figures 4 and 5 are calculated for the assumption of ideally straight linkers and are therefore only an average pattern of steric hindrance. Such pattern can only work in the case of linkers, like those studied in the present paper.

Small deviation of the DNA path in the nucleosome from the ideal superhelix have not been found to affect the hindrance diagrams to any noticeable degree.

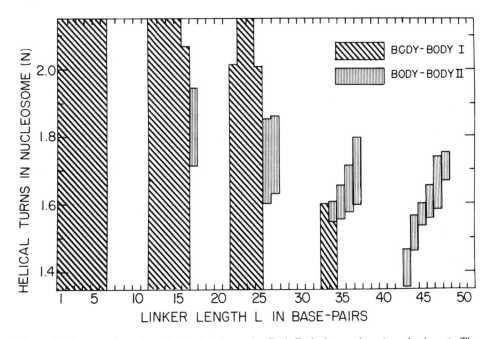

Figure 5. Diagram of steric exclusion in chromatin: Body-Body interactions (see the insert). The Linker-Body and Linker-Linker type exclusion areas are not shown for clarity. For notations, see Figure 4.

For elucidation of the higher order structure of chromatin the knowledge of the exact value of the DNA helical repeat P_n in the nucleosome is of more importance than that in linkers, P. Indeed, a small error dP_n in the value of P_n, accumulated over about 14 helical repeats, should result in a considerable turn of the end base-pairs in the nucleosome, thus dramatically changing the relative orientation of adjacent nucleosomes. For example an error of only 1% amounts to an appreciable change in the orientation, $14 \times 0.01 \times 360° = 50°$.

The most accurate estimate available of the average pitch of the DNA in the nucleosome N_p is 10.36 ± 0.02 base-pairs per turn, which is a weighted average of estimates obtained by five independent methods (20,21). The length of the nucleosomal DNA chosen for the calculations, 145 base-pairs, equals almost exactly to 14 double-helical periods, thus justifying the choice of an integral number of periods in the nucleosomal DNA. If, however, the number of the double-helical periods involved in the nucleosomal DNA eventually turns out to be equal to any other constant, not necessarily integer, it would result in an addition of a constant value to all the orientation angles, and, correspondingly, in a shift of the exclusion diagram as a whole, without affecting the pattern of the exclusion zones.

Satellite Chromatin

After the role of the linker length and of the nucleosomal DNA helical repeat is clarified, we could calculate specific types of higher order structure of chromatin.

Recent data on the nucleosome phasing in satellite DNA sequences (22,23) suggest that the nucleosomes in satellite chromatin are most probably phased with respect to the tandemly repeated DNA sequences. Of special interest are tandem sequence repeat lengths in the interval of 145 to 290 base-pairs which can contain one and only one nucleosome per repeat. Such satellite chromatin must have a very regular structure, all the linkers being of equal length. The available data about the tandem sequence repeat lengths on satellite DNA can be compared with Figure 5. Here the exclusion areas are shown for Body-Body II type hindrance resulting from the steric interaction between a nucleosome and its second neighbour with the two linkers between them being of equal length (see Figure 3). A well studied example of this kind is the green monkey alpha-satellite chromatin (23-25). This satellite DNA has been sequenced as well as related alpha-satellites of other primates (26-28). In two of these cases the repeats are strict in terms of length: green monkey alpha-satellite, 172 base-pairs, and bonnet monkey alpha-satellite, 171 base-pairs. These are the only examples known to us of repeating DNA sequences of nucleosomal size, such that their lengths are measured with sequencing accuracy. Assuming that the 14 full helical repeats of the nucleosomal DNA superhelix occupy 145 base-pairs, we obtian for these alpha-satellites two linker lengths: L = 26 and L = 27 base-pairs. Figure 5 shows exclusion areas for both these linker lengths covering together the interval of 1.60 to 1.86 turns of the nucleosomal superhelix. Therefore the tandem DNA repeat lengths of these alpha-satellites suggest that their nucleosomes should have the number N of superhelical turns either smaller than 1.6 or larger than 1.86. Since the nucleosomes of satellite chromatin seem to be standard in all respects studied (29,30), we believe that these limitations apply to nucleosomes in general.

It is worth noticing, that should we assume the nucleosomal DNA to contain 16 rather than 14 helical repeats (166 base-pairs rather than 145), the resulting linker lengths in these satellites would be forbidden: L = 5 and L = 6 rather than 26 and 27 base-pairs. Another reason to prefer 145 rather than 166 base-pairs as a standard nucleosomal DNA size is the fact that on many occasions the average "repeat" length of DNA in chromatin was found to be shorter than 166 base-pairs (see e.g. review 31).

The latest neutron diffraction measurements (3) give the estimate of about N = 1.8 for the number of superhelical turns in the nucleosome. The above steric limitation N > 1.86 turns agrees well with the neutron diffraction estimate. However, one can not exclude the possibility of the nucleosome being less compact in vivo than in crystal. Therefore the value of N < 1.6 can not be discarded altogether.

A point at the border of any of the hindrance exclusion areas shown in Figure 5 corresponds to a close contact between the nucleosomes, which might be important for chromatin packing in specific higher order structures. Interestingly, for the above alpha-satellite DNA with linker lengths L = 26 and L = 27 base-pairs the value N = 1.86 turns gives an extremely compact chromatin structure (see Figure 3), where each nucleosome is in close contact with four neighbouring nucleosomes— two first neighbours at the edge of the Body-Body I interaction and two second neighbours at the edge of the Body-Body II interaction. This combination of N = 1.86 turns and L = 26 (or L = 27) base-pairs is remarkable in that no other linker length can give such a compact satellite chromatin located simultaneously near the two different types of exclusion areas (see Figure 5). This is interesting in view of the recent experimental evidence of unusual compactness of satellite chromatins (32,33).

We could expect a priori that a regular chromatin structure would be essentially an array of stacked nucleosomes in parallel orientation (similar to the extreme left model of Figure 1). However, the alpha-satellite chromatin in Figure 3 appears to be rather like an antiparallel "zig-zag".

In conclusion, we believe that the higher order structure of cromatin has to obey certain rules imposed by steric exclusion, like the forbidden linker length intervals. These as well as all the other numerical results concerning steric exclusion effects discussed in this paper should not be viewed as absolute and final, because they are obtained under certain arbitrary though reasonable assumptions. They are rather meant to illustrate the effects of steric exclusion, which should exist under a much wider set of assumptions, in particular the effect of alternation of the forbidden and allowed linker length intervals.

Acknowledgments

The authors wish to express their gratitude to Dr. Z. Shakked and to Dr. H. Eisenberg for discussions and valuable comments.

References and Footnotes

1. Sussman, J. & Trifonov, E.N., *Proc. Natl. Acad. Sci. USA, 75,* 103-107 (1978).
2. Levitt, M., *Proc. Natl. Ac. Sci. USA, 75,* 640-644 (1978).
3. Bentley, G.A., Lewit-Bentley, A., Finch, J.T., Podjarny, A.D. & Roth, M. *J. Mol. Biol., 176,* 55-75 (1984).
4. Richmond, T.J., Finch, J.T., Rushton, B., Rhodes, D., & Klug, A., *Nature 311,* 532-537 (1984).
5. Noll, M., *Nucleic Acids Res., 1,* 1573-1578 (1974).
6. Sollner-Webb, B., Melchior, W. & Felsenfeld, G., *Cell, 14,* 611-627 (1978).
7. Lutter, L., *Nucleic Acids Res., 6,* 41-56 (1879).
8. Saragosti, S., Moyne, G. & Yaniv, M., *Cell, 20,* 65-73 (1980).
9. Lohr, D. & van Holde, K.E., *Proc. Natl. Acad. Sci. USA, 76,* 6326-6330 (1979).
10. Karpov, V.L., Bavykin, S.G., Preobrazhenskaya, O.V., Belyavsky, A.V., and Mirzabekov, A.D., *Nucl. Acids Res., 10,* 4321-4337 (1982).
11. Strauss, F. & Prunell, A., *EMBO Journal, 2,* 51-56 (1983).
12. Subirana, J.A., Munos-Guerra, S., Radermacher, M. & Frank, J., *J. Biomolec. Str. and Dyn. 1,* 705-714, (1983).
13. Borochov, N., Eisenberg, H. & Kam, Z., *Biopolymers, 20,* 231-235 (1981).
14. Strauss, F., Gaillard, C. & Prunell, A., *Eur. J. Bioch., 118,* 215-222 (1981).
15. Peck, L.J. & Wang, J.C., *Nature, 292,* 375-378 (1981).
16. Simpson, R., *Biochemistry, 17,* 5524-5531 (1978).
17. Trifonov, E.N., *Cold Spring Harbor Symp. Quant. Biol., 47,* 271-278 (1983).
18. Mengeritsky, G. & Trifonov, E.N., *Nucleic Acids Res., 11,* 3833-3851 (1983).
19. Kabsch, W., Sander, C. & Trifonov, E.N., *Nucleic Acids Res., 10,* 1097-1104 (1982).
20. Trifonov, E.N., in *Nucleic Acids: the Vectors of Life,* Ed. Pullman, B. & Jortner, J., Reidel Publ. Comp., pp. 373-385 (1983).
21. Wartenfeld, R., Mengeritsky, G. & Trifonov, E.N., *CODATA Bull. 56,* 14-16 (1984).
22. Igo-Kemenes, T., Omori, A. & Zachau, H.G., *Nucleic Acids Res., 8,* 5377-5390 (1980).
23. Musich, P.R., Brown, F.L. & Maio, J.J., *Proc. Nat. Acad. Sci. USA, 79,* 118-122 (1982).
24. Rosenberg, H., Singer, M. & Rosenberg, M., *Science, 2900,* 394-402 (1978).
25. Strauss, F., & Varshavsky, A., *Cell 37,* 889-901 (1984).
26. Donehower, L., Furlong, C., Gillespie, D. & Kurnit, D., *Proc. Natl. Acad. Sci. USA, 77,* 2129-2133 (1980).
27. Rubin, C.M., Deininger, P.L., Houck, C.M. & Schmid, C.W., *J. Mol. Biol., 135,* 151-167 (1980).
28. Wu, J.C. & Manuelidis, L., *J. Mol. Biol., 142,* 363-386 (1980).
29. Musich, P.R., Brown, F.KL. & Maio, J.J., *Proc. Nat. Acad. Sci. USA, 74,* 3297-3301 (1977).
30. Levinger, L. & Varshavsky, A., *Cell, 28,* 375-385 (1982).
31. Lilley, D.M.J., in *Topics in Nucleic Acid Structure,* Ed. Neidle, S., (ed.) Macmillan Publ. Ltd., pp. 141-176 (1981).
32. Zhang, X.Y. & Hörz, W., *Nucleic Acids Res., 10,* 1481-1494 (1982).
33. Reudelhuber, T.L., Ball, D.J., Davis, A.H. & Garrard, W.T., *Nucleic Acids Res., 10,* 1311-1325 (1982).

*Biomolecular Stereodynamics III, Proceedings of the Fourth Conversation in the
Discipline Biomolecular Stereodynamics, State University of New York,
Albany, NY, June 04-09, 1985, Eds., Ramaswamy H. Sarma & Mukti H. Sarma,
ISBN 0-940030-14-4, Adenine Press, ©Adenine Press 1986.*

Structural Studies on a DNA-*Eco*RI
Endonuclease Recognition Complex

Judith A. McClarin[1], Christin A. Frederick[1], John Grable[1], Cleopas T. Samudzi[1], Bi-Cheng Wang[2], Patricia Greene[3], Herbert W. Boyer[3] and John M. Rosenberg[4]

[1]Dept. of Biol. Sci., Univ. of Pittsburgh, Pittsburgh, PA 15260
[2]Biocrystallography Laboratory, Box 12055, VA Medical Center,
Pittsburgh, Pennsylvania 15260, USA
[3]Dept. of Biochem. and Biophys., University of Calif., SF.,
San Francisco, CA 94143

Abstract

The 3Å structure of a co-crystalline recognition complex between *Eco*RI endonuclease and the cognate oligonucleotide TCGCGAATTCGCG was solved by the ISIR method using a platinum isomorphous derivative. The complex possesses a common two-fold symmetry axis which relates both strands of the self-complementray oligonucleotide and the two identical subunits of the protein dimer. Each subunit is organized into an α/β domain based on a five stranded β-sheet and an extension, called the "arm", which wraps around the DNA. The primary β-sheet consists of anti-parallel and parallel segments which, repsectively, contain the sites of DNA strand scission and sequence specific recognition. (DNA hydrolysis was inhibited via omission of magnesium).

The DNA departs significantly from the conformations seen in the absence of protein, suggesting that binding of the enzyme is required for their stability. These are termed neo-conformations to distinguish them from those which are intrinsically stable in the absence of protein and include the torsional "type-1 neo-kink" which unwinds the DNA by approximately 25°. This separates the DNA backbones by approximately 3.5Å without unstacking the bases, and is a structural requirement for the recognition modules of the protein to gain access to the edges of the bases exposed at the bottom of the major groove. We suspect that there are a finite number of structurally feasible neo-conformations which are important for DNA-protein interactions in general.

Sequence specificity is determined by "modular" interactions based on the crossover α-helices, *ie.* those which connect the β-strands of the parallel segment of the principal β-sheet. They are pointing into the major groove of the DNA and amino acid side chains at the amino ends of these helices form bidentate interactions with the bases. The inner recognition module consists of two symmetry-related α-helices which recognize the inner tetranucleotide (AATT), while the two symmetry-equivalent outer recognition modules are single alpha-helices which recognize the GC base pairs.

[4]To whom inquiries should be addressed at: Dept. of Biol. Sci., Univ. of Pittsburgh, Pittsburgh PA 19260
(412) 628-4636

Introduction

The recognition by a protein of a specific sequence of bases along a strand of double helical DNA lies at the heart of many fundamental biological processes including the regulation of gene expression by repressors and activators, site specific genetic recombination and host dependent restriction and modification of DNA. The detailed molecular mechanisms of some of these interactions are begining to be understood due to efforts in many laboratories using genetic, biochemical and structural methods. One of the most intriguing questions in molecular biology today is whether the details of several of these recognition mechanisms will form a a small number of simple patterns which would lead to a general understanding of DNA recognition processes. In order to answer this question, structural data is required from representative DNA-protein co-crystalline complexes.

The structures of three sequence specific DNA-binding proteins, the *cro* and CI repressors from bacteriophage-λ and the *E. coli* catabolic gene activator protein (CAP) have been solved in the absence of DNA (1,2,3,4,5). These three proteins share a common "two-helical motif" at the putative DNA binding site which has led to model building of the recognition complexes (6,3,7). The 7Å structure of a co-crystalline complex between a tetradecanucleotide and bacteriophage 434 repressor supports the general features of these models (8,9). The common features in these structures suggests that they are examples of one class of DNA recognition proteins based on the two-helical motif.

Similarly, structural investigations are in progress on other systems, including the Klenow fragment of *E. coli* DNA polymerase I (10), nucleosome core particles (11,12), the histone octamer (13) and the "histone like" DNA-binding protein II of *B. stearothermophilus* (14). These analyses will facilitate the elucidation of the specific structural features of both protein and DNA in these specific protein-DNA interactions as well as general structural principals of DNA-protein interactions.

We recently reported (15) our observations based on a preliminary electron density map (see below) of a co-crystalline complex between *Eco*RI endonuclease and a cognate oligonucleotide. Here we report further analysis of this structure.

Methods

Crystallization conditions and methods of data collection were reported previously (16,15). It should be noted that Hydrolysis of the DNA was prevented by removing the required co-factor, Mg^{2+}, from the crystallization medium.

A platinum derivative was obtained by soaking crystals in 10 mM cis-$(NH_3)_2PtCl_2$, dissolved in 40 mM bis-tris-propane (BTP), pH 7.0 plus a mixture of 11% w/v polyethylene glycol (PEG) 6000 and 5% w/v PEG 400 for seven days at 25°C. A mercury derivative was likewise obtained by soaking crystals in 5 mM $Hg(SCH_2CH_2NH_3^+)_2$, dissolved in 40 mM BTP, pH 7.0 plus 16% w/v PEG 6000 for 3.5 days at 25°C. The heavy atom locations were independently determined by difference Patterson methods for both derivatives*. The platinum Difference Patterson Function is shown in Figure One. The Cullis R factors for the Pt and Hg derivatives to 3.5Å are 49.2% and 52.2%, respectively.

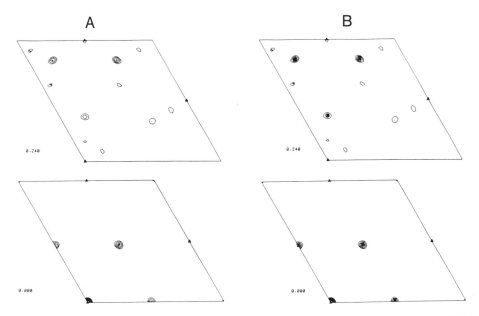

Figure 1. A). The w = 0 and w = 0.24 sections of the platinum difference Patterson function, which contains the Pt-Pt vectors. B). The indicated positions mark the expected positions for the Pt-Pt vectors based on the coordinates of the major platinum binding site.

An MIR electron density map was calculated from these two derivatives to 5Å resolution. In addition, Wang's ISIR method (see below) was independently applied to the Pt and Hg data. The general features, such as the solvent regions and the molecular outline, were similar in all three electron density maps; however the MIR and Hg-ISIR maps contained significant amounts of noise while the Pt-ISIR map was clear. We suspect that the noise is associated with non-isomorphism in the Hg derivative.

The statistics for the platinum derivative exhibited evidence of a slight non-isomorphism at high resolution, as can be seen in Figure Two, which is a plot of intensity differences as a function of resolution. The plot shows a minimum at 3.5Å, with a slight increase at higher resolution**. We felt that the Pt phase information

---

*Even though the platinum compound is a chemotherapeutic agent which appears to act at the DNA level, the location of the major and minor sites turn out to be on the surface of the protein, well away from the DNA.

**The source of this slight non-isomorphism was obvious once the structure was solved: The Pt was bonded to the sulphur atoms of two methionyl residues, which fortuitously were in close proximity on the surface of the protein. However, the bridging reaction appears to have required a small structural adjustment in a short segment of the polypeptide chain in the vicinity of the heavy atom substitution.

was dubious beyond 3.5Å and therefore did not utilize it for the calculations summarized below:

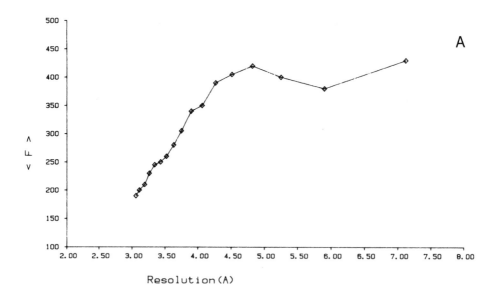

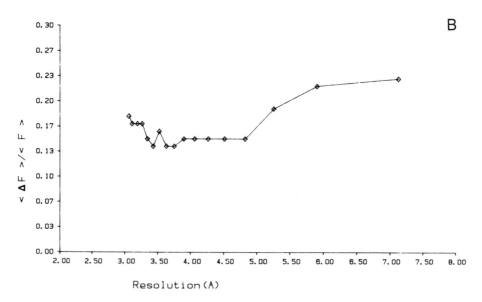

Figure 2. A). A plot showing the mean value of our observed structure factors as a function of resolution. B). A plot showing, as a function of resolution, the ratio of the mean of the absolute values of the differences between native and Pt-derivative structure factors to the mean of the structure factors.

The ISIR procedure was used to resolve the phase ambiguity in the Pt SIR data to 3.5Å as described earlier (15). It was then used to extend the data to 3.2Å and then 3.0Å resolution (15). The average figure-of-merit at the beginning of the process was 0.33 for those 4038 reflections which had both the native and the derivative information and at the end of the process it was 0.84 for all 5880 observed reflections including those 1842 reflections for which the derivative information had been rejected. At this stage, the R-factor (discrepancy index) based on the observed and the map-inversion structure factor amplitudes was 18%. An electron density map, based on these phases was the basis of our earlier report (15).

Although this electron density map was very clear in most places and allowed us to trace the entire DNA double helix and most of the polypeptide backbone, there were a few regions where the electron density was not easily interpretable. The problem at this point was that we were not able to obtain a continuous tracing of the entire polypeptide chain which that was unique and convincing. In hindsight, the difficulties centered primarily on two areas: The β-ribbon which forms two of the three strands in the subsidiary anti-parallel sheet in the "arm" (see discussion) was not clearly resolved in the initial map. A region surrounding a crystallographic three-fold symmetry axis is very densely packed with protein (see Figure Three), where three dimers form a tight complex. This region also, probably, contained some noise in the first map with the result that the molecular boundary was unclear in this localized area. In addition, several large ripples were present along some of the three-fold axes located in the solvent region, indicating some error. In summary, the first map was difficult to interpret in a region of the molecule which projected out into the solvent and in a second region around the three-fold axes where the protein molecules were densely packed.

A fraction of the data were missing from the original data sets and we concluded that the absence of this information was interfering with the ISIR procedure. Three factors led to the absence of data: First, a few reflections at very low resolution were obscured by the beam stop of our Arndt-Wonacott camera. Second, a few reflections were saturated even on the third film of our film packs and were deleted from the data sets by the computer programs we used to process our film data (17,18). Third, these programs also deleted a significant proportion of our weakly observed data because they were deemed statistically unreliable.

Efforts were then made to estimate the missing amplitudes and phases and incorporate these estimates in the electron density calculations. This was initiated for reflections within a 5Å resolution limit and iterated for four cycles (all the observed data to 3.0Å were used during this process). At each iteration, the structure factor amplitudes and phases were estimated for the missing reflections by Fourier inversion of the modified electron density map. These estimates were used in electron density calculations during the next cycle. At the end of the fourth cycle, a new solvent mask was calculated based on the 5880 originally observed reflections and the 298 estimates generated by this process. The process was repeated in a similar manner to estimate the missing reflectons to 4.0Å, then to 3.5Å and finally to 3.0Å. At the end of the process, the average figure-of-merit for the 5880 observed reflections increased from 0.84 to 0.87 and their R-factor dropped from 18% to 15%. This process produced 2394 estimated structure factor amplitudes and phases.

At this stage, an electron density map was calculated using all the observed and estimated reflections (8274 in total). This map showed considerable improvement over the original ISIR map based on the 5880 observed reflections only. This map however, still showed small ripples around some of the three-fold axes, although their magnitudes had been diminished considerably from the first map. These ripples around the three-fold were finally levelled by recalculating a solvent mask using 10Å radius in the masking function instead of the 5.1Å used earlier. After 12 cycles of iterations and two re-calculatations of the solvent mask, the final figure-of-merit and map inversion R-factor remained at 0.87 and 15%, respectively. The electron density based on the third set of phases exhibited a slight improvement in clarity and was used for the fitting of the chemical sequence of the enzyme, as described below.

We compared the electron density maps which preceeded and followed both of the "extension" steps (phase extension from 3.5Å to 3.0Å using observed amplitudes and estimation of missing intensities) and concluded in both cases that the extensions reduced noise and improved the clarity of the maps while maintaining the basic features which were present in the initial 3.5Å map. These features included the DNA, several prominent α-helices and strands of β-sheet. The improvements in detail were most noticeable in the regions described above, in some of the loops connecting secondary structure elements and in some of the side chains. They enabled us to distinguish possibilities which had been ambiguous before the extension.

The final electron density map was displayed on plexiglass sheets. The DNA and protein secondary structure elements were very clear, as noted above. Over two-thirds of the amino acid side chains were distinctly visible along with main chain density for all but four amino acid residues[*]. Almost all of the large hydrophobic side chains, tryptophans, phenylalanines and tyrosines were clearly recognizable. Many basic residues, especially arginines, which were located at the DNA-protein interface were also easily identifiable. Most of the poorly visualized side chains were located at the protein-solvent interface. Both the DNA-protein and protein subunit-subunit interfaces were well ordered and provided useful constraints when we assigned the known amino acid sequence to the electron density map. These assignments were made via inspection of the electron density map, aided by model building, distance measurements and the known stereochemistry of proteins. This process lead to a unique tracing of the polypeptide chain through the complex.

Coordinates for Cα and for Cβ, or a terminal side chain atom for larger amino acids were taken from the the ISIR map on plexiglass sheets and used to generate atomic

---

[*]The missing residues were in the immediate vicinity of the major heavy atom site, and it appears likely that their movement is associated with the small non-isomorphism noted previously.

coordinates for the entire molecule with the program FRODO (19,20,21). Electron density fitting continued with FRODO on an Evans and Sutherland PS300 computer graphics system. The coordinates were regularized to approximately ideal geometry alternately with improving the fit to the electron density. At this point, the model has been fit to all of the electron density features noted above.

It should be noted that the structure reported here represents an intermediate stage of a complete crystallographic structure determination that ultimately will include extensive refinement *ie.* adjustment of the molecular model to optimize the fit between the experimentally measured diffraction pattern and that calculated from the model. It is highly improbable that these adjustments will alter the basic conclusions reported here, however the fine details of the current model should be considered preliminary results.

Description and Discussion of the Structure

General Features of the Complex:

The co-crystalline asymmetric unit contains one protein subunit of 276 amino acid residues of known sequence (22,23) and one strand of the oligonucleotide TCGCGAATTCGCG (*Eco*RI site underlined). The complex is a two-fold symmetric dimer in which the protein-protein inter-subunit diad, the principal diad of the symmetric DNA double helix and the crystallographic two-fold axis all coincide (16). The molecular boundary, as seen in the Pt-ISIR electron density map, clearly encloses this complex in a well defined globular stucture, 50Å across, see Figure Three.

The DNA-protein complexes are packed within the crystalline lattice so that the DNA forms a continuous rod parallel to the *c*-axis. The unpaired 5' thymine residues at each end of the double helix appear to be stacked upon each other leading to a continuous series of stacked bases across a crystallographic two-fold axis[*]. Therefore, although the oligonucleotide is thirteen nucleotides long there are fourteen stacked nucleotides per unit cell. While this is consistant with a prediction by Harrison that oligonucleotides fourteen base pairs long should be particularly useful in the growth of DNA-protein co-crystals (8), it should be noted that Harrison's prediction was based on assumptions regarding the structure of the DNA which are violated in our case by the neo-kinking reported below (*ie.* Harrison assumed B-DNA with 10.5 base pairs/turn). The oligonucleotide is actually somewhat larger than the DNA-binding face of the protein. However, its length almost exactly matches the net width of the protein, which tapers slightly at the DNA interface leaving a solvent gap separating the ends of the oligonucleotide from the neighboring protein.

There are three major areas of protein-protein interaction which, together with the DNA-DNA interaction, form the crystalline lattice. First is the subunit-subunit interface within the dimeric complex which contains the determinants of dimer

---

[*]The space group is P321.

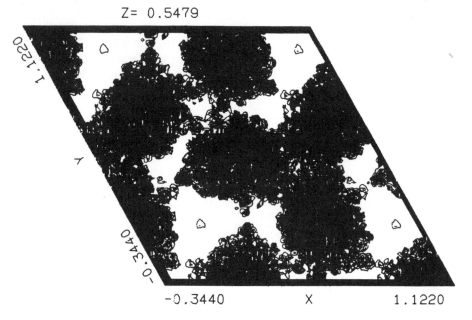

Z= 0.5479

1.1220

-0.3440

-0.3440   X   1.1220

Figure 3. A projection down the *c*-axis of the Pt-ISIR electron density map of the DNA-*Eco*RI endonuclease complex.

formation. (*Eco*RI endonuclease forms highly stable dimers in solution both in the presence and absence of DNA (24,25)). Second is the region around the threefold symmetry axis, where three dimers are tightly packed (see Figure Three). Third is a smaller region of limited protein-protein interactions along the direction of the *c*-axis. These involve contacts between loops at the molecular surface of the protein. The DNA-DNA interaction comprises a significant fraction of the net inter-molecular interactions along the *c*-axis. This confirms the concept that stablity in DNA-protein co-crystals requires compatiblity between the DNA-DNA and protein-protein contacts in the direction of the average DNA helix axis and suggests that the length and terminal sequence of the cognate oligonucleotide should be treated as a critical variable in future attempts to form sequence specific DNA-protein co-crystals.

The conformation of the DNA in this complex departs from classical B-DNA. There are three aburpt dislocations in the helix, termed neo-kinks (see below), which divide the DNA into four blocks of three base pairs each, as in Figure Four. In comparison to the neo-kinks, the helical parameters within each block are more regular. The recognition sequence is contained in the central two blocks of the structure and the flanking sequence, CGC/GCG, is contained in the terminal blocks.

Structural Organization of the Protein Subunit:

*Eco*RI endonuclease is an α/β protein consisting of a five stranded β-sheet surrounded on both sides by α-helices (see Figure Five). Four of the five strands are parallel, however the location of the single anti-parallel strand divides the sheet into parallel

$$\text{T C G C \overset{\lor}{G_*}A A \vdots T T C \overset{\lor}{G} C G}$$
$$\text{G C \underset{\land}{G_*}C T T \vdots A A \overset{*}{G_*}\underset{\land}{C} G C T}$$

Figure 4. The sequence of the synthetic oligonucleotide with the recognition sequence underlined. * denotes the location of phosphodiester bond hydrolysis resulting in a 5′ phosphate. The hydrolysis reaction requires Mg^{+2} as a cofactor. .... denotes the location of the type-1 neo-kink which is coincident with the crystallographic and molecular two-fold symmetry axis. The type-1 neo-kink unwinds the DNA by 25° and introduces a bend between the two central blocks, GAA and TTC, of 12° toward the minor groove. ∧ denotes the location of the asymmetric type-2 neo-kink, which separates the terminal blocks from the central blocks of nucleotide pairs. This kink bends the helical axis by 23°.

and anti-parallel three stranded segments (see Figure Six). Each of these segments forms a sizeable structural unit constructed on a simple three-dimensional pattern in which the physically adjacent elements of secondary structure are essentially contiguous within the primary sequence; *ie.* a sub-domain. It is interesting that the parallel sub-domain is the locale for the direct contacts between the protein and DNA bases as well as subunit-subunit interaction while the anti-parallel sub-domain contains the site of DNA strand scission. We also note that the parallel sub-domain is topologically very similar to one-half of the well known nucleotide binding domain (26). (The nucleotide binding domain is a six-stranded parallel β-sheet, constructed out of two three-stranded subdomains, which are very similar to each other).

Following the course of the polypeptide chain, we find the amino terminus of the polypeptide chain located in an extension of the principal α/β domain of the protein, refered to as the "arm", which wraps around the DNA. The polypeptide chain then forms a long α-helix on the surface of the molecule which is followed by a loop into the first strand of the β-sheet. This β-sheet is formed sequentially starting from the outside of the antiparallel sub-domain see Figure Six. The next loop, which connects the first and second β-strands, contains another α-helix situated on the surface of the molecule. The loop between the second and third (anti-parallel) β-strands projects somewhat into the solvent. The third β-strand is a common element of both the anti-parallel and parallel sub-domains. The parallel sub-domain is formed next, sequentially from the middle of the β-sheet to the fifth strand at the edge of the sheet. The α-helices found at the subunit interface are the crossover helices (27) of the parallel sub-domain, *ie.* those connecting the third β-strand to the fourth and the fourth to the fifth. After exiting the fifth β-strand, the polypeptide chain loops around the surface of the complex, placing the carboxy terminus in the proximity of the DNA backbone.

As can be seen in Figure Five, all the α-helices in the protein are aligned so that their amino terminal ends are pointing toward the DNA. This orients the α-helix dipoles (28) so that they interact favorably with the electrostatic field generated by the negatively charged phosphates on the DNA backbone. The two crossover helices of the parallel subdomains are actually oriented so that their amino terminal

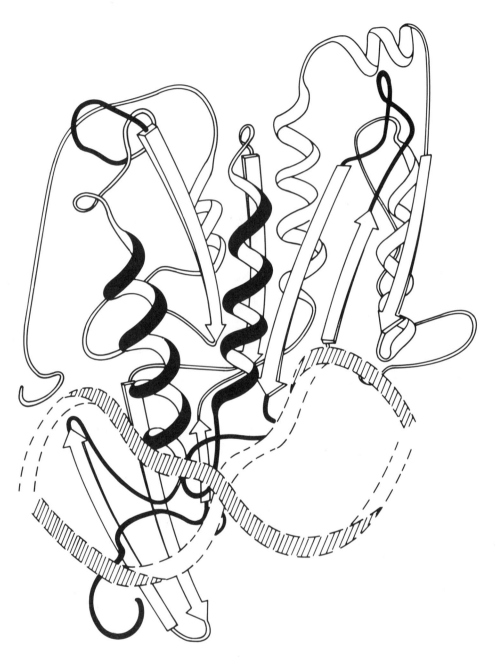

Figure 5. Schematic backbone drawing of one subunit of (dimeric) *Eco*RI endonuclease and both strands of the DNA in the complex. The arrows represent β-strands, the coils represent α-helices and the ribbons represent the DNA backbone. The helices in the forground of the diagram connect the third β-strand to the fourth and the fourth to the fifth They also interface with the other subunit. The amino-terminus of the polypeptide chain is in the arm near the DNA.

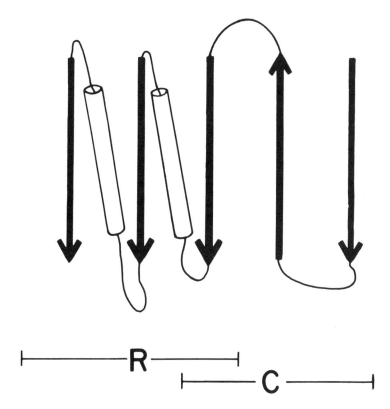

Figure 6. Topology diagram of the major α/β domain of *Eco*RI endonuclease. The β-sheet is divided into two overlapping topological segments, the parallel and anti-parallel sub-domains which corresponds to the functional division of the β-sheet into a sub-domain primarily responsible for recognition of the specific DNA sequence, R, and a subdomain primarily involved in catalytic activity, C.

ends project into the major groove of the DNA. The amino acid side chains which interact with the DNA bases are located at the ends of these helices.

The β-sheet exhibts the conventional twist (29,30,31,32,33,27) with the individual β-strands approximately perpendicular to the DNA helical axis.

The "arm" is an extension of the α/β domain (see Figure Seven) which wraps around the DNA partially encircling it, thereby clamping it into place on the surface of the enzyme. Due to the two-fold symmetry of the complex there are two arms, each of which interacts with the DNA directly across the double stranded helix from the scissile bond. Jen-Jacobson has demonstrated that these non-specific contacts between the arm and the DNA required for DNA cleavage by selective proteolysis in which portions of the arm are selectively removed (34). Many of the resulting "deleteion derivatives" retain sequence specific DNA binding but lack strand scission capability.

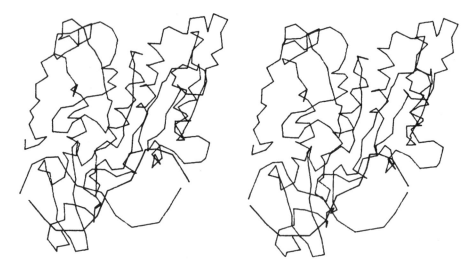

Figure 7. Stereo drawing of the C_α coordinates of one subunit of *Eco*RI endonuclease and the phosphorous coordinates of the double stranded DNA. This drawing was generated with the program FRODO.

Structurally, the arm has a structural "identity" of its own. It is composed of the amino terminus of the protein and a β-ribbon sequentially located between the fourth and fifth strands of the primary β-sheet. (A β-ribbon is a hairpin structure consisting of two anti-parallel β-strands connected by a tight turn.) Part of the amino terminal portion of the arm adds a third β-strand so that the structural foundation of the arm is a three stranded antiparallel β-sheet. Thus, there are two β-sheets in each *Eco*RI endonuclease subunit; the primary five-stranded sheet described above and the subsidiary three-stranded sheet described here. The first fourteen amino acid residues of the polypeptide chain form an irregular structure which is sandwiched in between the subsidiary β-sheet and the DNA; many of the non-specific DNA-protein contacts mediated by the arm are located here. Additional DNA backbone contacts are located in the short segment of polypeptide chain which connects the third subsidiary β-strand with the α-helix which follows it in the primary sequence (this α-helix is the "outer recognition module" described below).

Even though the arm has the structural features described above, it is not a domain, as defined by Richardson (27). It does not appear to have a fully developed hydrophobic core and it is composed of two passes of the polypeptide chain, rather than a single chain segment. Indeed, it is doubtful that it could assemble or maintain its tertiary structure in the absence of the principal domain. It is therefore an extension on the principal domain, but one which has an important functional role.

Structural Features of the DNA:

We reported (15) that the DNA conformation in the recognition complex departs from the B-motif in a way which suggests that the DNA is now adopting conformations which would be unstable in the absence of protein. The most striking of these departures are kinks which occur every three base pairs, as summarized in Figure Four.

The term "kink" has not been used consistently in the literature: We have been follwing the useage originally conceived by Crick and Klug (35), who defined a kink as an abrupt change in helical properties. They specifically included not only sharp bends (abrupt changes in the direction of the helix axis) but also lateral displacements (slip dislocations) of the helix axis, highly localized under- or over-windings (torsional dislocations) and combinations of these elements. They proposed a specific model for the wrapping of DNA around nucleosomes which invoked severe kinks in which the base pairs became unstacked at the kink. Subsequently, the term kink has also been used to describe localized unstacking (see the review by Saenger (36) for several examples). Our structure strongly suggests that highly localized changes in helical parameters are likely to be a significant feature of nucleic acid-protein interactions even though they may not involve actual unstacking of adjacent base-pairs. A word is therefore required to describe any abrupt change in helical parameters and we feel that "kink" is the most appropriate term. It should be noted that while there is no absolute descrimination between a "sharp bend" and a "kink" we would prefer to use the term "bend" to refer to a structure in which helical parameters are changed smoothly over several base-pairs and "kink" to refer to one in which the parameters change abruptly and/or irregularly over one or two base-pairs. In both cases, it is expected that the DNA (or RNA) be approximately helical on either side of the kink or bend.

It is now known that some DNA sequences are intrinsically bent, even in the absence of proteins (37) whereas the kinks we have noted in the DNA-*Eco*RI endonuclease complex only occur in the presence of protein (the oligonucleotide used in our co-crystals is virtually identical to that studied by Dickerson and colleagues (38,39,40), which was not kinked in their structures). The intrinsic bends and kinks are probably structurally different from those which require the binding of a specific protein and these two situations are certainly thermodynamically distinct. We feel that it is important that our terminology reflect these differences and refer to the intrinsically stable kinks and bends as such (or simply as kinks and bends). Those which require a specific binding protein are termed neo-kinks and neo-bends. Thus, Richmond *et al.* observed that the DNA in their nucleosome structure contained "sharp bends" and/or possible kinks (41) which would be neo-bends in our terminology.

We feel that one of the more intriguing observations to emerge from the *Eco*RI endonuclease structure is that the repertoire of conformational states intrinsically accessible to DNA has been expanded to include additional "neo-conformations" which are stabilized by the binding of a protein. Specifically, we define a neo-conformation as a structural distortion which is imposed on the double helix by a binding protein and which is not seen in the absence of protein[*]. (This should not

---

[*]This is based on the definitions of neo as "in a new or different form or manner" and "new chemical compound isomeric with or otherwise related to (such) a compound" (42).

be taken to exclude the possibility that thermally transient fluctuations in DNA structure would include neo-conformations in the absence of protein. Indeed, fluctuations of this sort may well be important intermediates in the formation of DNA-protein complexes. However the bulk of DNA molecules at any instant would not be in a neo-conformation according to our definition unless they were bound to a protein). We suspect that neo-conformations will be a general feature of many nucleic acid-protein interactions. We also suspect that there will be a finite number of well defined, structurally feasible neo-conformations, analogous to the set of tight turns enumerated for protein structure (see, for example, referece (27). Neo-conformations may also have a role in sequence specificity because some sequences may accept the distortion imposed by the protein more readily than other sequences.

The neo-kink which coincides with the crystallographic and molecular two-fold axis will be called a "type-1 neo-kink" (refered to in the previous work (15) as a neo-1 kink). It is located between the two (symmetry equivalent) central blocks of DNA ie. the GAA and TTC blocks (The blocks are refered to by the sequence along one strand of the DNA). The type-1 neo-kink effectively rotates the entire GAA double helical block with respect to the TTC block. It is possible to imagine the kink as a relative twisting of the GAA and TTC blocks about the average helix axis so as to unwind the DNA by about 25°. This corresponds very well with measurements of the unwinding of supercoiled plasmid DNA in solution by Kim and coworkers (43) who obtained a value of 25° per enconuclease dimer bound in the absence of Mg^{+2}. The principle effect of this twisting motion is that the major groove becomes wider even though the bases are not unstacked across the neo-kink (however, there are significant displacements in the base pair planes). The phosphate-phosphate distance across the major groove is increased by aproximately 3.5Å. The type-1 neo-kink also introduces a small bend of approximately 12° between the two central blocks toward the minor groove, away from the protein. The effect of a type-1 neo-kink on B-DNA can be seen in Figure Eight, where a single type-1 neo-kink was placed in between segments of DNA which have the helical parameters associated with "standard" B DNA ie. 10.3 residues/turn and 3.2Å/residue).

The type-2 neo-kinks are different from the type-1 in several respects. A type-2 neo-kink is asymmetric: The helix axis bends by 23°, however, the net bend consists of rotations about two axes. The first would coincide with the helix diad between the CGC and GAA blocks in unkinked DNA (rotation about this axis is responsible for the asymmetry). The second axis is perpendicular to both the first rotation axis and the average helix axis (which generates a small bend towards the minor groove). The asymmetry of the first rotation is indicated by the fact that the phosphate-phosphate distance on one side (CGC-GAA) of the neo-kink is significantly shorter than the comparable distance on the other side (TTC-GCG, see Figure Four). The type-2 neo-kink does not appear to induce substantial unwinding of the DNA.

Helical symmetry was determined for each block of three base pairs formed by the kinks, by analysis of the geometrical relationships of the centroids of the phosphate and internal deoxyribose moieties, (44). The central blocks exhibited B-type

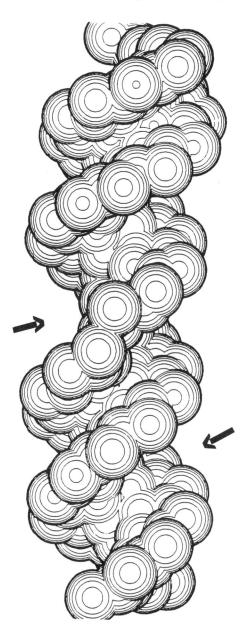

Figure 8. A drawing of double stranded B-DNA with a type-1 neo-kink inserted. The helical parameters used to generate the DNA double helix were the parameters determined for the central block, GAA, of the oligonucleotide bound to *Eco*RI endonuclease in the co-crystals. The arrows indicate the location of the major groove widened by the presence of the type-1 neo-kink and a standard width B-DNA major groove.

parameters with 10.3 residues per turn and and a helical diplacment of 3.2Å. The terminal blocks exhibit A-type parameters of 11.3 residues per turn and a helical displacement of 2.7Å per residue. The 10.3 bp per turn helical repeat of the central blocks is within the values observed for DNA in solution, bound to crystalline surfaces and in nucleosomes (45,46,47,48,49). The A-type parameters are reminiscent of the A-like features of the CGC segment in the dodecamer structure of the same sequence (38). The tendency to wind the double helix with A-like parameters in the terminal blocks should introduce an additional unwinding on the order of 10° into a long DNA molecule (depending on how quickly the DNA reverted to B-type winding).

DNA-Protein Interactions:

The major groove of the recognition hexanucleotide (GAATTC) appears to be filled with protein, forming a large, complementary interface. All the base-amino acid interactions appear to be located in the major groove of the DNA while the edges of the bases which are exposed in the minor groove are open to the solvent. Thus, the direct interaction of complementary surfaces in the major groove is a major determinant of the recognition specificity. However, there is the additional possibility that the energy required to drive DNA into the neo-conformations noted above depends on the base sequence. This would provide an additional, indirect mechanism of sequence recognition which may also contribute to *Eco*RI endonuclease specificity.

A crucial component of the DNA-protein interface is formed by the amino terminal ends of the crossover α-helices in the parallel subdomain. Since the DNA-protein complex possesses two-fold symmetry, α-helices from both subunits participate in the formation of a four-helix bundle which inserts into the major groove of the DNA. One of the roles of the neo-kinks is to make room for this bundle, which would not fit into the major groove of conventional B-DNA.

The α-helix which connects the third and fourth β-strands makes an angle of approximately 60° with the average DNA helix axis as shown in Figure Nine. The poly-peptide chain turns sharply at the end of the α-helix so that the amino acid residues at the amino-terminal end of the helix, those in the bend and in the adjacent stretch of chain are in close proximity to the DNA. This α-helix is also adjacent to the molecular two-fold axis and the amino terminus of the helix is physically adjacent to the amino terminus of the symmetry related helix from the other subunit. These two α-helices form a symmetric module which is responsible for the direct interactions between the endonuclease and the bases of the inner tetranucleotide (AATT). This structural unit will be refered to as the inner recognition module. Two symmetry related pairs of amino acid side chains make contact with two symmetry related pairs of sequential adenine residues. Interestingly, each pair of adjacent adenines interact with one amino acid from each subunit, see Figure Ten. In our current interpretation of the electron density map, these residues are glutamic acid 144 and arginine 145. The position of the arginine side chain in our current model is consistent with a hydrogen bonding interaction with the N7 moieties of adjacent adenines while the glutamic acid is probably hydrogen bonding to the exocyclic N6 groups of both adenines.

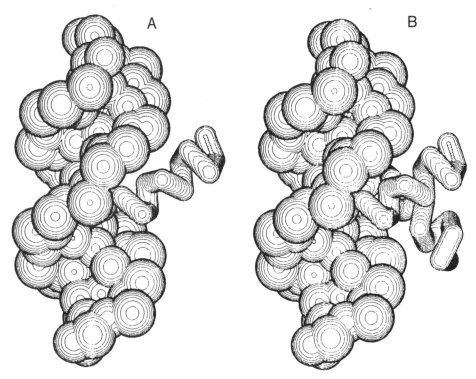

Figure 9. A). The double stranded DNA in the *Eco*RI endonuclease-DNA complex and the α-helix which connects the third and fourth β-strands. B). The same view as in A) with the symmetery related α-helix from the other subunit included. The two α-helices form the inner module, which recognizes the inner tetranucleotide, AATT.

Seperate α-helical modules are responsible for the direct contacts between the protein and the two outermost G-C pairs of the canonical hexanucleotide, GAATTC. These modules are called the outer recognition modules. They are identical by virtue of the two-fold symmetry and each independent module consists of the crossover α-helix which connects the fourth and fifth strands of the principal β-sheet. This helix has many interactions with both α-helices of the inner recognition module and is thereby positioned so that it projects its amino terminus into the major groove of the DNA. At this stage of our analysis, it appears likely that arginine 200 interacts with the guanine in a manner predicted by Seeman, Rosenberg and Rich (50).

In addition to the recognition of and precise requirement for the canonical site, GAATTC, *Eco*RI endonuclease also exhibits a dependence on the sequence of the nucleotides flanking this hexanucleotide (51,52,53). These workers noted that the flanking sequence environment can effect the overall rate of hydrolysis by at least an order of magnitude. Furthermore, Modrich and coworkers showed that *Eco*RI endonulcease can dissociate from DNA after making only a single strand nick and that the frequency of such nicking depended on the flanking sequences (54). It is unlikely that these effects are due to direct contacts between the enzyme and bases

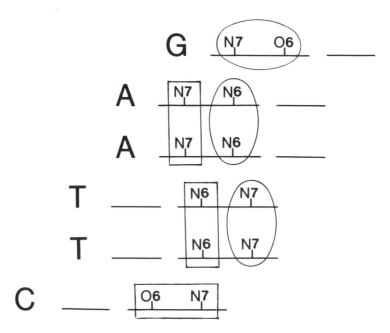

Figure 10. A schematic drawing depicting the interactions between base pairs in the *Eco*RI recognition site and amino acid side chains of *Eco*RI endonuclease. Rectangles denote interactions from one subunit and ovals denote interactions from the symmetery related subunit. The proposed interactions are arginine 200 hydrogen bonding to the N7 and O6 of the GC base pairs, arginine 145 hydrogen bonding to the two N7 moieties of adjacent adenines and glutamic acid 144 hydrogn bonding to the N6 groups of adjacent adenines.

outside of the canonical hexanucleotide because we have not noted any such interactions in our electron density maps. However, there are extensive contacts between the enzyme and the DNA backbone which extend well beyond the hexanucleotide. This suggests that the conformational free energy of the type-2 neo-kink and/or the A-like terminal block depends on the sequence of oligonucleotides immediately flanking the *Eco*RI site.

There appear to be interactions between the protein and the backbone of the DNA from the second through the nineth phosphates on each strand. Counting from the 5′ end of the oligonucleotide, TpCpGpCpGpApApTpTpCpGpCpGp, the third, fourth and seventh phosphates are buried in the protein, *ie.* they appear to be inaccessible to solvent. The remaining phosphates in the indicated region interact with the enzyme even though they are partially exposed to solvent. The third, fourth and fifth phosphates are bound in a large cleft in the protein which forms the active site for DNA hydrolysis. This cleft is partially open to solvent in the vicinity of the fifth phosphate, where the scissile bond is located. It is through this channel that magnesium probably enters the active site[*].

[*]We have recently demonstrated that Mg^{+2} can be perfused into the crystals and the hydrolytic reaction carried out *in situ*. (J. Picone, manuscript in preparation). The enzyme-product co-crystals still diffract X-rays and their structure analysis is in progress.

The cleft surface contains many basic amino acid residues which interact electro-statically with the phosphates. These interactions contribute to the overall stability of the complex. There are, of course, two identical clefts in the surface of the two-fold symmetric enzyme. They are too far apart to fit regular B-DNA and we suspect that this separation promotes the formation of the type-1 neo-kink via long range electrostatic attraction between the basic clefts and phosphates on the incoming DNA molecule.

The contacts noted above in the X-ray structure of the tridecamer-protein complex probably contain all of the major DNA-protein contacts between the endonuclease and larger natural DNA substrates. The association constant measured for the dodecamer, CGCGAATTCGCG, is within experimental error of that measured for plasmid DNA (25,23,55,56). The lower association constant for an octanucleotide substrate as compared with dodecameric or larger substrates (57,25) suggests strongly that interactions between the enzyme and the flanking regions of the DNA backbone make sigificant contributions to the net stability of the complex.

Inferred and observed contact points:

There is good general agreement between the results of ethylation interference experiments (58) and the phosphates contacts described previously. The largest effects observed by Lu *et al.* correspond to the third, fourth and seventh phosphates, which as noted above are buried in the protein and protected from solvent. The next largest effect is observed for the reactive phosphate at the fifth position. Small effects are noted for the sixth phosphate, which is probably forming interactions to the protein even though it is partially exposed to the solvent. (We suspect that a stronger ethylation interference would have been observed at lower protein concentrations where the equilibrium is more sensitive to smaller reductions in the protein-DNA association constant.

Methylation protection and interference experiments implicated both the major groove and the minor groove as points of DNA-protein contact. It is clear that the predicted contacts in the major groove of the recognition sequence is in good agreement with the X-ray structural data. The implications of the minor groove data must be reevaluated in light of the structural results. The N3 positions on the central adenine, in the minor groove, were protected from dimethylsulphate by endonuclease and prior methylation at the N3 blocked subsequent binding of the enzyme. Since there is no density observed in the minor groove of the DNA, and there are no sections of polypeptide chain left unaccounted for in the chain tracing of the protein, the observed effects at the N3 are probably related to the conformational changes induced in the DNA by protein binding.

Another approach used to identify contacts on the DNA is based on the hydrogen bonding degeneracy rules of Seeman *et al.* (50,59,60). In *Eco*RI endonuclease such degeneracy can be induced by conditions of elevated pH (8-9.5), Mn^{+2}, low ionic strength and by the addition of organic compounds such as glycerol or ethylene glycol (59,61,62,63,64,65). This has been called the *Eco*RI* reaction (61) and the non-canonical sites so recognized are referred to as *Eco*RI* sites. We previously showed that hierarchies of base preferences could be discerned by qualitative estimations of the rate of cleavage at *Eco*RI* sites in plasmid DNA molecules. These observed hierarchies are consistant with a recognition model which correctly predicted the specific contacts between the protein and DNA we subsequently observed in the X-ray structure. The original model assumed two major groove hydrogen bonds per base pair under standard conditions (60), which turned out to be those identified above. An additional assumption was that one or more of these hudrogen bonds are randomly replaced by a water-DNA hydrogen bond under *Eco*RI* conditions. The position of any *Eco*RI* sequence within the appropriate hierarchy is determined by the number of remaining protein-DNA hydrogen bonds (60). Woodbury *et al.* (59) reached similar conclusions from a somewhat different perspective. This method of "degeneracy analysis" therefore appears to be a promising method for the elucidation of DNA-protein contact points.

Structural Suggestions for Conformational Mobility:

The formation of the *Eco*RI endonuclease-DNA complex requires conformational changes in both substrate DNA and the enzyme in order to acheive specific binding. The neo-kinks have already been discussed. The protein must also alter it's conformation during the binding event because the arms encircle the DNA to such an extent that it appears unlikely that substrate DNA could enter the active site in the absence of some movement. There are three possibilities: The arms may be "rigidly" mobile, retaining their structure structure while they move with respect to the rest of the molecule; the arms, or part of them may undergo an order-disorder transition in which they are disordered in the absence of DNA and condense on it during complexation and/or the dimeric endonuclease could undergo a quaternary conformational change in which the subunits move with respect to each other. We also have crystals of the protein in the absence of DNA and that structure is in progress.

There is probably an intermediate conformational state, *viz.* that of the non-specific DNA-protein complex, which presumably appears as the uncomplexed DNA and protein and progress to the specific DNA-protein complex seen here. *Eco*RI endonuclease certainly binds non-specifically to DNA of random sequence and it has been suggested that this binding accelerates formation of the specific complex via facilitated diffusion of the protein along the DNA.

Conclusions

*Eco*RI endonuclease specifically binds the canonical sequence, GAATTC, through DNA-protein inteactions in the major groove of the DNA. The minor groove of the canonical sequence is not directly involved in sequence specificity. Upon binding to the endonuclease, the DNA adopts new (neo) conformational states not previously seen in protein-free DNA. We have defined a neo-conformation as a structural distortion which is imposed on DNA by a binding protein and which is not seen in the absence of protein. Two neo-kinks have been observed in this structure, the type I neo-kink and the type II neo-kink. The type I neo-kink unwinds the DNA by 25° and renders the major groove accessible to the protein. The type II neo-kink's major effect is to bend the DNA which also is required for protein accessibility and may have a role in the hydrolytic activity of the protein.

EcoRI endonuclease is a dimer with identical, symmetry related subunits. Each subunit is an α/β domain with an extension, a pseudo-domain called the "arm". The α/β domain is organized into topological sub-domains which have identifiable functional roles. The 3-stranded parallel subdomain of the β-sheet is associated with sequence recognition and the subunit interface. The 3-stranded anti parallel subdomain of the β-sheet is associated with phosphodiester bond cleavage. The two segments overlap to form a 5-stranded β-sheet. The "arms" which embrace the DNA and clamp it into place are based on a subsidiary three-stranded anti-parallel sheet.

DNA sequence recognition is broken down into modular elements based on the cross-over α-helices in the α/β domains. The inner recognition module which recognizes the sequence AATT consists of two symmetry related α-helices, one from each subunit. This inner recognition module appears to hydrogen bond with the adenines in the four adjacent AT base pairs. The outer bases pairs at either end of the canonical sequence are recognized by the two outer recognition modules. Each outer module consists of one α-helix which appears to hydrogen bond to the guanines.

Acknowledgements

This work was supported by NIH grant GM25671 (JMR).

References and Footnotes

1. Andreson, W.F., Ohlendorf, D.H., Takeda, Y., and Matthews, B.W., "Structure of the *cro* Repressor from Bacteriophage λ and its Interaction with DNA", *Nature, 290,* 754-758 (1981).
2. Anderson, W.F., Takeda, Y., Ohlendorf, D.H. and Matthews, B.W., "Proposed α-helical Super-secondary Structure Associated with Protein-DNA Recognition", *J. Mol. Biol., 159,* 745-751 (1982).
3. Pabo, C.O. and Lewis, M., "The Operator-Binding Domain of λ Repressor: Structure and DNA Recognition", *Nature, 298,* 443-447 (1982).
4. McKay, D.B. and Steitz, T.A., "Structure of Catabolite Gene Activator Protein at 2.9 Å Resolution Suggests Binding to Left-Handed B-DNA", *Nature, 290,* 744-749 (1981).
5. Steitz, T.A., Ohlendorf, D.H., McKay, D.B., Anderson, W.F. and Matthews, B.W., "Structural Similarity in the DNA-binding Domains of Catabolite Gene Activator and *cro* Repressor Proteins", *Proc. Natl. Acad. Sci. USA, 79,* 3097-3100 (1982).
6. Ohlendorf, D.H., Anderson, W.F., Fisher, R.G., Takeda, Y. and Matthews, B.W., "The Molecular Basis of DNA-Protein Recognition Inferred From the Structure of *cro* Repressor", *Nature, 298,* 718-723 (1982).
7. Sauer, R.T., Yocum, R.R., Doolittle, R.F., Lewis, M. and Pabo, C.O., "Homology Among DNA-Binding Proteins Suggests Use of a Conserved Super-Secondary Structure", *Nature, 298,* 447-451 (1982).
8. Anderson, J., Ptashne, M., and Harrison, S.C., "Cocrystals of the DNA-binding domain of phage 434 repressor and a synthetic phage 434 operator", *Proc. Natl. Acad. Sci. USA, 81,* 1307-1311 (1984).

9. Anderson, J.E., *The 7Å Structure of a 434 Repressor-Operator Complex,* PhD dissertation, Harvard University (1984).

10. Ollis, D.L., Brick, P., Hamlin, R., Xuong, N.G. and Steitz, T.A., "Structure of large fragment of *Escherichia coli* DNA polymerase I complexed with dTMP", *Nature, 313,* 762-766 (1985).

11. Richmond, T.J., Finch, J.T., Rushton, B., Rhodes, D. and Klug, A., "Structure of the Nucleosome Core Particle at 7Å Resolution", *Nature, 311,* 532-537 (1984).

12. Bentley, G.A. and Lewit-Bentley, A., "Crystal Structure of the Nucleosome Core Particle at 16Å Resolution", *J. Mol. Biol., 176,* 55-75 (1984).

13. Burlingame, R.W., Love, W.E., Wang, B-C., Hamlin, R., Xuong, N.H., and Moudrianankis, E.N., "Crystallographic Structure of the Octameric Histone Core of the Nucleosome at a Resolution of 3.3Å", *Science, 228,* 546-553 (1985).

14. Tanaka, I., Appelt, K., Dijk, J., White, S.W., and Wilson, K.S., "3Å resolution structure of a protein with histone-like properties in prokaryotes", *Nature, 310,* 376-381 (1984).

15. Frederick, C.A., Grable, J., Melia, M., Samudzi, C., Jen-Jacobson, L., Wang, B.-C., Greene, P.J., Boyer, H.W. and Rosenberg, J.M., "Kinked DNA in Crystalline Complex with *Eco*RI Endonuclease", *Nature, 309,* 327-331 (1984).

16. Grable, J., Frederick, C.A., Samudzi, C, Jen-Jacobson, L., Lesser, D., Greene, P.J., Boyer, H.W., Itakura, K. and Rosenberg, J.M., "Two-Fold Symmetry of Crystalline DNA-*Eco*RI Endonuclease Recognition Complexes", *Journal of Biomolecular Structure and Dynamics, 1,* 1149-1160 (1984).

17. Rossmann, M.G., "Processing Oscillation Diffraction Data for Very Large Unit Cells with an Automatic Convolution Technique and Profile Fitting", *J. Appl. Crystallogr., 12,* 225-238 (1979).

18. Rossmann, M.G., Leslie, A.G.W., Abdel-Meguid, S.S. and Tsukihara, T., "Processing and Post-Refinement of Oscillation Camera Data", *J. Appl. Crystallogr., 12,* 570-581 (1979).

19. Jones, T.A., "A Graphics Model Building and Refinement System for Macromolecules", *J. Appl. Crystallogr., 11,* 268-272 (1978).

20. Jones, T.A., "FRODO: A Grahics Fitting Program for Macromolecules", in *Computational Crystallography,* Sayre, D., ed., Clarendon Press, 303-317 (1982).

21. Pflugrath, J.W., Saper, M.A., and Quiocho, F.A., "New generation graphics system for molecular modeling", in *Papers presented at the International Summer School on Crystallographic Computing, Kyoto Japan,* ?, (1983).

22. Greene, P.J., Gupta, M., Boyer, H.W., Brown, W.E. and Rosenberg, J.M., "Sequence Analysis of the DNA Encoding the *Eco*RI Endonuclease and Methylase", *J. Biol. Chem., 256,* 2143-2153 (1981).

23. Newman, A.K., Rubin, R.A., Kim, S.-H. and Modrich, P., "DNA Sequences of Structural Genes for *Eco*RI DNA Restriction and Modification Enzymes", *J. Biol. Chem., 256,* 2131-2139 (1981).

24. Modrich, P. and Zabel, D., "*Eco*RI Endonuclease: Physical and Catalytic Properties of the Homogeneous Enzyme", *J. Biol. Chem., 251,* 5866-5874 (1976).

25. Jen-Jacobson, L., Kurpiewski, M., Lesser, D., Grable, J., Boyer, H.W., Rosenberg, J.M. and Greene, P.J., "Coordinate Ion Pair Formation Between *Eco*RI Endonuclease and DNA", *J. Biol. Chem., 258,* 14638-14646 (1983).

26. Rossman, M.G., Liljas, A., Branden, C.-I., and Banaszak, L.J., "Evolutionary and Structural Relationships among Dehydrogenases", in *The Enzymes,* Boyer, P., ed., 3, Vol. 11, 61-102 (1975).

27. Richardson, J.S., "The Anatomy and Taxonomy of Protein Structure", *Adv. in Protein Chem., 34,* 167-339 (1981).

28. Hol, W.G.S., "The Role of The α-Helix Dipole in Protein Function and Structure", *Prog. Biophys. Molec. Biol., 45,* 149-195 (1985).

29. Chothia, C., "Conformation of Twisted β-Pleated Sheets in Proteins", *J. Mol. Biol., 75,* 295-302 (1973).

30. Quiocho, F.A., Gilliland, G.L., and Phillips, G.N., "The 2.8Å Resolution Structure of the L-Arabinose-binding Protein from Escherichia coli", *J. Biol. Chem., 252,* 5142-5149 (1977).

31. Shaw, P.S. and Muirhead, H., "Crystallographic Structure Analysis of Glucose 6-Phosphate Isomerase at 3.5Å Resolution", *J. Mol. Biol., 109,* 475-485 (1977).

32. Weatherford, D.W. and Salemme, F.R., "Conformations of Twisted Parallel β-Sheets and the Origin of Chirality in Protein Structures", *Proc. Natl. Acad. Sci. USA, 76,* 19-23 (1979).

33. Schulz, G.E., Elzinga, M., Marx, F., and Schirmer, R.H., "Three dimensional structure of adenyl kinase", *Nature, 250,* 120-123 (1974).

34. Jen-Jacobson, L., Lesser, D., and Kurpiewski, M., "The Mobile Arms of *Eco*RI Endonuclease: Role in DNA Binding and Cleavage", *Nature,* 000-000 Submitted.

35. Crick, F.H.C., and Klug, A., "Kinky Helix", *Nature, 255,* 530-533 (1975).

36. Saenger, W., *Principles of Nucleic Acid Structure,* Springer-Verlag (1984).

37. Wu, H., and Crothers, D.M., "The locus of sequence-directed and protein-induced DNA bending", *Nature, 308,* 509 (1984).

38. Dickerson, R.E., and Drew, H.R., "Structure of a B-DNA Dodecamer: II. Influence of Base Sequence on Helix Structure", *J. Mol. Biol., 149,* 761-786 (1981).

39. Dickerson, R.E., "Base Sequence and Helix Structure Variation in B- and A-DNA", *J. Mol. Biol., 166,* 419-441 (1983).

40. Dickerson, R.E., and Drew, H.R., "Kinematic Model for B-DNA", *Proc. Natl. Acad. Sci. USA, 78,* 7318-7322 (1981).

41. Richmond, T.J., Finch, J.T., Rushton, B., Rhodes, D., and Klug, A., "Structure of the nucleosome core particle at 7 Å resolution", *Nature, 311,* 532-537 (1985).

42. Gove, P.B., *Webster's Seventh New Collegiate Dictionary,* G. & C. Merrian Co., (1963).

43. Kim, R., Modrich, P. and Kim, S.-H., "Interactive Recognition in *Eco*RI Restriction Enzyme-DNA Complex", *Nucl. Acids. Res., 12,* 7285-7292 (1984).

44. Rosenberg, J.M., Seeman, N.C., Day, R.O. and Rich, A., "RNA Double Helices Generated from Crystal Structures of Double Helical Dinucleoside Phosphates", *Biochem. Biophys. Res. Commun., 69,* 979-987 (1976).

45. Wang, J.C., "Helical Repeat of DNA in Solution", *Proc. Natl. Acad. Sci. USA, 76,* 200-203 (1979).

46. Peck, L.J. and Wang, J.C., "Sequence Dependence of the Helical Repeat of DNA in Solution", *Nature, 292,* 375-378 (1981).

47. Rhodes, D. and Klug, A., "Sequence-Dependent Helical Periodicity of DNA", *Nature, 292,* 378-380 (1981).

48. Rhodes, D. and Klug, A., "Helical Periodicity of DNA Determined by Enzyme Digestion", *Nature, 286,* 573-578 (1980).

49. Lutter, L.C., "Precise Location of DNase I Cutting Sites in the Nucleosome Core Determined by High Resolution Gel Electrophoresis", *Nucl. Acids. Res., 6,* 41-56 (1979).

50. Seeman, N.C., Rosenberg, J.M. and Rich, A., "Sequence-Specific Recognition of Double Helical Nucleic Acids by Proteins", *Proc. Natl. Acad. Sci. USA, 73,* 804-808 (1976).

51. Thomas, M. and Davis, R.W., "Studies on the Cleavage of Bacteriophage Lambda DNA with *Eco*RI Restriction Endonuclease", *J. Mol. Biol., 91,* 315-328 (1975).

52. Halford, S.E. and Johnson, N.P., "The *Eco*RI Restriction Endonuclease With Bacteriophage λ DNA", *Biochem. J., 191,* 593-604 (1980).

53. Alves, J., Pingoud, A., Haupt, W., Langowski, J., Peters, F., Maass, G. and Wolff, C., "The Influence of Sequences Adjacent to the Recognition Site on the Cleavage of Oligonucleotides by the *Eco*RI Endonuclease", *Eur. J. Biochem., 140,* 83-92 (1984).

54. Jack, W.E., Terry, B.J. and Modrich, P., "Involvement of Outside DNA Sequences in the Major Kinetic Path by Which *Eco*RI Endonuclease Locates and Leaves its Recognition Sequence", *Proc. Natl. Acad. Sci. USA, 79,* 4010-4014 (1982).

55. Lillehaug, J.R., Kleppe, R.K. and Kleppe, K., "Phosphorylation of Double-Stranded DNAs by T4 Polynucleotide Kinase", *Biochemistry, 15,* 1858-1865 (1976).

56. Jack, W.E., Rubin, R.A., Newman, A. and Modrich, P., "Structures and Mechanisms of *Eco*RI DNA Restriction and Modification Enzymes", in *Gene Amplification and Analysis Volume I: Restriction Endonucleases,* J.G. Chirikjian, ed., Elsevier/North-Holland, 165-179 (1981).

57. Greene, P.J., Poonian, M.S., Nussbaum, A.L., Tobias, L., Garfin, D.E., Boyer, H.W. and Goodman, H.M., "Restriction and Modification of a Self-complementary Octanucleotide Containing the *Eco*RI Substrate", *J. Mol. Biol., 99,* 237-261 (1975).

58. Lu, A-L., Jack, W.E., and Modrich, P., "DNA Determinants Important in Sequence Recognition by *Eco*RI Endonuclease", *J. Biol. Chem., 256,* 13200-13206 (1981).

59. Woodbury, C.P., Jr., Hagenbuchle, O. and von Hippel, P.H., "DNA Site Recognition and Reduced Specificity of the *Eco*RI Endonuclease", *J. Biol. Chem., 255,* 11534-11546 (1980).

60. Rosenberg, J.M. and Greene, P.J., "*Eco*RI* Specificity and Hydrogen Bonding", *DNA, 1,* 117-124 (1982).

61. Polisky, B., Greene, P., Garfin, D.E., McCarthy, B.J., Goodman, H.M. and Boyer, H.W., "Specificity of Substrate Recognition by the *Eco*RI Restriction Endonuclease", *Proc. Natl. Acad. Sci. USA, 72,* 3310-3314 (1975).

62. Hsu, M. and Berg, P., "Altering the Specificity of Restriction Endonuclease: Effect of Replacing Mg^{+2} with Mn^{+2}", *Biochemistry, 17,* 131-138 (1978)

63. Malyguine, E., Vannier, P. and Yot, P., "Alteration of the Specificity of Restriction Endonucleases in the Presence of Organic Solvents", *Gene, 8,* 163-177 (1980).

64. Woodhead, J.L., Bhave, N. and Malcolm, A.D.B., "Cation Dependence of Restriction Endonuclease *Eco*RI Activity", *Eur. J. Biochem., 115,* 293-296 (1981).

65. Gardner, R.C., Howarth, A.J., Messing, J, and Shepherd, R.J., "Cloning and sequencing of restriction fragments generated by *Eco*RI*", *DNA, 1,* 109-115 (1982).

*Biomolecular Stereodynamics III, Proceedings of the Fourth Conversation in the
Discipline Biomolecular Stereodynamics, State University of New York,
Albany, NY, June 04-09, 1985, Eds., Ramaswamy H. Sarma & Mukti H. Sarma,
ISBN 0-940030-14-4, Adenine Press, ©Adenine Press 1986.*

Recognition of Nucleic Acid Structure
by Monoclonal Autoantibodies Reactive With B-DNA
and a Monoclonal Anti-GMP Antibody

B. David Stollar, Bao-Shan Huang (1), **and Michael Blumenstein** (1)
Department of Biochemistry and Pharmacology
Tufts University Health Science Campus
Boston, MA 02111, USA

Abstract

Native B-DNA is not an effective immunogen in normal animals, but antibodies that react with native B-DNA do occur in sera of patients with autoimmune disease, (systemic lupus erythematosus), in both humans and certain strains of mice. Monoclonal examples of these antibodies have been isolated. Most react preferentially with denatured DNA, but some examples show preference for the native form. They distinguish surprisingly among helices of different base composition. One monoclonal antibody, H2, reacts with poly(dA-dT) and only very poorly with poly(dG-dC), whereas another, H241, shows the reverse selectivity. Electron microscopy confirmed that H241 binds to closed circular native DNA. Its order of reactivity with helical nucleic acids is: poly(dG-dC)(B-form) $\gg$ Native DNA = br-poly(dG-dC)(Z-DNA) > poly(dG-dT)•poly(dA-dC) = poly(dG)•poly(dC) $\gg$ poly (dG-dA)•poly(dT-dC); it did not react detectably with poly(dA-dT). This monoclonal antibody also shows some preference in binding for certain sites within closed circular DNA, as reflected in protection of these sites from restriction nuclease cleavage. As H241 shows selectivity that depends on base sequence, it must either recognize subtle differences in the shapes and outer features of these helices or make contact with portions of the bases, probably in the major groove of the DNA. Competitive immunoassays with phospholipids and oligonucleotides indicate that binding may involve both the backbone and the bases.

To probe the interactions between antibody and nucleic acid more precisely, we are exploring the application of NMR spectroscopy. As a simple system to test for possible interactions of the base, sugar and phosphate with antibody, a monoclonal antibody to the nucleotide GMP was prepared. Free nucleotide or antibody and mixtures of GMP and antibody were studied with both phosphorus and proton NMR. Results indicated that the phosphate of GMP did not interact directly with this antibody's combining site. Binding of GMP did lead to a slight chemical shift change for the C-8 proton, but a significant base stacking between the purine ring and an aromatic amino acid on the antibody was ruled out.

Introduction

The functions of a large number of proteins require their recognition of specific sites in DNA. Some of this specificity arises from contacts of proteins with portions of bases accessible in the major groove of DNA (2). Other interactions may depend on recognition of local conformational variations in the helical nucleic acid structure. Increasing interest in these mechanisms has arisen from descriptions of dramatic structural variations such as the B-DNA to Z-DNA transition (reviewed recently by Rich et al. (3)), fluctuations in B-helix geometry in fibers of synthetic polydeoxyribo-nucleotides (4,5); and more subtle non-uniformities of adjacent base pairs within crystal structures of helical oligonucleotides (6).

Anti-nucleic acid antibodies serve as non-destructive models for examining protein-nucleic acid interactions. They can distinguish almost qualitatively among members of different helical families (reviewed in 7-10) and quantitatively among different members of one helical class (7,11-13). As their specificities are defined these antibodies become useful analytical probes, serving as sensitive and specific reagents for detecting particular structures in complex biological materials. For example, experimentally induced antibodies to left-handed Z-DNA (14-17), double-stranded RNA (18,19) or DNA-RNA hybrids (20-22) have been able to detect corresponding structures without interference from a large excess of other nucleic acids.

In contrast to these other helical structures, native DNA has not been an effective immunogen in normal animals, but spontaneously arising antibodies characteristic of the autoimmune disease systemic lupus erythematosus (SLE) do react with native DNA (reviewed in 8,9). Monoclonal examples of such antibodies have been obtained from hybridomas derived from autoimmune MRL-1pr/1pr or NZB/NZW mice (24-27) or humans (28). They often cross-react with several polynucleotides, presumably by recognition of features common to all reactive forms (24-28). The question arises whether a homogeneous autoantibody reactive with helical DNA can recognize base sequence dependent structural variations in the helix. This article describes a monoclonal autoantibody that discriminates nearly completely between synthetic poly(dG-dC) and poly(dA-dT) and protects certain restriction nuclease cleavage sites preferentially. A second antibody did not show this selectivity. We also describe NMR experiments with a simpler anti-nucleic acid antibody, a monoclonal anti-GMP. This system is being developed to establish a basis for determining how nucleic acid-protein interactions are reflected in ^{31}P and ^{3}H NMR spectra.

Materials and Methods

Monoclonal autoantibodies. H241 and H143 were monoclonal antibodies derived from autoimmune MRL-1pr/1pr mice, which spontaneously produce anti-DNA antibodies (24,29). Hybridomas were grown in tissue culture in medium supplemented with 15% fetal calf serum, in a 5% CO_2 atmosphere. Antibodies were purified from culture supernatant by affinity chromatography on columns of Sepharose 4B to which purified goat anti-mouse Ig antibody was conjugated. Culture fluid was applied to the column, which was then washed with PBS, borate-buffered saline and PBS again before elution of antibody with 0.1M glycine pH 3. Eluted antibody was dialyzed against PBS.

Preparation of monoclonal anti-GMP antibody and Fab fragments. BALB/c mice were immunized with GMP-hemocyanin conjugates prepared with periodate-oxidized GMP (30). When a strong serum antibody response was established, spleen cells of an immunized animal were fused with Sp2/0-Ag14 non-secreting mouse myeloma cells (24). Hybridomas were grown in selective medium and screened with an enzyme-linked immunoassay (31) for production of antibody that reacted with denatured DNA and was competed specifically by GMP. Cells were subcloned twice by limiting dilution. A positive clone was grown in several lots of 500 ml of tissue culture medium. Antibody was absorbed from medium to a GMP-Agarose affinity column (Sigma Chemical Co.) and eluted with 30 mM GMP. The GMP was removed by dialysis, and antibody was concentrated by membrane filtration. The Fab fragment was prepared by digestion of the antibody at pH 3.5 with pepsin (1% of the amount of antibody) for 8 hr at 37°, neutralization with 3M Tris buffer, and incubation with 20 mM cysteine for 1 hr. The Fab fragment was then concentrated and dialyzed against 0.1 M NaCl. For proton NMR experiments, the Fab fragment was lyophilized and redissolved in D_2O twice.

Nucleic acids and polynucleotides. Calf thymus DNA (Worthington Biochemicals) was purified as described (32). *E. coli* and *Pseudomonas* DNA samples were purified from bacteria in log phase growth as described by Marmur (33). ΦX174 RFI DNA was purchased from New England Biolabs (Beverly, MA). MP8 phage DNA was isolated from phage-infected *E. coli* strain JM 103. ^{3}H-labeled *E. coli* DNA was prepared from a partial thymine-requiring mutant, B-3, labeled with ^{3}H-thymidine during log phase growth (32). Synthetic polynucleotides purchased from P-L Labs were: poly(A), poly(U), poly(C), poly(I), poly(G), poly(A)•poly(U), poly(I)•poly(C), poly(dG-dC), poly(dA-dT), poly(dC-dA)•poly(dT-dG), poly(dG-dA)•poly(dT-dC), and poly(dA)•poly(dT). Poly(A)•poly(dT) was from Collaborative Research. Triple-helical polymers were prepared by mixing of homopolymers in appropriate ratios under annealing conditions (34,35). The left-handed Z-DNA form of poly(dG-dC) was prepared by bromination of poly(dG-dC) in 4M NaCl (36-37).

Radioimmunoassays. Binding of antibody to radioactively labeled DNA was measured with goat anti-mouse Ig as a second step (32). Competitive assays with unlabeled nucleic acids were as described (32).

Analytical nuclease digestion. Closed circular MP8 phage DNA was digested with HaeIII nuclease in 10 mM Tris, 140 mM NaCl with 16 mM $MgCl_2$ in the absence or presence of normal mouse Ig or purified H241 or H143 monoclonal antibody. The digestion products were extracted with phenol-chloroform-isoamyl alcohol (50:48:2), precipitated with ethanol, redissolved in PBS and analyzed by gel electrophoresis in 8% polyacrylamide gels.

Electron microscopy of antigen-antibody complexes. 10 μg of ΦX174 RFII DNA and 120 μg of antibody H241 or H43 or normal mouse IgG were incubated in 750 μ1 of TBS (0.01 M Tris-HCl, 0.14 M NaCl, pH 7.4) for 30 min at room temperature and then applied to a 1.5 $\times$ 27 cm column of Sepharose 4B and washed through with

TBS. A peak appearing just after the void volume contained DNA and bound antibody. A sample of 1-1.5 μg of this DNA was examined by electron microscopy as described (38).

NMR spectroscopy. ^{31}P and ^{1}H NMR spectra were measured with a Brüker AM400 MHz wide bore NMR spectrometer. For ^{31}P NMR, GMP and the monovalent Fab fragment of antibody were mixed in a 10 mm tube in a 1:1 molar ratio at a concentration of 0.22 mM in 2 ml of 0.1 M NaCl in water. The solution contained 0.2 mM EDTA, and D_2O was added to 10% to provide a field frequency lock signal. For ^{1}H NMR, Fab samples were lyophilized, redissolved in D_2O and incubated at 37°C for 1 hr; they were then re-lyophilized and redissolved in 99.996% D_2O (Stohler Isotope Chemicals, Waltham, MA). Fab was used at a concentration of 0.3 mM in 0.1 M NaCl, with 0.4 mM EDTA; GMP in D_2O was added to yield molar ratios of 0.5:1 to 4:1 of GMP:Fab. Spectral measurements were generally made at 23°, 29°, and 35°C. In a saturation transfer experiment, samples were pre-irradiated for 0.2 sec. at frequency intervals of 100 Hz spanning a range of 1000 Hz on either side of the signal from the free GMP. An additional experiment was performed with frequency intervals of 20 Hz spanning 8.65 to 8.05 ppm. The acquisition time for these spectra was 0.819 sec., and the total recycle time was 3 sec.

Results and Discussion

H241 is a murine monoclonal autoantibody that binds well to native DNA but also displays several cross-reactions. It reacts with high concentrations of cardiolipin, as do several other monoclonal autoantibodies (39), probably by recognizing repeated and appropriately spaced phosphates of the phospholipid. In its reaction with polynucleotides, H241 shows more cross-reactivity than do most induced anti-

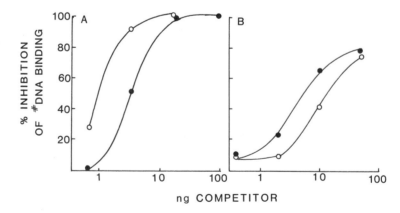

Figure 1. Competitive binding immunoassay to measure reactions of native and denatured DNA with monoclonal autoantibodies H241 (A) and H143 (B). Purified antibodies (5 μg) were incubated with varying amounts of native (○) or denatured (●) calf thymus DNA and then added to 50 ng of ^{3}H-labeled *E. coli* native DNA. Immune complexes were precipitated with goat anti-mouse Ig.

nucleic acid antibodies, as it combines with native DNA, denatured DNA, and both the B-DNA and Z-DNA forms of poly(dG-dC). On the other hand, it shows no reactivity, in either direct binding or competitive binding assays, with several single-stranded homopolymers, [poly(A), poly(C), or poly(U)]; double-stranded RNA, [poly(A)·poly(U) or poly(I)·poly(C)]; DNA-RNA hybrid, [poly(A)·poly(dT)]; or triple-stranded helices [poly(A)·poly(I)·poly(I) or poly(A)·poly(U)·poly(U)].

Because a mixed picture, with aspects of wide cross-reactivity and aspects of selectivity, were observed in these preliminary studies, a more detailed analysis was performed to probe the specificity of H241.

Reactions with native and denatured DNA. Most monoclonal autoantibodies to DNA studied in several laboratories have shown a marked preference for denatured DNA over native DNA (24,27), but some examples have been found for which binding of native DNA is equal to or greater than that of denatured DNA (25,26). H241 and H143 are two examples, in our series, of antibodies that bind S1 nuclease-treated ^{3}H-thymidine-labeled native DNA in radioimmunoassay. With H241 this binding was inhibited by lower concentrations of native than denatured DNA; with H143 the reverse was true (Fig. 1). Binding of H241 to native DNA was also observed by electron microscopy, which revealed its binding to either linear (not shown) or circular native DNA (Fig. 2).

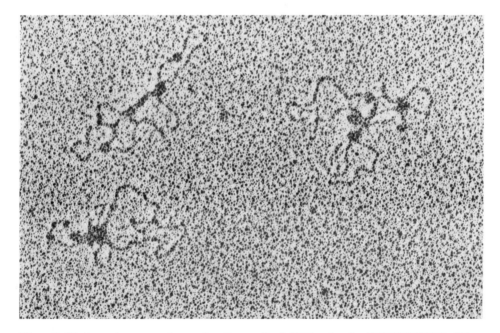

Figure 2. Electron microscopy of monoclonal autoantibody H241 and native ϕX174 RF II DNA. 120 μg of purified antibody was incubated for 30 min with 10 μg of the DNA. The mixture was passed through a 1.5 × 27 cm column of Sepharose 4B and material emerging at the void volume was prepared for electron microscopy as described (38).

Discrimination among polydeoxyribonucleotides. In spite of a tendency to cross-react with several polynucleotides, H241 distinguished among double-helical polydeoxyribonucleotides of varying base composition. The presence of guanine was common to the most reactive forms. Most striking was the nearly complete discrimination between poly(dG-dC) and poly(dA-dT) (Fig. 3a). This was characteristic of H241 but not of the other monoclonal antibody that bound native DNA H143 (Fig. 3b).

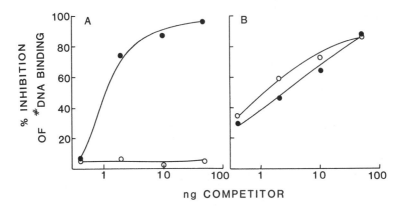

Figure 3. Competitive binding immunoassay to measure selectivity of monoclonal antibodies H241 (A) and H143 (B) for helical DNA of varying base composition. Purified antibodies were incubated with varying amounts of poly(dG-dC) (●) or poly(dA-dT) (○) and then added to 50 ng of ^{3}H-labeled *E. coli* native DNA. immune complexes were precipitated with goat anti-mouse Ig.

The role of the neighboring bases was explored more fully with a series of polymers (Table I). The B-form of alternating poly(dG-dC) was the most reactive, inhibiting DNA binding by H241 at the lowest concentration. Native DNA of mixed base composition, the Z-DNA form of poly(dG-dC), poly(dG)•poly(dC) and poly(dG-dT)•poly(dA-dC) were similar to each other and less reactive than poly(dG-dC). Poly(dG-dA)•poly(dT-dC) was significantly less effective a competitor and poly(dA-dT) or poly(dA)•poly(dT) were very ineffective even at higher concentrations. Again, H143 did not make these distinctions among these polymers of varying base sequence (Table I).

Protection of MP8 DNA against Hae III nuclease digestion. A test of whether the antibody could recognize and protect specific regions within native DNA was performed with the circular phage DNA MP8, a double-stranded derivative form of M13 with a known base sequence and known sites of cleavage by restriction nucleases (40). HaeIII makes 15 cuts in this DNA molecule. Small amounts of H241 protected selective sites in this DNA, as evidenced by the appearance of larger DNA fragments than are present in the uninhibited digestion products. In the presence of small amounts of H241, only a few new bands were observed and certain of the control bands disappeared while others were preserved, and certain of the control bands

Table I
Fine Specificities of H241 and H143

Competitor	Relative inhibition of H241	Binding of ^{3}H-native DNA H143
$(dG-dC)_n (dG-dC)_n$	1.0	1.0
native DNA	.084	4.5
$Br-(dG-dC)_n (dG-dC)_n$	.06	1.0
$(dG-dT)_n (dA-dC)_n$	.028	
$(dG-dG)_n (dC-dC)_n$	.014	
$(rG-rG)_n (rC-rC)_n$	.0028	
$(dG-dA)_n (dT-dC)_n$	<.002	
$(dA-dT)_n (dA-dT)_n$	$-_a$	2.4
$(dA-dA)_n (dT-dT)_n$	$-_a$	

The concentration of each polynucleotide causing 50% inhibition of the binding of ^{3}H-native DNA was determined. Relative inhibition is the ratio of the concentration required for poly(dG-dC) to the concentration required for each of the polynucleotides.

[a]no competition at the highest concentration tested.

disappeared while others were preserved (Fig. 4, lanes 1-4). In contrast, H143, which does not show preference for given base sequences, caused the simultaneous appearance of many new bands with no selective disappearance of any of the control digest fragments (Fig. 4, lanes 5-8). These results indicate that, at low concentrations, H241 binds preferentially to particular regions of the MP8 DNA (Fig. 5), whereas H143 binds randomly. At high concentrations, H241 can bind to many sites and it can precipitate all the fragments produced by the HaeIII digestion.

Monoclonal antibody H241 (unlike H143) has a striking ability to recognize helical structural features that depend on base sequence (Fig. 4, Table I). One interpretation of this is that the antibody binds to the outer surface of the helix, requiring that the dG-dC (or dC-dG) sequence presents a unique local shape on the outside of the helix. This antibody would then be a very precise yardstick of the backbone geometry. Physical measurements have revealed differences in backbone geometry among different members of the B-DNA class of helices. In poly(dA-dT) the adjacent ApT and TpA segments are not identical, as revealed by the presence of two distinct peaks in the ^{31}P-NMR spectrum (42); the repeating unit in this polymer is a dinucleotide rather than a mononucleotide (41,42). Arnott et al. (5) have described variations in helical geometry as presentation of a '"wrinkled"' backbone surface, which is different in poly(dA-dT) and poly(dG-dC), both of which differ in precise dimensions from average B-DNA. Furthermore, experimentally induced antibodies to double-helical polydeoxyribonucleotides have been able to detect variations from average B-DNA; for example, antibodies induced by poly(dG)·poly(dC), or poly (dC-dG) react with the immunizing polymer but not with native DNA (43-45); they probably make the distinction on the basis of the precise helical shape. These immunogens have been weak, inducing formation of small amounts of antibody.

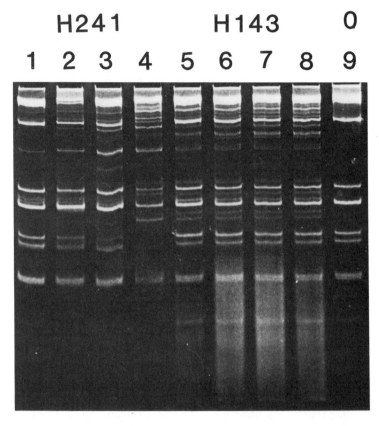

Figure 4. Protection of circular MP8 phage DNA from digestion by HaeIII nuclease. Protection was measured in mixtures of DNA with monoclonal autoantibody H241 (Lanes 1-4) or H143 (Lanes 5-8), in comparison with DNA plus normal mouse IgG (Lane 9). 1.5 μg of MP8 DNA was incubated with enzyme and 3.2 μg (lanes 1,5), 4.8 μg (lanes 2,6), 6.5 μg (lanes 3,7) or 8 μg (lanes 4,8,9) of immunoglobulin and enzyme. This provided Ab:DNA ratios of 1:110 base pairs to 1:44 base pairs.

When the specific populations are isolated by hybridoma selection and monoclonal antibody production, they and autoantibodies such as H241 may provide a library of reagents for the detection of varying local conformation.

The alternative explanation for the specificity of H241 would have the antibody obtain sequence-dependent specificity by interaction with portions of the bases. The reactivity of H241 with several different guanine-containing polymers (of varying conformation) but not adenine-containing polymers would suggest this view. An examination of three-dimensional model of B-DNA reveals that this would require that a portion of the antibody binding surface protrude substantially into the major groove. This same choice of possibilities was suggested by the ability of oligo-nucleotides to inhibit H241 as well as other monoclonal anti-DNA autoantibodies studied by Ballard and Voss (25). Antibody combining sites that have been defined by X-ray crystallography have, so far, been grooves or clefts in the tip of the Fab

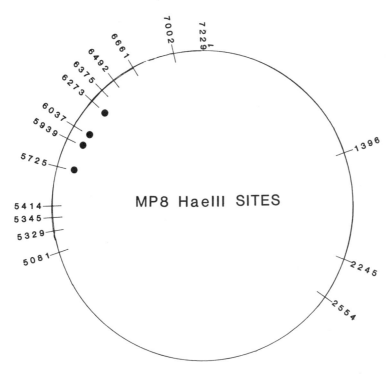

Figure 5. Map of region of MP8 phage DNA preferentially protected by small amounts of antibody H241. Analysis of the sizes of new fragments and diminishing bands in the gel shown in Figure 4 indicated that the regions marked by closed circles were preferentially protected by small amounts of antibody H241. Because of the large size of the fragments between nucleotides 1 and 5081, the experiment does not provide information on binding to this portion of the molecule.

arm of the immunoglobulin (reviewed in 46). For interaction of the antibody with the C7 or C8 of guanine, or the C5 of cytosine, an edge of the antibody's binding site would have to protrude into the antigen structure to a depth of several angstroms. Experiments with defined oligonucleotides (G. Zon and B.D. Stollar, unpublished data) support the conclusion that portions of the bases are important parts of the determinants. H241 also reacts with cardiolipin, which contains phosphate groups spaced by 3 carbon atoms but does not have purines or pyrimidines. Although this is not a strong cross-reaction, it indicates that some of the binding site also interacts with the phosphate backbone.

NMR studies of monoclonal anti-GMP antibodies. We have begun to explore the use of NMR spectroscopy to identify more precisely the nucleic acid sites that interact with antibody. In preliminary experiments with oligonucleotides, difficulties in interpretation arose from the complexity of the spectra. To establish principles for analysis of these interactions, we established a simpler system by preparing monoclonal antibodies to the mononucleotide GMP. Monovalent Fab fragments of one monclonal antibody were prepared by pepsin digestion of the IgG, followed by

Stollar, et. al.

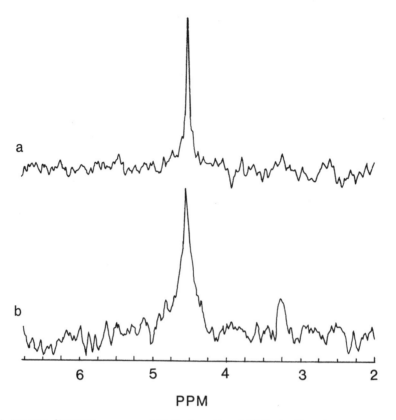

a

b

6   5   4   3   2

PPM

Figure 6. 162 MHz ³¹P NMR spectrum of GMP alone (a) and GMP plus Fab of monoclonal antibody AD8 (b). Fab and GMP were both at a concentration of 0.22 mM in NaCl, buffered at pH 8. NMR spectra were obtained with a 90° pulse, 0.819 sec. acquisition time, 2.819 sec recycle time, 1950 transients and 5Hz line broadening. Chemical shifts are referenced to 85% phosphoric acid. The small signal at 3.2 ppm is from inorganic phosphate.

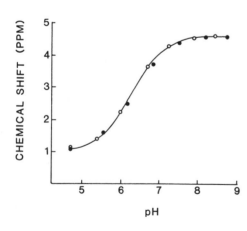

Figure 7. Titration of phosphate of free GMP (a) or GMP in the presence of Fab of monoclonal anti-GMP antibody AD8 (b). Chemical shifts were measured under conditions described in the legend to Figure 6.

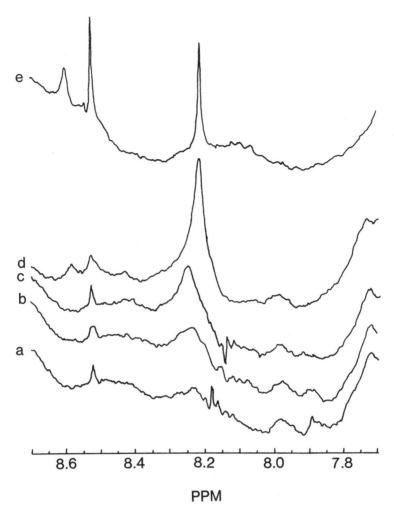

Figure 8. 400 MHz ^{1}H NMR spectrum of the Fab of monoclonal anti-GMP antibody alone (a) and of mixtures of GMP and Fab in molar ratios of 1:1 (b); 2:1 (c); and 4:1 (d); and of GMP with Fab of irrelevant non-binding immunoglobulin in a molar ratio of 1:1 (e). The figure shows the region of the spectrum of the C8 proton of guanine. Fab was at a concentration of 0.3 mM in D_2O. Sweep width was 5000 Hz; acquisition time was 0.819 sec and recycle time 2.819 sec; 2800 transients were accumulated; line broadening of 2 Hz was applied. The peak at 8.56 ppm in the spectra is of unknown origin; its size in (e) was the same as in the spectrum of that Fab preparation alone.

incubation of the digestion mixture in 20 mM cysteine, and then removal of small peptide fragments by dialysis. On SDS-polyacrylamide gel electrophoresis in the presence of 2-mercaptoethanol, the resulting fragments yielded a single band of protein fragments of approximately 22,000 molecular weight, as expected for Fab. The Fab reacted with denatured DNA and this reaction was inhibited specifically by guanosine or GMP.

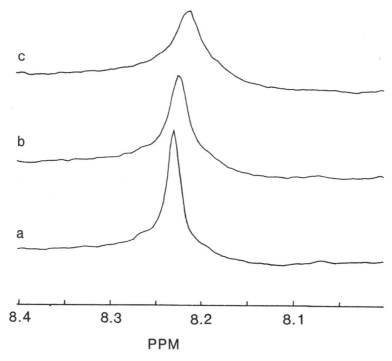

Figure 9. Temperature dependence of the linewidth of the ^{1}H NMR spectrum of the C8 proton of GMP in a 4:1 (molar ratio) mixture of GMP and Fab of monoclonal anti-GMP antibody AD8. Spectra were measured under conditions indicated in the legend to Figure 8, at 25°C (a), 29°C(b) and 35°C(c).

In comparison with GMP alone, the Fab-GMP (in a 1:1 molar ratio) mixture yielded a somewhat broadened ^{31}P NMR peak, (Fig. 6), reflecting the binding of GMP. The phosphorus itself, however, did not appear to be involved in direct contact with amino acid residues in the antibody combining site, as the titration of the bound GMP was identical to that of free GMP (Fig. 7). This result was consistent with the finding that guanosine was 50-fold more effective than GMP in competing for binding of this antibody to denatured DNA (data not shown).

Fab was lyophilized and dissolved in D$_2$O for proton NMR experiments. In the presence of a non-binding protein (the Fab of normal immunoglobulin), the C8 proton of guanine was clearly identifiable when the molar ratio of GMP to Fab was 1:1 (Fig. 8e) and the width of the peak was the same as that of GMP alone (not shown). With the same 1:1 ratio of GMP to Fab of the specific monoclonal antibody, there was a barely visible broad peak in this region at 35°C and this was even less evident at 23°C. It became more clearly visible as excess GMP was added to the mixture (Fig 8 b,c,d), but remained broad in comparison with that of free GMP (Fig 8e). As the temperature was increased from 23° to 29° to 35°, this peak became broader (Fig. 9), suggesting that the ligand was undergoing slow exchange. No additional peak was visible in the spectrum, but difference spectra subtracting Fab

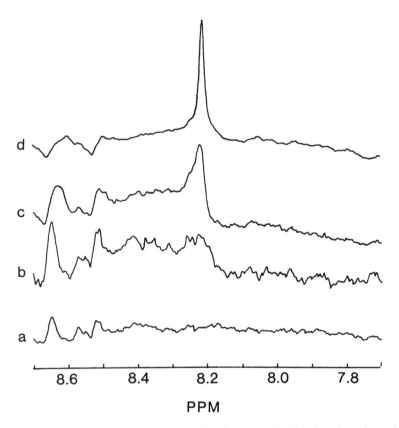

d

c

b

a

8.6   8.4   8.2   8.0   7.8

PPM

Figure 10. Difference spectra obtained by subtraction of spectrum for Fab alone from those of GMP + Fab with molar ratios of GMP:Fab of 0.5:1 (a); 1:1 (b); (2:1) (c); and 4:1 (d).

alone from (Fab + GMP) suggested there may be a broad signal slightly downfield from that of free GMP (Fig. 10).

To test further whether the observed signal represented only free GMP, with broadening due to exchange, another resonance for a bound GMP form with a different chemical shift was sought by use of a saturation transfer experiment. When a mixture of GMP and Fab (molar ratio 4:1) was irradiated at varying frequencies, diminution of the major peak of GMP at 8.23 ppm was observed upon irradiation at 8.4 ppm. This effect showed a sharp maximum at 8.4 ppm and was not observed when the sample was irradiated at 8.1 ppm, a similar distance upfield from the major signal peak. The spectrum resulting from irradiation at 8.4 ppm was subtracted from that caused by irradiation at a distant frequency. In the difference spectrum, peaks were visible at both 8.4 ppm and 8.23 ppm (Fig. 11b). The 8.23 ppm peak was not present in difference spectra calculated from spectra caused by irradiation at 8.1 ppm (Fig. 11e) (on the other side of the free GMP signal) or at more distant downfield frequencies (Fig. 11a). This result suggested there was a different chemical shift for the C8 proton of the bound form, though the difference

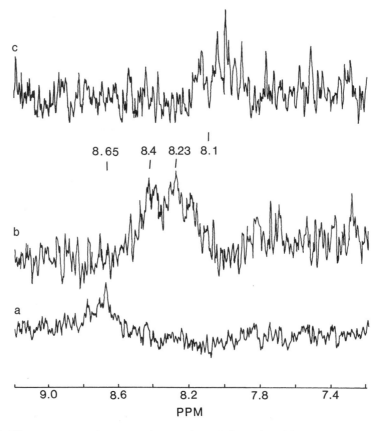

Figure 11. Ditterence spectra from saturation transfer experiment in which a 4:1 (molar ratio) mixture of GMP and Fab of monoclonal anti-GMP antibody were irradiated at varying frequencies for 600 transients for each spectrum. The spectra obtained with irradiation at 8.65 ppm (a), 8.4 ppm (b) and 8.1 ppm (c) were subtracted from that obtained with irradiation at 1 ppm. Spectra were obtained at 29°C under conditions described in Figure 8.

from the position of free form was very small. These findings suggest that the C8 proton is involved in interaction with the protein residues, though perhaps to a limited extent. A significant base stacking between the purine ring of GMP and an aromatic amino acid of the Fab fragment, which would be expected to cause a significant upfield chemical shift (greater or equal to 0.5 ppm) is ruled out. Some involvement of the C8 position in interaction with antibody was also indicated by the finding that 8-bromoguanosine was 50-fold less effective a competitor than guanosine for binding of antibody to denatured DNA (data not shown). That this region was not the center of the major interaction with antibody was suggested by the fact that higher concentrations of the bromoguanosine, with a bulky substitution on C8, could completely inhibit the reaction of antibody with DNA. In contrast, 7-methylguanosine reacted with antibody only very weakly and did not inhibit the DNA binding completely even at much higher concentrations. Inosine was about 100-fold weaker than guanosine, but like 8-bromoguanosine, it could completely block the antibody-DNA interaction. The combined results of competitive immunoassays and NMR spectroscopy indicate that major interaction with antibody probably involves the purine base in the region of N1, C6, and C7; this is consistent with findings of Munns et al (47). Definitive information on the role of N1 and N7 may be determined by 15N NMR with suitably enriched nucleotides. Additional studies revealing the criteria by which strong interactions can be measured in this relatively simple system of monoclonal anti-GMP antibody will establish a basis for progressing to NMR analysis of antibodies that react with oligonucleotides and short helical nucleic acids.

Acknowledgements

This research was supported by grants AM 27232 and GM32275 from the National Institutes of Health. We are thankful for the excellent technical assistance of Beverly Esielonis and Irene Dougas.

References and Footnotes

1. Present address of Bao-Shan Huang: Department of Biochemistry, Beijing University Medical School, Beijing, China. Present address of M. Blumenstein: Chemical Research Department, McNeil Pharmaceuticals, Springhouse, Pa, 19477.
2. Y. Takeda, D.H. Ohlendorf, W.F. Anderson and B.W. Matthews, *Science 221,* 1020 (1983).
3. A. Rich, A. Nordheim and A.H.-J. Wang, *Annu. Rev. Biochem., 53,* 791 (1984).
4. H. Shindo, R.T. Simpson and J.S. Cohen, *J. Biol. Chem. 254,* 8125 (1979).
5. S. Arnott, R. Chandrasekaran, I.H. Hall, L.C. Puigjaner, J.K. Walker and M. Wang, *Cold Spring Harbor Symp. Quant. Biol. 47,* 53 (1983).
6. A.V. Fratini, M.L. Kopka, H.R. Drew and R.E. Dickerson, *J. Biol. Chem. 257,* 14686 (1982).
7. F. Lacour, E. Nahon-Merlin and M. Michelson, *Curr. Top. Microbiol. Immunol. 62,* 1 (1973).
8. B.D. Stollar, in *The Antigens,* Ed. M. Sela, Academic Press, New York p.1 (1973).
9. B.D. Stollar, *Crit. Rev. Biochem. 3,* 45 (1975).
10. B.D. Stollar, *Crit. Rev. Biochem.* in press (1985).
11. M. Guigues and M. Leng, *Eur. J. Biochem. 69,* 615 (1976).
12. M.I. Johnston and B.D. Stollar, *Biochemistry 17,* 1959 (1978).
13. E. Nahon-Merlin, A.M. Michelson, F. Lacour and G.Camus, *Ann. Immunol. 131C,* 279 (1980).
14. A. Nordheim, M.L. Pardue, E.M. Lafer, A. Möller, B.D. Stollar and A. Rich, *Nature 294,* 417 (1981).
15. F. Lemeunier, C. Derbin, B. Malfoy, M. Leng and E. Taillandier, *Exp. Cell Res. 141,* 508 (1982).
16. D.J. Arndt-Jovin, M. Robert-Nicoud, D.A. Zarling, C. Greider, E. Weimer and T.M. Jovin, *Proc. Natl. Acad. Sci. U.S.A. 80,* 4344 (1983).
17. E.B. Kmiec and W.K. Holloman, *Cell 36,* 593 (1984).
18. B.D. Stollar and V. Stollar, *Virology 42,* 276-280 (1970).
19. J.R. Miller, L.A. Caligiuri and I. Tamm, *J. Virol. 16,* 290 (1975).
20. G.T. Rudkin and B.D. Stollar, *Nature 265,* 472 (1977).
21. W. Büsen, J.M. Amabis, O. Leoncini, B.D. Stollar and F.J.S. Lara, *Chromosoma 87,* 247 (1982).
22. A. Alcover, M. Izquierdo, B.D. Stollar, Y. Kitagawa, M. Miranda and C. Alonso,*Chromosoma 87,* 263 (1982).
23. W.D. Stuart, J.G. Bishop, H.L. Carson and M.B. Frank, *Proc. Natl. Acad. Sci. U.S.A. 78,* 3751 (1981).
24. C. Andrzejewski, J. Rauch, E. Lafer, B.D. Stollar and R.S. Schwartz, *J. Immunol. 126,* 226 (1981).
25. D.W. Ballard and E.W. Voss Jr., *Mol. Immunol. 19,* 793 (1982).
26. F. Tron, D. Charron, J.-F. Bach and N. Talal, *J. Immunol. 125,* 2805 (1980).
27. J.S. Lee, D.F. Dombroski and T.R. Mosmann, *Biochemistry 21,* 4940 (1982).
28. Y. Shoenfeld, J. Rauch, H. Massicotte, S.K. Datta, J. Andre-Schwartz, B.D. Stollar and R.S. Schwartz, *New Eng. J. Med. 308,* 414 (1983).

29. B.S. Andrews, R.S. Eisenberg, A.N. Theofilopoulos, S. Izui, C.B. Wilson, P.J. McConahey, E.D. Murphy, J.B. Roths and F.J. Dixon, *J. Exp. Med. 148,* 1198 (1978).
30. Erlanger, B.F., *Methods in Enzymology 70,* 85 (1980).
31. Engvall, E., *Methods in Enzymology 70,* 419 (1980).
32. M. Papalian, E. Lafer, R. Wong and B.D. Stollar, *J. Clin. Invest. 65,* 469 (1980).
33. J. Marmur, *J. Mol. Biol. 3,* 208 (1961).
34. B.D. Stollar and V. Raso, *Nature 250,* 231 (1974).
35. L.C. Rainen and B.D. Stollar, *Biochemistry 16,* 2003 (1977).
36. E.M. Lafer, A. Möller, A. Nordheim, B.D. Stollar and A. Rich, *Proc. Natl. Acad. Sci. U.S.A. 78,* 3546 (1981).
37. A. Möller, A. Nordheim, S.A. Kozlowski, D.J. Patel and A. Rich, *Biochemistry 23,* 54 (1984).
38. A. Nordheim, E.M. Lafer, L.J. Peck, J.C. Wang, B.D. Stollar and A. Rich, *Cell 31,* 309 (1982).
39. E.M. Lafer, J. Rauch, C. Andrzejewski Jr., D. Mudd, B. Furie, B. Furie, R.S. Schwartz and B.D. Stollar, *J. Exp. Med. 153,* 897 (1981).
40. J. Messing and J. Vieira, *Gene 19,* 269 (1982).
41. A. Klug, A. Jack, M.A. Viswamitra, O. Kennard, Z. Shakked and T. Steitz, *J. Mol. Biol. 131,* 669 (1979).
42. D. Patel, S.A. Kozlowski, J.W. Suggs and S.D. Cox, *Proc. Natl. Acad. Sci. U.S.A. 78,* 4063 (1981).
43. J.S. Lee, M.L. Woodsworth and L.J.P. Latimer, *Biochemistry 23,* 3277 (1984).
44. E.M. Lafer and B.D. Stollar, *J. Biomolec. Structure Dynamics, 2,* 487 (1984).
45. E.M. Lafer, Doctoral Thesis, Tufts University (1982).
46. D.R. Davies and H.M. Metzger, *Ann. Rev. Immunol. 1,* 87 (1983).
47. T.W. Munns, M.K. Liszewski, and B.H. Hahn, *Biochemistry 23,* 2958 (1984).

Biomolecular Stereodynamics III, Proceedings of the Fourth Conversation in the Discipline Biomolecular Stereodynamics, State University of New York, Albany, NY, June 04-09, 1985, Eds., Ramaswamy H. Sarma & Mukti H. Sarma, ISBN 0-940030-14-4, Adenine Press, ©Adenine Press 1986.

Probing Nucleic Acid Structure with Metallointercalators

Challa V. Kumar, Adrienne L. Raphael, and Jacqueline K. Barton*

Department of Chemistry, Columbia University,
New York, NY 10027

Abstract

Metallointercalators are useful reporters of the local structure of nucleic acids. Their rich spectroscopy and reactivity may be tailored to follow the telltale events that occur along the DNA helix. Chiral octahedral complexes have been designed which recognize the local asymmetry in the DNA helix and by coupling redox activity in the metal center to the recognition, conformation-specific DNA cleaving agents are obtained.

Introduction

Understanding the nature and structure of complexes of small molecules with nucleic acids can be important and useful to experiments in chemistry and biology (1-7). The shapes, charges and sizes of these small molecules make them excellent probes for the local structure and conformation of biomolecules in solution. Several such small molecular reagents have been successfully employed and have proven to be very valuable in studying biological macromolecules. A simple example which we will focus on in this article is the intercalative binding of aromatic cations to double helical DNA (3-5). In particular we will examine intercalators containing a metal center.

Intercalation is a non-covalent binding mode (8) where the aromatic planar moiety stacks in between the adjacent base pairs of the DNA and causes unwinding of the double helix. Intercalators often exhibit biological activity. Many are frame shift mutagenic agents, resulting presumably from their distortions of the native DNA structure. Several natural antitumor antibiotics, such as daunomycin and echino-mycin, contain planar aromatic moieties and their pharmacological activity stems in part from their ability to intercalate into DNA (6). Importantly, the spectroscopic and chemical properties of intercalators can be exploited for practical purposes in reporting about the nature of the nucleic acid to which it was bound. For example, the fluorescence intensity of ethidium bromide, an intercalator, is enhanced 20 times upon binding to DNA and is used even today as a common DNA stain in

---

*to whom reprint requests should be addressed.

electrophoresis experiments (7). Fluorescence depolarization experiments using ethidium have been successfully employed to measure the torsional rigidity of nucleic acids (9,10).

Inorganic metal complexes as well which possess planar aromatic ligands bind to nucleic acids by intercalation and these are increasingly being used as molecular probes of nucleic acid structure and conformations in solution. The metal center, owing either to its high electron density, rich charge transfer characteristics, or redox activity, becomes a unique probe for x-ray diffraction experiments, spectroscopic studies or in site-specific chemical reactions. The coordination about the metal center, with easily varied ligands and stereochemically well-defined geometries, further permits one to match the structure of the anchored intercalator to the local shape of the DNA helix. In this article we describe recent efforts made in our laboratory and by others to develop and employ metallointercalators as probes of nucleic acid structure.

Planar Metallointercalators

Lippard and coworkers were the first to discover that metal complexes could intercalate into DNA (11). Shown in figure 1 is the planar (2-hydroxyethanethiolate)-terpyridylplatinum(II), (terpy)Pt(HET)$^+$, a well-characterized cationic complex which binds to DNA and RNA by intercalation. Increases in DNA melting temperature as a function of binding, viscosity changes indicating the lengthening of the helix, and superhelical DNA unwinding experiments all were consistent with intercalation of the square planar platinum(II) species in between the base pairs of the helix. A crystal structure of (terpy)Pt(HET)$^+$ intercalated in d(CpG) revealed the geometry of the bound complex (12). A clear advantage to using a *metallo*intercalator as a probe was demonstrated early in fiber x-ray diffraction experiments where the electron dense metal center of Pt(terpy)(HET)$^+$ served as a unique marker for intercalation (13). The regular distribution of platinum atoms seen in the diffraction patterns at 10.2 Å intervals, that is at every other interbase pair site (3x3.4 Å) along the helix, proved the neighbor exclusion principle (14) for intercalative drug binding to DNA.

An important criterion for intercalation by a small molecule is the presence of a planar aromatic moiety. This was amply demonstrated by the binding studies of [(phen)Pt(en)]$^{2+}$; [(bpy)Pt(en)]$^{2+}$ and [(py)$_2$Pt(en)]$^{2+}$ to DNA (15). These studies involving measurements of helix unwinding and x-ray fiber diffraction experiments clearly showed that [(phen)Pt(en)]$^{2+}$ and [(bpy)Pt(en)]$^{2+}$, which contain flat ligands intercalate whereas the bis(pyridyl) complex, with rings rotated out of the coordination plane, does not. Subsequent studies (16) were used to characterize the binding of various platinum(II) complexes to DNA, with variations both of the intercalating, aromatic ligand and the free ligand (the ethanethiolate tail in (terpy)Pt(HET)$^+$). Hydrogen bonding donors could be changed in the tail, or the tail could even be replaced by a labile chloride ion, (terpy)PtCl$^+$ leading to a reagent that binds to DNA by direct coordination, following initial intercalative binding. Binding of these complexes to DNA further showed ionic strength dependent affinities for DNA which were somewhat lower than that of ethidium and a linear dependence in binding on the guanine-cytosine content of the DNA.

Dervan and co-workers devised an important method for functionalizing various DNA binding molecules by tethering a metal ion to an organic drug. Methidiumpropyl EDTA-Fe(II) (MPE-Fe(II)), also shown in Figure 1, has the metal chelator EDTA tethered to methidium, a well characterized planar intercalator (17). This complex causes single strand scission of DNA in the presence of iron(II) and oxygen. The DNA cleavage becomes catalytic with the addition of a reducing agent. MPE-Fe(II) is perhaps not a standard metallointercalator but it illustrates simply the notion of coupling intercalative binding to a redox active metal center, and in so doing targeting the chemical reactivity of a metal to a DNA site. Essentially the methidium group, an ethidium analogue, binds to the helix, bringing with it a high local concentration of ferrous ion, which presumably through Fenton chemistry delivers hydroxyl radicals to the helix, thereby cleaving the strand.

Intercalative binding of MPE-Fe(II) has been demonstrated by a number of methods (18). Upon binding to calf thymus DNA, the visible absorption spectrum of MPE-Fe(II) undergoes a metachromic shift. Studies of the helical unwinding of the supercoiled plasmid pM2 DNA that is characteristic of intercalation showed that MPE-Ni(II) or MPE-Mg(II) unwinds the helix by 11.2° per molecule bound. The binding affinity of the MPE-Ni(II) or Mg(II) is approximately 5 times that of MPE alone. DNA cleavage is thought to occur via production of diffusable hydroxyl radical species. The scission is inhibited by exogenous iron chelators, high dithiothreotol concentrations, catalase and superoxide dismutase, all of which can quench the radical species. The DNA products indicate an oxidative degradation of the deoxyribose ring (free base, 5′-phosphate, 3′-phosphate and 3′-phosphoglycolic acid) (18).

The small size and lack of sequence-specific binding of MPE-Fe(II), due to the non-specific intercalative binding of methidium, has made it ideal for footprinting the sites where other small drug molecules bind along the DNA strand. Using MPE-Fe(II) the specific binding sites and sizes of several drugs including actinomycin, netropsin and mithramycin have been determined (19-24). After limited cleavage by MPE-Fe(II) of the bound drug-DNA complex, Maxam-Gilbert sequencing is performed. Since the DNA is protected wherever the drug is bound, what results is a direct visualization of the "footprint" of drug binding sites in DNA sequencing gels. MPE-Fe(II) has also been useful in investigations of chromatin structure (25).

Metalloporphyrins represent another family of essentially planar aromatic metal complexes that have been shown to intercalate into DNA (26-28). The meso-tetra(4-N-methylpyridyl)porphine cation, through Scatchard binding and hypochromic analyses, viscometry, thermal denaturation and circular dichroism studies, appears to intercalate into DNA. DNA helical unwinding experiments have been performed on the metal-free porphyrins. Intercalation of these complexes is somewhat surprising given

(2-hydroxyethanethiolate)terpyridylPt(II)

methidiumpropylEDTA-Fe(II)

bisphenanthrolineCu(I)

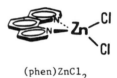

(phen)ZnCl$_2$

Figure 1. Metallointercalators.

their very large size, and the high charge on the metal complex. The N-methyl-pyridyl groups must bend into the plane of the porphyrin and the helix must become very unwound to permit complete insertion of the complex into the helix. Once there, dissociation must be very slow. Not surprisingly, the iron(II) derivatives of this porphyrin induce strand scissions in DNA. However, no inhibition of this activity was observed with ethidium bromide, and other non-DNA binding Fe(II) porphyrins have similar nicking efficiencies. With several derivatives of porphyrins containing different metal ions, it was demonstrated that complete insertion of the metal complex requires GC rich DNA whereas with AT rich DNA either partial intercalation or outside binding occurs.

Metallointercalators with Tetrahedral Coordination

Direct intercalation of metal complexes is not restricted to those metal centers having a square planar geometry. Non-planar metal complexes which possess planar aromatic ligands can bind to DNA by intercalation. A good example is the tetrahedral zinc complex (phen)ZnCl$_2$ (Fig. 1) (29). Bound intercalators unwind supercoiled closed circular DNA (5). If one examines the electrophoretic mobilities rapidly with increasing (phen)ZnCl$_2$, with no effect on the nicked circular form. At greater than 0.05 mM concentration of zinc, after comigration with the nicked species, the mobility of the closed circular form increases once again. As unwinding continues beyond the relaxed circular form, additional intercalation leads to positive supercoils with the concomitant increase in its mobility through the gel. Thus intercalative binding of this simple tetrahedral complex containing the planar phenanthroline ligand is likely. In order to check any interference from hydrolysis products from this dichloride, binding of (phen)$_2$Zn^{2+} and (phen)$_3$Zn^{2+} were examined. These coordinatively saturated complexes also appear to bind to DNA through intercalation as shown through the electrophoretic assay. Thus we demonstrated that a planar ligand, not necessarily a planar coordination geometry is required for intercalation, that is for at least partial insertion into the helix.

The analogous tetrahedral complex of cuprous ion, (phen)$_2$Cu$^+$, generated *in situ,* causes efficient DNA strand scission in the presence of H$_2$O$_2$ and a reducing agent (30-31). Here again intercalation is coupled to a redox active metal. Intercalative binding of one of the ligands was suggested to be necessary for this cleavage reaction, as the cleavage occurs only on double stranded DNA or self-complementary single stranded DNAs. Binding competition experiments with well known intercalators such as ethidium bromide supported this suggestion. Additionally, the analogous bipyridyl or terpyridyl Cu(I) complexes do not show any reactivity and 2,9-dimethyl-1,10-phenanthroline, which competes with phenanthroline in chelating the metal, inhibits the cleavage reaction (32).

A very interesting finding by Sigman and coworkers was that the cleaving efficiencies of bis(phen)Cu(I) with A, B, and Z form DNAs differ greatly (33). DNA/RNA heteroduplexes (A structure) are cleaved approximately one third as efficiently as B-form duplexes (poly d(AT) or poly d(GC)), and poly d(GC) in 3 M NaCl (Z form) is not cleaved at all. It is not clear whether the low cleavage efficiency is due to low binding affinity or if the bound complex is improperly oriented for cleavage in these various conformations. Perhaps additionally the formation of (phen)$_2$Cu$^+$ *in situ* is somewhat inhibited in the presence of Z-DNA where coordination sites along the strand are easily accessible to the copper.

The bis(phen)Cu(I) has also been used in chromatin structural studies (34,35). The nucleosomal array of *Drosophila* embryo chromatin is cut in a similar manner by bis(phen)Cu(I) and micrococcal nuclease in its linker regions. This implies that the two very disparate cleavage agents are "recognizing" a localized variation in secondary structure. Perhaps these regions of altered conformation help align histones or serve as recognition sites for other proteins as well.

Octahedral Complexes as Chiral Probes for DNA

A significantly different but perhaps fruitful approach to study the conformations of nucleic acids is to design a probe that exploits the inherent handedness of nucleic acid duplexes. DNA is an asymmetric molecule and greater specificity of binding

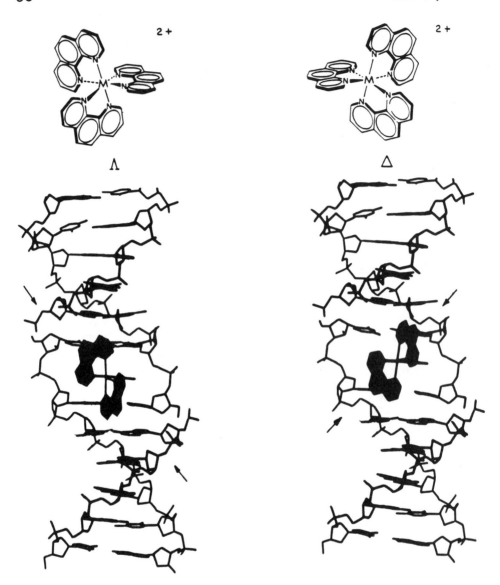

Figure 2. Δ and Λ isomers of trisphenanthroline metal complexes and illustration of chiral constraints on intercalation into right-handed DNA.

may be accomplished by the matching of stereochemistries. Octahedral tris(phen-anthroline) complexes form optical isomers and these may be resolved for several transition metal ions (Figure 2). In our first experiments (29), when calf thymus DNA was dialyzed against racemic [Zn(phen)$_3$]Cl$_2$, the dialysate was found to be optically enriched, the obvious result of the preferential binding of one isomer to DNA. At the time we did not know the absolute configuration of the favored iso-mer, but the result could be readily explained based on the intercalative mode of binding, where one of the phenanthroline ligands is sandwiched between the adjacent base pairs. Importantly, since the metal complexes are quite rigid, and intercalative binding clearly positions the complex in the helix, the orientation of all portions of the molecule becomes easily defined. As illustrated in Figure 2, the Δ isomer would intercalate preferentially into right-handed DNA. The vast expanse of the two non-intercalated ligands for the Δ isomer are not sterically hindered by the DNA backbone; in fact they lie along the right-handed groove of the helix. Under similar conditions for the Λ isomer, however, with one ligand intercalated, the remaining ligands are involved in severe steric interactions with the right-handed phosphate backbone. Here the disposition of non-intercalated ligands opposes the right-handed groove. Thus binding is favored only when the right handed screw symmetry of the helix matches that of the metal complex. Chiral discrimination is found based upon the relative handedness of the metal complex and that of the DNA helix.

In addition to the ground state properties of these chiral intercalators, their excited state behavior can be very conveniently exploited to extract the telltale events that occur in the interaction of these probes with nucleic acids. The stereoselective binding characteristics of octahedral metal centers have therefore been further examined by taking advantage of the spectroscopic properties of tris(phen)Ru(II) complexes. The polypyridyl complexes of d^6 ruthenium(II) have been well characterized and are inert to racemization (36-39). The stereoselective binding of tris(phen)Ru(II) complexes was first observed in our laboratory using these methods (40,41). For ruthenium the absolute configurations had been assigned independently (38) and indeed it was the Δ isomer that was preferred. Binding could be assayed by equilibrium dialysis, by superhelical DNA unwinding, or in particular by several spectroscopic assays. The intense charge transfer transitions of these complexes provide spectroscopic methods to follow the binding of these complexes to DNA. Shown in Figure 3 are the emission spectra of $Ru(phen)_3^{2+}$ and tris(diphenylphenanthroline)ruthenium(II) ($Ru(DIP)_3^{2+}$) in the absence and presence of DNA (42). Emission is clearly enhanced with intercalative binding to DNA. Changes in the absorption spectra are also seen, a small shift to lower energies and hypochromism (41). The excited state lifetimes of the ruthenium species are significantly lengthened with DNA binding, supporting an intimate association with the helix, with decreased vibrational relaxation and perhaps with effects in part a result of the altered water structure about the metal.

Spectroscopic evidence for intercalative binding is obtained also from luminescence quenching and polarization measurements (42). Certain molecules can act as energy sinks for excited states and by appropriately choosing these quenchers, it is possible to distinguish sensitively between the bound and free ruthenium cations. For example, a negatively charged quencher, such as ferrocyanide $[Fe(CN)_6]^{4-}$ can effectively quench the emission from $Ru(phen)_3^{2+}$ (quenching rate = 45×10^9 M^{-1}s^{-1}). The Stern Volmer constant, a quantity that is proportional to the rate of quenching of the excited state, is dramatically reduced from 4700 M^{-1} to 475 M^{-1} in the presence of DNA. The bound ruthenium cations are protected from the negatively charged quencher, ferrocyanide, by the array of negative charges of the phosphate backbone of DNA. Thus a reduction in quenching constant reflects the binding of the ruthenium complex to the helix.

That this binding is intercalative, or at least rigid, is apparent in polarization measurements. Intercalation of one of the ligands would immobilize the metal complex and, if fixed on the timescale of the emission, the emitted photon will be polarized (9,10). Indeed when DNA is added to these metal complexes, the binding to DNA results in increased polarization for $Ru(phen)_3^{2+}$ and $Ru(DIP)_3^{2+}$ (42). That there is polarization in aqueous solution, without slowing the DNA motions, is remarkable given the long timescales (2 μsec) for ruthenium emission. No such polarization is observed for $Ru(bpy)_3^{2+}$, consistent with its low binding affinity. In fact these various methods can be used to compare and contrast binding for the three ruthenium complexes with differing ligands. As shown in Table I, $Ru(phen)_3^{2+}$ and $Ru(DIP)_3^{2+}$ appear to bind to DNA by intercalation. The increasing hydrophobicity for the ligands bpy, phen, and DIP seems to correlate with binding ability to the DNA.

Table I
Binding Characteristics of Ru(II) Complexes Bound to Calf Thymus DNA

	$Fe(CN)_4^{-4}$ Quenching	τ_1	τ_2 (nsec)	Polarization	Intercalation	Ionic
$Ru(bpy)_3^{2+}$	linear	420		0.002	−	?
$Ru(phen)_3^{2+}$	biphasic	733	2645	0.036	+	+
$Ru(DIP)_3^{2+}$	biphasic	855	3348	0.012	+	+

The stereoselectivity associated with intercalative binding can also be examined using these techniques. Table II shows the comparison of Λ- and Δ-$Ru(phen)_3^{2+}$ (43). In particular it is apparent that the chiral discrimination is a result of the differential *intercalation* of enantiomers. Δ-$Ru(phen)_3^{2+}$ shows greater retention of emission polarization than Λ-$Ru(phen)_3^{2+}$ in the presence of DNA, consistent with the fact that intercalative binding by the right-handed propeller-like Δ isomer is preferred into the right-handed helix.

Table II
Binding Characteristics of $Ru(phen)_3^{2+}$ Isomers to Calf Thymus DNA

	$K_{sv}(M^{-1})$	Lifetimes(ns)	Polarization
Δ-$Ru(phen)_3^{2+}$	250	559; 2137	0.0367 ± 0.002
Λ-$Ru(phen)_3^{2+}$	370	583; 1999	0.0258 ± 0.002

How might the stereoselectivity be increased so as to provide a more specific probe? Based on the intercalative model, the enantiomeric selectivity depends upon the diameter of the metal complex relative to the width of the helical groove.

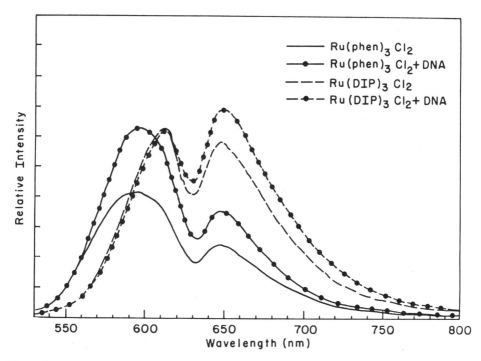

Figure 3. Emission spectra of Ru(phen)$_3$$^{2+}$ and Ru(DIP)$_3$$^{2+}$ in the absence and presence of B-form calf thymus DNA.

With larger ligands, where the steric exclusion size is increased in comparison to the size of a DNA groove, the discrimination is therefore enhanced. Binding of Ru(DIP)$_3$$^{2+}$ to DNA was found to be stereospecific; only the Δ isomer intercalates into the B-DNA helix (44). This stereospecific binding was seen in absorption titrations but made very clear in ferrocyanide quenching experiments (42), as shown in Figure 4. The quenching plot for the Δ isomer, a biphasic curve, is characteristic of two distinct forms, the intercalated and surface bound or free ruthenium. In contrast, the quenching plot for the Λ isomer indicates no intercalated ruthenium; it is a simple one component system showing a linear Stern-Volmer plot. Also the quenching constant for the Λ isomer is much larger than Δ, indicating that the intercalated ruthenium is well protected from the ferrocyanide quencher by the negatively charged backbone of DNA.

Based on this stereospecificity for the binding of Ru(DIP)$_3$$^{2+}$ to DNA, it is possible to design probes for left-handed DNA in solution. If the DNA can discriminate between metal complex enantiomers, then analogously one could use one metal enantiomer to discriminate between DNAs of differing helicity. Although Λ-Ru(DIP)$_3$$^{2+}$ does not intercalate into DNA, its binding to the left hand Z-DNA was evident from absorption experiments (44). It is interesting that little or no stereoselectivity was observed in binding experiments with Z-DNA. Both Δ- and Λ-Ru(DIP)$_3$$^{2+}$ bind to this conformation. The shallow and wider major groove provides a rather

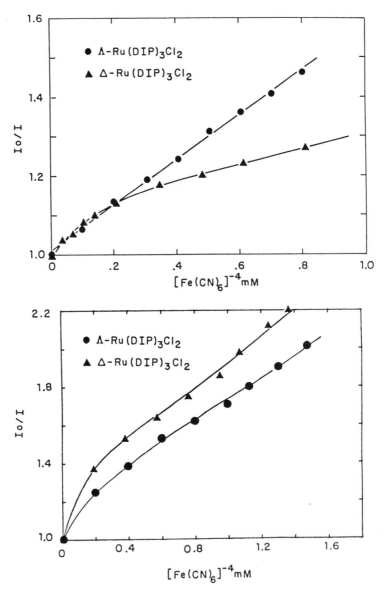

Figure 4. Quenching of emission from $\varDelta$ and $\varLambda$ isomers of Ru(DIP)$_3^{2+}$ bound to, (top) calf thymus DNA and (bottom) Z-form poly d(GC).

nondiscriminatory helix for the chiral intercalators. Similar results were observed in ferrocyanide quenching experiments with the optical isomers of Ru(DIP)$_3^{2+}$ and Z-form poly d(GC), as shown also in Figure 4. Indeed here a slight stereoselective preference is observed for the $\varLambda$ isomer in binding to the left-handed helix, given its quenching curve of reduced slope. Thus, the Ru(DIP)$_3^{2+}$ isomers serve as excellent probes to distinguish between the right- and left-handed helices in solution. By measuring the degree of stereoselectivity, be it for right or even left-handed helices, unique information about DNA conformations may be obtained in solution. These probes may therefore be helpful to characterize in solution the Z-DNA conformation, but also perhaps to describe any left-handed non-Z-conformations (45), for example in the extreme those which would be stereospecific for Λ-Ru(DIP)$_3^{2+}$.

Measurements of relative stereoselectivities have been useful to us as well in characterizing subtle variations in conformations of right-handed B-like helices. The stereoselectivities are easily measured in either a dialysis or spectroscopic experiment. We set out to determine the base specificity of the intercalative binding of these probes to DNA and any variations we might detect in chiral preferences as a function of the guanine-cytosine (GC) content of the DNA (43). We examined several synthetic and natural DNAs of varying GC content. The overall binding affinity, as measured in a dialysis experiment, showed no clear dependence on GC content. Unlike for the platinum(II) intercalators (16), the octahedral species showed no base specificity in binding. Interesting variations in stereoselectivity with GC content were apparent, however, as can be seen in Figure 5. With increasing %GC, increases in the preference for the Δ isomer are found. The stereoselectivity for intercalative binding as measured through emission polarization can be contrasted to that for total binding as measured through equilibrium dialysis. The AT-rich regions appear to show little or no preference for one isomer for intercalative

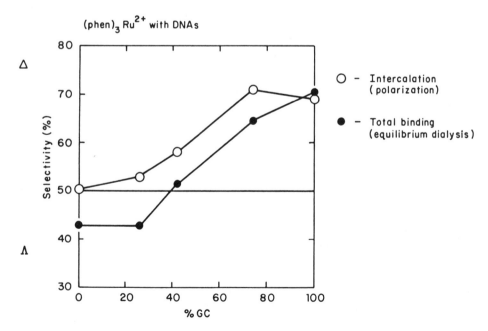

Figure 5. Stereoselectivity, S, associated with binding of Ru(phen)$_3^{2+}$ to DNAs of varying GC content determined by differences in steady state polarization (○) and through equilibrium dialysis (●). The polarization results indicate enantioselectivity for intercalation and the dialysis results, selectivity for total binding, (43).

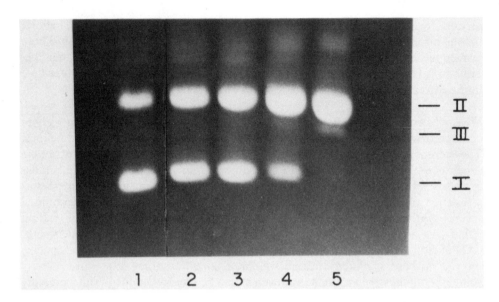

Figure 6. Strand scission of pColE1 DNA in the presence of Co(phen)$_3^{3+}$ and light. pColE1 (100 μM nucleotide) (lane 1), or pColE1 and 10 μM Co(phen)$_3^{2+}$ (lanes 2-5), irradiated at 254 nm (4W Mercury lamp) for 0, 0, 20, 40, 60 min (lanes 1-5, respectively). A complete conversion of form I to form II DNA is found in the 1% agarose gels.

binding, whereas the GC-rich DNAs show significant preference for the Δ isomer. Thus the AT sequences are less discriminatory than the GC sequences. One explanation could be that GC sequences are packed more closely together than AT sequences, hence providing a smaller groove for high chiral discrimination. Similar variations are found with increasing ionic strength where the helix surely is compressed owing to neutralization of phosphate-phosphate repulsions along the strand with increased salt. It is also evident from these experiments that the preference for the Λ isomer seen with AT rich DNAs is clearly not due to intercalative binding and here we introduce a new binding mode termed surface binding. The surface binding involves the strong hydrophobic interactions between the ligands and the helical groove. From models it is very clear that such type of binding would be highly stereoselective for the Λ isomer. Binding against the groove requires a *complementary* symmetry. A similar complementarity was found for direct coordination of phen$_2$RuCl$_2$ isomers to DNA (46). Certainly these alternate binding modes should provide the basis for designing new chiral probes for DNA conformations.

Metal Complexes as Site-Specific DNA Cleaving Molecules

By coupling the recognition characteristics of metal complexes for DNA conformations, which we learned about using the spectroscopy of ruthenium complexes, to the redox characteristics of cobalt(III) polypyridyl species, we actually are able to design a probe that both recognises and brings chemistry to local sites along the

DNA strand, in some sense mimicking DNA site-specific enzymes. The redox properties of the metal center can thus be appropriately exploited to carry out site specific and conformation-specific chemistry on the DNA template.

Other site-specific metal complexes as cleavers had been described earlier. Bleomycin, a glycopeptide antibiotic perhaps easily illustrates how nature has taken advantage of this approach (47-49). A tripeptide S, part of the bleomycin A_2 molecule binds to DNA with the same affinity as the complete molecule but does not result in strand scission or base release as does the complete metal complex (50). This tripeptide, which causes partial unwinding of DNA contains bithiazole and dimethyl sulfonium moieties and is thought to intercalate. Bleomycin complexes with Fe(II) (47-49,51), Mn(II) (52,53) or Cu(I) (54,55) in the presence of O_2 and reducing agents or Co(III) (56,57) upon irradiation, lead to DNA cleavage. Thus bleomycin is apparently a natural example to illustrate bifunctionality, where the ability exists to direct the redox active metal to a specific site (bleomycins much prefer 5'-GT-3' sites for cleavage). A comparable "affinity-cleaver" distamycin-EDTA-Fe(II) has been synthesized by Dervan and coworkers (58). These cleaving molecules are sequence specific. The octahedral trisphenanthroline complexes, however, are *structurally* directed; they recognize the DNA helicity and local conformation. Hence coupling to a redox active metal would yield a *conformation-specific* cleaving agent.

Actually the photoactivated nicking properties of the Co(III) bleomycin derivative (57) led us to investigate the chiral $Co(phen)_3^{3+}$ complex. $Co(phen)_3^{3+}$ causes single stranded cleavage of DNA upon irradiation (59). This is sensitively detected by agarose gel electrophoresis as seen in Fig. 6. With increasing irradiation of a 10:1 nucleotide:$Co(phen)_3^{3+}$ mixture, the conversion of compact supercoiled form I to nicked form II pColE1 DNA can be seen. This photoactivated cleaving reaction could be of use for *in vivo* experiments.

Interestingly the $Co(DIP)_3^{3+}$ scission reaction is stereospecific. Extensive cleavage of low superhelical densities of pColE1 occurs using the $\varDelta$-$Co(DIP)_3^{3+}$ enantiomer but none is evident using $\varLambda$-$Co(DIP)_3^{3+}$. This result suggests that *intercalation* of $\varDelta$-$Co(DIP)_3^{3+}$ into the right-handed helix is necessary for the scission reaction; electrostatic binding of the left-handed bulky trication is not sufficient for cleavage.

Significantly, while the $\varLambda$ isomer will not cleave pColE1 of low superhelix density, *both* $\varDelta$ and $\varLambda$-$Co(DIP)_3^{3+}$ cause strand scission of pBR322 ($\sigma = -0.058$). This is shown in Figure 7 plotted as the % loss of supercoiled (form I) DNA versus time of irradiation of 10:1 nucleotide:$Co(DIP)_3^{3+}$ solutions. $\varLambda$-$Co(DIP)_3^{3+}$ also cleaves pLP32, a pBR322 derivative with a Z-form $d(GC)_{16}$ insert (60). Thus the reagent is able to recognize small segments of varying local helicity. $\varLambda$-$Co(DIP)_3^{3+}$ will not cleave right-handed DNAs but will cleave left-handed DNAs and we can locate small sites in a left-handed conformation by virtue of cleavage in large segments of B-DNA.

In order to locate the discrete cleavage sites of $\varLambda$-$Co(DIP)_3^{3+}$ on a plasmid the experimental scheme detailed in Figure 8 may be used (61). pBR322 or pLP32 nicking by

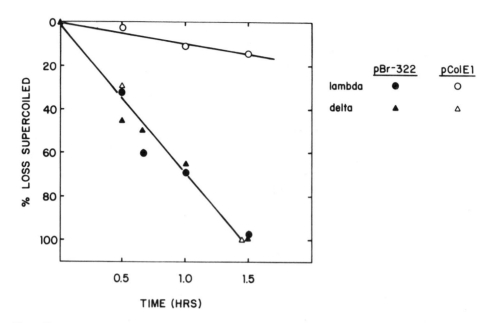

Figure 7. Stereoselective cleavage by Λ-Co(DIP)$_3^{3+}$. This plot shows the percent reduction in supercoiled (form 1) DNA intensity for pColE1 and pBR322, incubated with either Δ ($\triangle$) or Λ-Co(DIP)$_3^{3+}$ (O) (5 μM) as a function of time irradiated at 254 nm. pColE1, here at low superhelical densities, is only cleaved by the Δ isomer whereas pBR322 shows cleavage by both Δ and Λ-Co(DIP)$_3^{3+}$.

Λ-Co(DIP)$_3^{3+}$ (315 nm, 90 sec irradiation) is followed by single site restriction cleavage (with EcoRI, BamHI, Ava I, or Nde I) using an enzymatic excess. S1 nuclease is then used to cleave opposite the nick, and the resulting double stranded fragments are run out on a 1% agarose gel, stained with ethidium bromide and sized by comparing with molecular weight markers. A representative gel is shown in Figure 8. In fact double stranded fragments result that sum to 4.4 kilobase pairs (the full length of pBR322) indicating discrete recognition sites for cleavage.

The Λ-Co(DIP)$^3_{3+}$ cleavage sites map to the 32 base pair d(GC)$_{16}$ insert in pLP32 but also to four additional sites in pBR322 and pLP32. Results with pLP32 indicated to us that we were indeed able to cleave at a Z-DNA region on the plasmid, since the d(GC)$_{16}$ insert in pLP32 had been shown already to form a Z-DNA structure under these conditions (60). Perhaps then the four additional sites also were in a left-handed form. Our conditions used included low sodium concentrations which we believe favor the left-handed form. These four sites furthermore correspond to long alternating purine/pyrimidine sequences in the plasmid, and purine/pyrimidine stretches favor the left-handed Z-form under a variety of conditions (62). One last intriguing result is the association of these sites with the ends of genetic coding regions along the plasmid (see Table III). Perhaps Λ-Co(DIP)$_3^{3+}$ recognizes left-handed helices that are conformational punctuation marks designating the ends of genes.

Table III
Λ-Co(DIP)$_3$$^{3+}$ Cleavage Sites

	Alternating Purine-Pyrimidine Sequences*	Corresponding Regions of Biological Function
1.45 ± 0.05 kb	1447-1460 CACGGGTGCGCATG	The 3'-end of the tetracycline resistance gene is near 1425.
2.25 ± 0.07 kb	2315-2328 CGCACAGATGCGTA	The essential region for replication (origin of replication) extends from 2536-2360.
3.32 ± 0.11 kb	3265-3277 GTATATATGAGTA	The 3'-end of the β-lactamase gene is at 3295.
4.24 ± 0.02 kb	4254-64 TCCGCGCACAT	The 5'-start of the β-lactamase gene is at 4201, and the −35 consensus region for the promoter ends at 4236.

*The underlined residue identifies the base out of register.

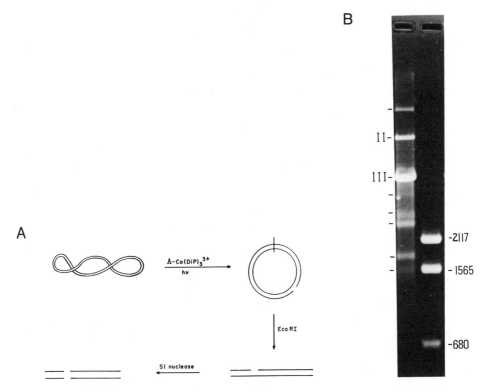

Figure 8. (a) Schematic of experimental procedure for coarse mapping of cleavage sites of Λ-Co(DIP)$_3$$^{3+}$ on plasmid DNAs. (b) 1% agarose gel of (left) double stranded fragments resulting from Λ-Co(DIP)$_3$$^{2+}$ and S1 nuclease cleavage of pBR322 and (right) a Rsa I digest of pBR322.

In sum, employing this simple octahedral complex, Λ-Co(DIP)$_3^{3+}$, we are able to locate small segments of left-handed Z regions within a long stretch of B-structure, in particular four alternating purine-pyrimidine sequences within pBR322. The power of this structural discrimination is clear; it leads to conformation specific cleavage of double helical DNA, based on the handedness of the metal complex. Moreover the results hint at a possible role for the left-handed conformation in gene expression.

Conclusion

It is thus possible to design small molecular probes, using chiral metal complexes that can detect local conformations of nucleic acids and can direct redox chemistry to specific locations along the double helical DNA. Based on these general principles we are currently developing chiral probes that can differentiate the range of conformations that nucleic acids can adopt. We would like to target these reagents "to various sites" along the strand and use the chemistry of the metal to report back to us the local polynucleotide structure. We must refine our techniques of recognition and take still greater advantage of DNA polymorphism, but using these small molecular probes, with well-defined structure, stereochemistry, and reactivity, we might be able to obtain unique information about local nucleic acid structure and its importance to gene expression both *in vitro* and even perhaps *in vivo*.

Acknowledgements

We are grateful to the National Institute of General Medical Science and the National Science Foundation for their financial support. We also thank Prof. N.J. Turro for his advice and collaborations.

References and Footnotes

1. Spiro, T.G., ed., *Nucleic Acid-Metal Ion Interactions* (John Wiley and Sons Inc., New York, 1980).
2. Barton, J.K., *Comments Inorg. Chem. 3,* 321 (1985).
3. Berman, H.M. and Young, P.R., *Ann. Rev. Biophys. Bioeng., 10,* 87 (1981).
4. Gale, E.F., Cundliffe, E., Reynolds, P.E., Richmond, M.H. and Waring, M. *The Molecular Basis of Antibiotic Action* (Wiley, London, 1972); Neidle, S. and Waring, M.J. ed., *"Molecular Aspects of Anti-Cancer Drug Action,"* Macmillan Press, Ltd., London (1983).
5. Waring, M., *J. Mol. Biol. 54,* 247 (1970).
6. Neidle, S., *Prog. Med. Chem. 16,* 151 (1979).
7. LePecq, J.B. and Paoletti, C., *J. Mol. Biol. 27,* 87 (1967).
8. Lerman, L.S., *J. Mol. Biol. 3,* 18 (1961).
9. Millar, D.P., Robbins, R.J. and Zewail, A.H., *J. Chem. Phys. 76,* 2080 (1982).
10. Fujimoto, B.S., Shibata, J.H., Schurr, R.L. and Schurr, J.M., *Biopolymers, 24,* 1009 (1985).
11. Lippard, S.J., *Acc. Chem. Res. 11,* 211 (1978); Jennette, K.W., S.J. Lippard, Vassiliades, G.A. and Bauer, W.R., *Proc. Natl. Acad. Sci., USA 71,* 3839 (1974).
12. Wang, A.H-J., Nathans, J., van der Marel, G., van Boom, J.M. and Rich, A., *Nature, 276,* 471 (1978).
13. Bond, P.J., Langridge, R., Jennette, K.W. and Lippard, S.J., *Proc. Natl. Acad. Sci. USA 72,* 4825 (1975).
14. Cairns, J., *Cold Spring Harbor Symp. Quant. Biol. 27,* 311 (1962); Crothers, D.M., *Biopolymers 6,* 595 (1968).
15. Lippard, S.J., Bond, P.J., Wu, K.C. and Bauer, K.C., *Science 194,* 726 (1976).
16. Howe-Grant, M., Wu, K.C., Bauer, W.R. and Lippard, S.J., *Biochemistry 15,* 4339 (1976).

17. Herzberg, R.P. and Dervan, P.B., *J. Am. Chem. Soc. 104*, 313, (1982).
18. Hertzberg, R.P. and Dervan, P.B., *Biochemistry 23*, 3934 (1984).
19. Van Dyke, M.W. and Dervan, P.B., *Biochemistry 22*, 2373 (1983).
20. Van Dyke, M.W. and Dervan, P.B., *Cold Spring Harbor Symp. Quant. Biol. 47*, 347, (1982).
21. Van Dyke, M.W. and Dervan, P.B., *Nucleic Acids Res. 11*, 5555 (1983).
22. Van Dyke, M.W. and Dervan, P.B., *Science, 225*, 1122 (1984).
23. Van Dyke, M.W., Hertzberg, R.P. and Dervan, P.B., *Proc. Natl. Acad. Sci., USA 79*, 5470 (1982).
24. Harshman, K.D. and Dervan, P.B., *Nucleic Acids Res. 13*, 4825, (1985).
25. Cartwright, I.L., Hertzberg, P.B., Dervan, P.B., Elgin, S.C.R., *Proc. Natl. Acad. Sci., USA 80*, 3213 (1983).
26. Pasternack, R.F., Gibbs, E.J., Villafranca, J.J., *Biochemistry 22*, 2406 (1983).
27. Fiel, R.J. and Munson, B.R., *Nucleic Acids Res. 8*, 2835 (1980).
28. Fiel, R.J., Beerman, T.A., Mark, E.M. and Datta-Gupta, N., *BBRC 107*, 1067 (1983).
29. Barton, J.K., Dannenberg, J.J. and Raphael, A.L., *J. Am. Chem. Soc. 104*, 4967 (1982).
30. Graham, D.R., Marshall, L.E., Reich, K.A. and Sigman, D.S., *J. Am. Chem. Soc. 102*, 5419 (1980).
31. Reich, K.A., Marshall, L.E., Graham, D.R. and Sigman, D.S., *J. Am. Chem. Soc. 103*, 3582 (1981).
32. Marshall, L.E., Graham, D.R., Reich, K.A. and Sigman, D.S., *Biochemistry, 20*, 244 (1981).
33. Pope, L.E. and Sigman, D.S., *Proc. Natl. Acad. Sci., USA 81*, 3, (1984).
34. Cartwright, I.L. and Elgin, S.C.R., *Nucleic Acids Res. 10*, 5835 (1982).
35. Jessee, B., Gargiulo, G., Razvi, F. and Worcel, A., *Nucleic Acids Res. 10*, 5823 (1982).
36. Brandt, W.W., Dwyer, F.P. and Gyarfas, E.C., *Chem. Rev., 54*, 959 (1954).
37. Gillard, R.D., Hill, R.E., *J. Chem. Soc., Dalton Trans. 1217* (1974).
38. McCaffery, A.J., Mason, S.F. and Norman, B.J., *J. Chem. Soc., A, 1428* (1969).
39. Sutin, N. and Creutz, C., *Pure Appl. Chem. 52*, 2717 (1980).
40. Barton, J.K., *J. Biomol. Struct. Dyn. 1*, 621 (1983).
41. Barton, J.K., Danishefsky, A.T. and Goldberg, J.M., *J. Am. Chem. Soc. 106*, 2172 (1984).
42. Kumar, C.V., Barton, J.K. and Turro, N.J., *J. Am. Chem. Soc. 107*, 5518 (1985).
43. Barton, J.K., Goldberg, J.M., Kumar, C.V. and Turro, N.J., *J. Am. Chem. Soc.*, in press.
44. Barton, J.K., Basile, L.A., Danishefsky, A.T. and Alexandrescu, A., *Proc. Natl. Acad. Sci., USA 81*, 1961 (1984).
45. Fueurstein, B.G., Marton, L.J., Keniry, M.A., Wade, D.L. and Shafer, R.H., *Nucleic Acids Res. 13*, 4133 (1985).
46. Barton, J.K. and Lolis, E., *J. Am. Chem. Soc. 107*, 708 (1985).
47. D'Andrea, A.D. and Haseltine, W.A., *Proc. Natl. Acad. Sci., USA 75*, 3608 (1978).
48. Uesugi, S., Shida, T., Ikehara, M., Kobayashi, Y. and Kyogoku, Y., *Nucleic Acids Res. 12*, 1581 (1984).
49. Giloni, L., Takeshita, M., Johnson, F., Iden, C. and Grollman, A.P., *J. Biol. Chem. 256*, 8608 (1981).
50. Takeshita, M., Grollman, A.P., Uhtsubo, E. and Ohtsubo, H., *Proc. Natl. Acad. Sci., USA 75*, 5983 (1978).
51. Sugiyama, H., Xu, C., Murugesan, N., Hecht, S.M., *J. Am. Chem. Soc., 107*, 4104 (1985).
52. Ehrenfeld, G.M., Murugesan, N., Hecht, S.M., *Inorg. Chem. 23*, 1498 (1984).
53. Burger, R.M., J.M. Freedman, Horwitz, S.B. and Peisach, J., *Inorg. Chem. 23*, 2215 (1984).
54. Murugesan, N., Ehrenfeld, G.M. and Hecht, S.M., *J. Biol. Chem. 257*, 8600 (1982).
55. Freedman, J.H., Horwitz, S.B. and Pesach, J., *Biochemistry 21*, 2203 (1982).
56. Chang, C-H., Meares, C.F., *Biochemistry 23*, 2268 (1984).
57. Chang, C-H., Meares, C.F., *Biochemistry 21*, 6332 (1982).
58. Schultz, P.G., Taylor, J.S. and Dervan, P.B., *J. Am. Chem. Soc. 104*, 6861 (1984).
59. Barton, J.K. and Raphael, A.L., *J. Am. Chem. Soc. 106*, 2466 (1984).
60. Peck, L.J., Nordheim, A., Rich, A., and Wang, J.C., *Proc. Natl. Acad. Sci., USA 79*, 4560 (1982).
61. Barton, J.K. and Raphael, A.L., *Proc. Natl. Acad. Sci., USA 82*, 6460 (1985).
62. Rich, A., Nordheim, A., and Wang, A.H.J., *Ann. Rev. Biochem., 53*, 791 (1984).

Biomolecular Stereodynamics III, Proceedings of the Fourth Conversation in the
Discipline Biomolecular Stereodynamics, State University of New York,
Albany, NY, June 04-09, 1985, Eds., Ramaswamy H. Sarma & Mukti H. Sarma,
ISBN 0-940030-14-4, Adenine Press, ©Adenine Press 1986.

The Interactions of Drugs and Carcinogens with Left-Handed and Right-Handed DNA

Thomas R. Krugh, G. Terrance Walker, David G. Sanford, John M. Castle and John A. Alley

Department of Chemistry
University of Rochester
Rochester, New York 14627

Abstract

The interactions of the drugs actinomycin D, adriamycin, and ethidium with poly(dG-dC) and poly(dG-m^{5}dC) have been studied under both B- and Z-form conditions. Each of the drugs exhibit highly cooperative binding under Z-form conditions. Circular dichroism spectra provide evidence that the conformation of the polynucleotide at the intercalation site is right-handed. Further support for the marked preference of the drugs for the right-handed conformation comes from an analysis of the binding isotherms in terms of an allosteric binding model. A covalent adduct of the carcinogen AAAF (N-acetoxy-2-acetylaminofluorene) with the single guanine in the deoxyoligonucleotide (CCACGCACC) was prepared, and mixed with the complementary strand (GGTGCGTGG) to form a stable deoxyoligonucleotide carcinogen-modified duplex. Comparison of the circular dichroism spectra of the unmodified duplex to the modified duplex suggests that the conformation of the modified duplex is left-handed, even in 0.1 M NaCl buffer. The present experiments provide information on the interaction of these drugs with Z-DNA and show that these drugs are convenient probes for studying conformational transitions in DNA.

Introduction

Left-handed (Z) DNA is a clear example of the structural flexibility of DNA. The salt-induced conformational transition of poly(dG-dC) as observed by Pohl & Jovin in 1972 (1) and the elucidation of the three dimensional structure of a left-handed duplex by Wang et al. in 1979 (2) have spurred a tremendous interest in the relationship between the structure and function of DNA (1-27). Poly(dG-dC) assumes a left-handed conformation in 4.4 M NaCl to which the intercalating ligand ethidium (Figure 1) does not bind efficiently until the concentration of unbound ethidium reaches ~20 μM (4). Circular dichroism (CD) spectroscopy shows that the binding of ethidium to poly(dG-dC) in 4.4 M NaCl is accompanied by a left-handed to right-handed conformational transition of the polynucleotide.

The observation that ethidium binds very poorly until the free ethidium concentration reaches a minimum value led to the question of whether other antibiotics would

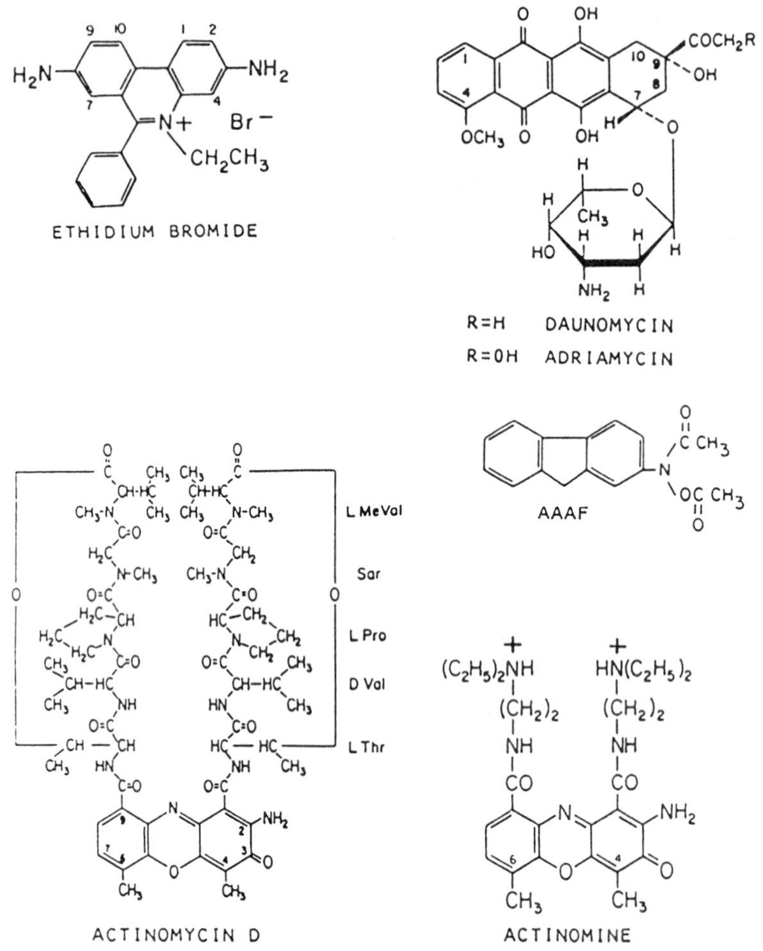

Figure 1. Molecular structures of ethidium bromide, daunomycin, adriamycin, actinomycin D, N-acetoxy-2-acetylaminofluorine (AAAF), and actinomine. Abbreviations: Thr = threonine; Val = valine; Pro = proline; Sar = sarcosine; MeVal = methylvaline. The tertiary nitrogens of actinomine are protonated at pH 7.

The observation that ethidium binds very poorly until the free ethidium concentration reaches a minimum value led to the question of whether other antibiotics would exhibit similar behavior. Our experiments to address this question began in the summer of 1981 with the study of the binding of actinomycin D to poly(dG-dC) in 4.4 M NaCl. Actinomycin D is a particularly interesting antibiotic which combines an intercalating chromophore with two cyclic pentapeptides that bind in the minor groove of the helix (28). We learned quickly that actinomycin D behaves differently than ethidium in binding to poly(dG-dC) in 4.4 M NaCl; actinomycin D exhibits efficient binding at free actinomycin concentrations lower than 1 μM. We extended these studies with actinomine and adriamycin. Actinomine interactions with DNA provide an important contrast to actinomycin D because the cyclic pentapeptides are replaced by charged substituents (Figure 1). Adriamycin is an important antitumor drug which binds cooperatively to calf thymus (29) DNA and has been shown to inhibit the B to Z transition due to the preferential binding of adriamycin to right-handed DNA (6,7). Initial experiments led to an extended investigation of the interaction of ethidium, actinomycin D and actinomine with the left-handed forms of poly(dG-dC) and poly(dG-m⁵dC) in which it was considered essential that the degree of drug binding be correlated with the conformational state of the poly-nucleotide. The results of these studies are reported in this manuscript and else-where (12-14).

The carcinogen AAAF (N-acetoxy-2-acetylaminofluorene) reacts at the C-8 position of guanine to form a covalent adduct (38), with a concomitant distortion in the structure of the helix at the site of adduct formation. Modification of poly(dG-dC) by AAAF facilitates conversion of the polynucleotide from B-form to a left-handed form (17-20). Detailed conformational information relevant to AAF-guanine adducts in double helical DNA requires a model system of an AAF adduct with a stable deoxyoligonucleotide. We present below the CD spectra of AAAF modified and unmodified duplexes of a oligonucleotide, which suggest that modification results in the adoption of a left-handed conformation.

Although the role of left-handed (Z) DNA in biology is not yet certain, the present experiments provide new insights into the interaction of important ligands with both B- and Z-DNA and provide a means of studying the B-Z transitions of nucleic acids.

Materials and Methods

Buffers. A high salt buffer consisting of 4.4 M NaCl, 10 mM Na_2PO_4, 10 mM Na_2EDTA, pH 7, will be referred to as 4.4 M NaCl Buffer. A low salt buffer consisting of 50 mM NaCl, 5 mM TRIS, pH 8, will be referred to as 50 mM NaCl Buffer. Other buffers consisted of 50 mM NaCl, 5 mM TRIS, pH 8, and varying amounts of magnesium chloride or cobalt hexamine chloride as indicated; these buffers will be referred to as 2 mM $MgCl_2$ Buffer, 25 mM $MgCl_2$ Buffer, or 40 μM $[Co(NH_3)_6]Cl_3$ Buffer. A buffer consisting of 0.1 M NaCl, 10 mM $NaHPO_4$, 0.1 mM Na_2EDTA, pH 7, was used in the oligonucleotide experiments.

Drug Solutions. Ethidium bromide (Sigma) was recrystalized from methanol. Actinomycin D (NSC-3053) was a gift of Merck, Sharp and Dohme. Actinomine was a gift of Dr. Sisir Sengupta of the Children's Cancer Foundation. Actinomycin D and actinomine were checked for purity by TLC and HPLC as previously described (14). Adriamycin (NSC-123127), which was obtained from the Natural Products Branch of the National Cancer Institute, was used without purification. Drug solutions were prepared immediately before each experiment.

DNA Solutions. Poly(dG-dC) and poly(dG-m⁵dC) were purchased from P. L. Biochemicals/Pharmacia and were used without purification. Establishment of the left-handed forms of the polynucleotides and preparation of calf thymus DNA were as described previously (13).

Optical Titrations. Optical titrations were performed on a Varian CARY 219 spectrophotometer at 25° C using 1, 5 or 10 cm pathlength cells. A detailed description of the experimental procedure and methods of data analysis were previously described (13,14).

Fluorescence Spectroscopy. Spectra were recorded on a Perkin-Elmer MPF44A spectrofluorometer at ambient temperature. Excitation and emission slit widths were 10 nm. Ethidium spectra were recorded with excitation at 510 nm, the isosbestic point between the absorption spectra of free and DNA-bound ethidium.

Circular Dichroism (CD) Spectroscopy. Spectra were recorded on a Jasco J-40 spectropolarimeter at ambient temperature using 1, 2 and 5 cm pathlength cells. The spectra were recorded at 5 nm intervals and were signal averaged and baseline corrected using a Digital PDP 11/34 computer. Sample preparation was as previously described (13,14). Molar ellipticity, [θ], values are reported in terms of DNA base pairs, except for the AAF modified strands where the molar ellipticity is reported in terms of nucleotide concentration for comparison to previous CD spectra of modified polymers.

Delta epsilon values at 320 nm for bound ethidium ($\Delta\epsilon_b^{320}$) were calculated from the equation:

$$\Delta\epsilon_b^{320} = \epsilon_L - \epsilon_R = \theta/(32.98 \cdot C_b \cdot l) \tag{1}$$

where θ is ellipticity (degrees); C_b is the molar concentration of bound ethidium; and l is the pathlength (cm). Bound ethidium concentrations were determined directly from absorption spectroscopy.

Results and Discussion

Ethidium Binding and B-Z Equilibrium

The effect of the B-Z transition upon ethidium binding is illustrated through parallel binding experiments in 4.4 M NaCl Buffer with calf thymus DNA and poly(dG-dC). Addition of calf thymus DNA to a 5 μM ethidium solution in 4.4 M NaCl results in bathochromic and hypochromic shifts in the visible absorption spectrum and a large increase in fluorescence intensity (Figure 2). These spectral changes are characteristic of ethidium intercalation and are the basis of absorption and fluorescence titration techniques (30,31). Nearly identical spectral changes are observed immediately after addition of B-form poly(dG-dC) to a solution of 5 μM ethidium in 4.4 M NaCl. Similar spectral properties immediately after mixing are expected for the two samples because ethidium intercalation into B-DNA occurs on a much faster timescale than the B to Z transition (9). However, differences between the calf thymus DNA and poly(dG-dC) samples become apparent after B-Z equilibrium is established. After heating the poly(dG-dC) sample, which hastens B-Z equilibration, the absorption and fluorescence spectra revert back toward the spectrum characteristic

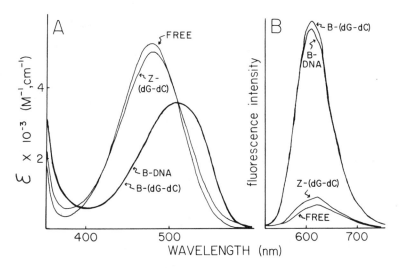

Figure 2. Absorbance (A) and fluorescence spectra (B) for bound and free ethidium in 4.4 M NaCl Buffer. The spectra correspond to free ethidium [Free], ethidium in the presence of calf thymus DNA [B-DNA], ethidium in the presence of B-form poly(dG-dC) [B-(dG-dC)], and ethidium in the presence of Z-form poly(dG-dC) [Z-(dG-dC)]. B-Z equilibrium in the poly(dG-dC) sample was established by heating for 10 minutes at 60° C. The fluorescence spectra are uncorrected. Further explanation as to experimental conditions are provided in the text.

of free ethidium, which is consistant with the dissociation of significant amounts of bound drug. A comparison of these samples clearly illustrates that ethidium binding to right-handed DNA is much more favorable than binding to a left-handed poly(dG-dC) helix in 4.4 M NaCl Buffer. It should be noted that the B to Z conversion and the concomitant dissociation of ethidium occur without heating, but very long equilibration times are required, consistent with the observation by Mirau & Kearns (9) that the presence of ethidium slows the rate of the B to Z transition.

Cooperative Binding with Z-DNA

The binding preference of ethidium for right-handed DNA and the highly cooperative B-Z equilibrium results in highly cooperative binding under Z-form conditions. The equilibrium binding isotherm for the interaction of ethidium with poly(dG-dC) in 4.4 M NaCl is shown in Figure 3. Positive cooperative binding is evident by the positive slope and near zero r/C_f intercept in the Scatchard (32) plot (Figure 3A) as well as the sigmoidal shape of the r versus C_f plot (Figure 3B) where r is defined as the bound drug to DNA base pair ratio and C_f is the concentration of free drug. The near zero r/C_f intercept results from the free drug concentration reaching approximately 20 μM before significant binding occurs, as illustrated in Figure 3B and as originally reported by Pohl et al. (4).

Shafer et al. (11) have performed ethidium-poly(dG-dC) experiments in 4.4 M NaCl similar to those described by Figures 2 and 3, but the much greater binding affinity

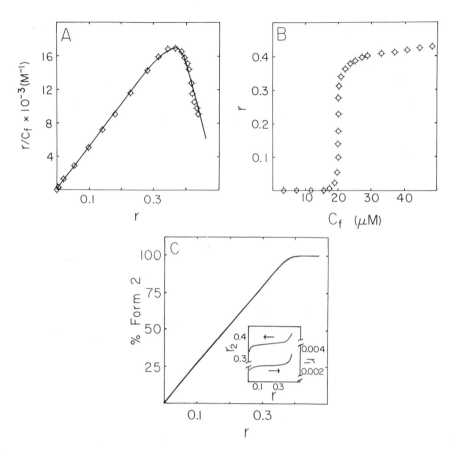

Figure 3. Equilibrium binding isotherm and binding model fit for the interaction of ethidium and Z-form poly(dG-dC) in 4.4 M NaCl Buffer. The data were obtained by optical titration methods (13). (A) Scatchard plot with the solid line representing a fit to the allosteric transition binding model (33). (B) Plot of r versus the free drug concentration, C_f. (C) Calculated percent of the DNA in Form 2 as a function of r according to the allosteric transition binding model (33). The inset contains plots of the average r value for binding to Forms 1 and 2 (r_1 and r_2, respectively) as a function of the overall r value for the polynucleotide. Note the separate scales for r_1 and r_2. The best fit of the model estimates that binding to right-handed DNA occurs with a cooperativity parameter (τ) of 3 and a binding constant which is ~300 times that for binding to left-handed DNA.

of ethidium for right-handed DNA was not readily apparent due, at least in part, to the high ethidium and poly(dG-dC) concentrations used in their experiments. At ethidium concentrations much greater than 20 μM and with millimolar poly(dG-dC) concentrations, a significant fraction of ethidium is expected to bind to poly(dG-dC) in 4.4 M NaCl due to the highly cooperative binding nature under these Z-form conditions, thus making it more difficult to observe the marked binding preference of ethidium for the right-handed conformation.

As shown in Figure 3A, the binding isotherm can be satisfactorily represented by an allosteric transition binding model (33) in which the polynucleotide can adopt two conformations, Form 1 (left-handed) and Form 2 (right-handed). In 4.4 M NaCl, poly(dG-dC) exists as Form 1 DNA at the start of the experiment but changes to Form 2 because of the preferential binding of ethidium to Form 2. The allosteric model describes a sequential conversion of Z-DNA to a right-handed form as ethidium binds (Figure 3C). The model confirms the clustering of ethidium into regions of right-handed DNA, as deduced from the experimental CD data described below. For example, r_2, the r value for ethidium binding to right-handed DNA, rises sharply to a value of ~0.35 at low overall values of r, and then remains constant until saturating levels of ethidium are reached (inset of Figure 3C). In contrast, binding of ethidium to left-handed DNA is calculated to occur at very low densities ($r_1 \approx 0.0025$).

Ethidium Binding Induces a Left- to Right-Handed Conversion in DNA.

The binding of ethidium to B-form poly(dG-dC) produces significant changes in the CD spectrum of the sample (Figure 4A). Comparison of the CD spectra in Figure 4A with fluorescence detected circular dichrosim (FDCD) spectra of ethidium bound to poly(dG-dC) (15) and to dinucleotides (34) reveals that the majority of the change in CD under B-form conditions may be attributed to contributions from bound ethidium. CD spectrocopy provides a convenient method for monitoring the left- to right-handed transition of Z-form poly(dG-dC) in the presence of ethidium. The initial Z-form CD spectrum in the absence of ethidium is converted to a spectrum characteristic of ethidium bound to right-handed poly(dG-dC) by the end of the titration in 4.4 M NaCl Buffer (Figure 4). At saturating levels of bound ethidium, the resultant CD spectra in Figures 4A & 4B are quite similar, suggesting that the conformation of the ethidium saturated polynucleotide is relatively independent of B- or Z-form conditions. The occurance of the isoelliptical point at

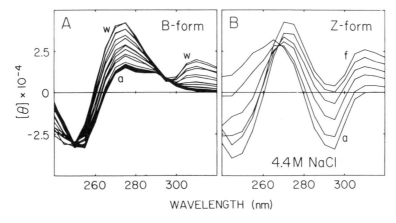

Figure 4. Circular dichroism (CD) spectra of ethidium:poly(dG-dC) samples in 50 mM NaCl Buffer (B-form) (A) and 4.4 M NaCl Buffer (Z-form) (B). In each panel the initial spectrum of poly(dG-dC) in the absence of ethidium has been denoted by an 'a', while the final spectrum of each titration, which corresponds to the largest [ethidium]/[base pair] ratio, has been denoted with a 'w' (A) and an 'f' (B).

~268 nm in Figure 4B is consistant with a systematic left- to right-handed conversion without the formation of major components of alternate conformation. This observation supports the existance of short interfaces between right-handed drug-bound DNA and left-handed DNA (B'-Z interfaces), in agreement with the observation of short B-Z interfaces in bacterial plasmids (35).

CD spectra of ethidium bound to poly(dG-dC) and to poly(dG-m⁵dC) have been recorded under a variety of Z-form conditions with the r value corresponding to each spectrum determined from optical titration and phase partition techniques (13). This approach enables determination of the extent of the left- to right-handed conversion per bound ethidium. Changes in ellipticity were monitored at 295 nm as a function of r. The CD data at 295 nm were chosen because the spectrum of bound ethidium approaches a minimum at this wavelength (15,34). Data are shown in Figure 5 for ethidium and poly(dG-dC) under B-form conditions, Z-form conditions

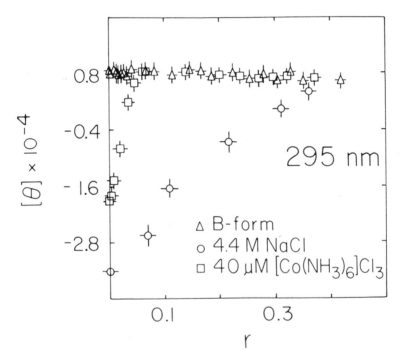

Figure 5. Molar ellipticity values at 295 nm as a function of r for ethidium titrations of poly(dG-dC) under B- and Z-form conditions. The buffers were: 50 mM NaCl (B-form), (△); 4.4 M NaCl (Z-form), (○); and 40 23μM [Co(NH₃)₆]Cl₃ (Z-form), (□). The values of r were determined from optical titration techniques (13).

in 4.4 M NaCl, and for the Z-form stabilized by 40 μM cobalt hexamine chloride. The CD data at 295 nm remain relatively constant as a function of r for the B-form titration, thereby providing an upper limit for the left- to right-handed conversion which produces large changes in [θ] at 295 nm. The reciprocal of the r value where each set of Z-form data intersect the B-form data provides an indication of the bound ethidium level at which the left-handed form of the polynucleotide is converted to a right-handed (drug-bound) form under the particular Z-form conditions. The reciprocal of this r value represents the average number of left-handed base pairs converted per bound ethidium. These values are listed in Table I.

Table I
Number of Left-Handed Base Pairs Converted Per Bound Drug

| | Z-poly(dG-dC)·poly(dG-dC) | | Z-poly(dG-m⁵dC)·poly(dG-m⁵dC) | |
	4.4 M NaCl	40 μM [Co(NH₃)₆]Cl₃	2 mM MgCl₂	25 mM MgCl₂
actinomycin D	4-6	20-25		
ethidium¹⁺	2-3	20-25	6-8	4-5
actinomine²⁺		10-15	6-8	

Is Cooperative Binding with Z-DNA Unique to Ethidium?

We have studied the binding of adriamycin, actinomycin D and actinomine under Z-form conditions through a variety of phase partition and optical titration techniques. All four drugs exhibit cooperative binding isotherms under Z-form conditions, although the exact shape of a binding isotherm under a given set of conditions is unique to each drug.

The binding isotherm for adriamycin and Z-form poly(dG-dC) in 4.4 M NaCl is shown in Figure 6. As described for ethidium, the sigmoidal shape of the r versus C_f plot is indicative of positive cooperative binding. Chaires (16) has performed similar binding experiments with adriamycin and daunomycin and has obtained similar results.

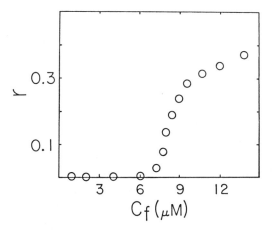

Figure 6. Equilibrium binding isotherm for the interaction of adriamycin and Z-form poly(dG-dC) in 4.4 M NaCl Buffer. The data are plotted as r versus the free drug concentration, C_f. The data were obtained by optical titration methods using molar extinction coefficients (M⁻¹, cm⁻¹) at 477 nm of 5350 and 9680 for the bound and free forms of adriamycin, respectively.

Note that very little adriamycin binds to poly(dG-dC) in 4.4 M NaCl until the free drug concentration reaches approximately 7 μM (Figure 6). This compares to values of ~20 μM for ethidium (Figure 3) and ~0.3 μM for actinomycin D (14); these values are influenced by the binding affinity of the ligand to the right-handed form, which in turn is related to the charge of the ligand. In 4.4 M NaCl Buffer the binding affinity of actinomine to the right- and left-handed forms is so small that no binding to poly(dG-dC) is observable by absorption spectroscopy even at 100 μM free drug concentrations.

The extent of the left- to right-handed conversion per bound drug has been determined for ethidium, actinomycin D, and actinomine under a variety of Z-form conditions (Table I). The number of base pairs converted per bound drug does not vary significantly among the drugs for a given set of Z-form conditions. Slight differences in the values listed in Table I primarily reflect differences in the site exclusion parameters, which are defined as the number of DNA base pairs per bound ligand at saturation levels of drug (36,37). The polynucleotide and the particular Z-form conditions employed are the dominant factors controlling the number of left-handed base pairs converted per bound drug.

Effects of Z-DNA Stability and Ionic Strength on Drug Binding

Factors which influence the B-Z equilibrium alter the shape of the binding isotherms, as illustrated by a comparison of ethidium binding to left-handed poly(dG-m^{5}dC) in 2 and 25 mM MgCl$_2$. The stability of the left-handed form of poly(dG-m^{5}dC) increases with higher magnesium concentrations under these buffer conditions. Consequently, the free ethidium concentration (C$_f$), at which half saturation of the polynucleotide occurs, increases from ~0.2 to ~3.5 μM in going from 2 to 25 mM MgCl$_2$ Buffer, respectively (Figure 7). Likewise the number of left-handed base pairs converted per bound ethidium decreases from ~7 to ~4 base pairs. Analogous behavior is observed with ethidium and poly(dG-dC) as the salt concentration is varied from 2 to 4.4 M NaCl (unpublished results), as well as with adriamycin and daunomycin (16).

Under conditions where the Z-form of poly(dG-dC) is only slightly lower in energy than the B-form, more than 20 base pairs of left-handed polynucleotide switch to a right-handed conformation for each bound ethidium. The shape of the binding isotherms also reflects the type of binding. When the Z-form is much more stable than the B-form, the isotherms exhibit a near zero intercept on the r/C$_f$ axis which results from inefficient binding of the drug until a minimum concentration is reached.

Ethidium Clustering Under Z-form Conditions

The CD bands above 315 nm for an ethidium-polynucleotide sample arise solely from bound ethidium because the polynucleotide has negligible absorption above 315 nm. It has been observed that the magnitude of the ellipticity at 320 nm, when normalized per bound ethidium ($\Delta\epsilon_b^{320}$), is a function of the average distance between

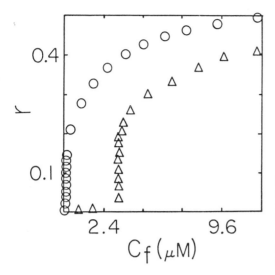

Figure 7. Equilibrium binding isotherms for the interaction of ethidium and Z-form poly(dG-m⁵dC) in 2 mM MgCl₂ (○) and 25 mMᶠ MgCl₂ Buffer (△). The data are plotted as r versus the free drug concentration, C_f. The data were obtained by optical titration methods (13).

ethidiums intercalated along the polymer (34). Under B-form conditions, the average distance between bound ethidiums decreases with increasing r as drug binding occurs at random positions along the polynucleotide. This is reflected in an increase in $\Delta\epsilon_b^{320}$ as a function of r (Figure 8). In contrast, for the ethidium titration of poly(dG-dC) in 4.4 M NaCl Buffer, the magnitude of $\Delta\epsilon_b^{320}$ remains essentially constant as a function of r at a value consistent with nearly saturated polymer (Figure 8). We interpret this observation as evidence for the formation of nearly saturated regions of right-handed ethidium-bound DNA (i.e., clusters), even at low overall levels of saturation where the majority of the polynucleotide exists in a left-handed helix.

In the case of ethidium and Z-form poly(dG-m⁵dC) in 25 mM MgCl₂ Buffer, approximately 5 left-handed base pairs are converted per bound ethidium in comparison to ~3 base pairs with Z-form poly(dG-dC) in 4.4 M NaCl. Consequently, the values of $\Delta\epsilon_b^{320}$ correspond to an average distance of ~5 base pairs between bound ethidiums for the r value range over which the left- to right-handed conversion occurs in 25 mM MgCl₂ (0<r<0.2). At r values greater than 0.2, the left-handed form of poly(dG-m⁵dC) is eliminated, and the values of $\Delta\epsilon_b^{320}$ increase with r as ethidium binds to the right-handed form of poly(dG-m⁵dC). Verification of ethidium clustering has also been obtained through fluorescence detected circular dichroism (FDCD) studies (15).

A representation of the clustering model is shown in Figure 9 in which left-handed and right-handed DNA coexist on the same polynucleotide, separated by a B'-Z interface. While Figure 9 was designed to represent ethidium binding to poly(dG-dC) in 4.4 M NaCl, it may be taken as a general depiction of drug clustering under Z-form conditions. The tendency of ligands to form nearly saturated regions of

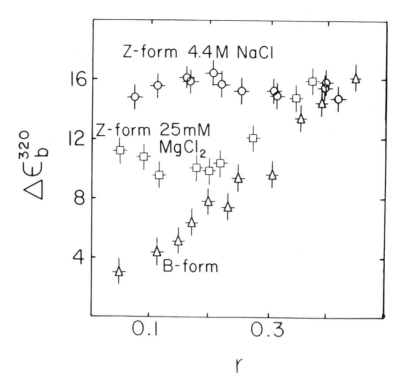

Figure 8. Plot of ϵ_L-ϵ_R at 320 nm per bound ethidium as a function of r for ethidium titrations of poly(dG-dC) under B-form conditions (50 mM NaCl Buffer), ($\triangle$); poly(dG-dC) under Z-form conditions (4.4 M NaCl Buffer), ($\bigcirc$); and poly(dG-m⁵dC) under Z-form conditions (25 mM MgCl₂ Buffer), ($\square$). The values of $\Delta\epsilon_b^{320}$ were calculated as described in the text. The values of r were obtained by optical titration techniques (13).

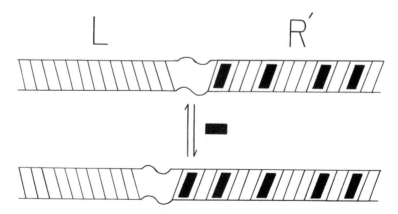

Figure 9. Illustration of the clustering model for drug binding under Z-form conditions. The left-handed form (L) is converted to a drug-bound right-handed form (R′) with a B′-Z interface separating the two forms of the DNA.

right-handed DNA (i.e., clusters) in 4.4 M NaCl Buffer may be understood in terms of the relative stability of B- and Z-DNA, and by the presence of energetically unfavorable B-Z (and B'-Z) interfaces, the number of which are expected to remain at a minimum during the titration.

The binding isotherms and CD data obtained with adriamycin and actinomycin D strongly suggest that these drugs bind in clusters also, especially in 4.4 M NaCl, following the same general binding model elaborated above for ethidium.

Ethidium Induces a Conformational Change in a Carcinogen-Oligonucleotide Adduct

The carcinogen N-acetoxy-2-acetylaminofluorene (AAAF) binds covalently to guanine (38) and facilitates the transition from B- to Z-form DNA in alternating pyrimidine-purine sequences (17-20). The carcinogen was reacted with d(CCACGCACC) to form a covalently modified oligonucleotide. A modified duplex was formed when the complementary strand, d(GGTGCGTGG), was added (Figure 10). The

```
        C           G                           C · G
        C           G                           C · G
        A           T                           A · T
        C           G                           C · G
AAF - G   +   C          ⇌            AAF - G   C
        C           G                           C · G
        A           T                           A · T
        C           G                           C · G
        C           G                           C · G
```

Figure 10. Schematic representation of the covalent modification of an oligonucleotide and the formation of a modified duplex. AAAF reacts with the single guanine in d(CCACGCACC) to form a covalent adduct. The purified adduct is then mixed 1:1 with the complementary strand d(GGTGCGTGG) to form a modified duplex.

unmodified duplex (no AAAF) has a CD spectrum characteristic of B-DNA, whereas the CD spectrum of the AAAF modified duplex supports formation of a left-handed duplex conformation even in 0.1 M NaCl (Figure 11). Titration of the modified duplex with ethidium results in conversion of the CD spectrum to one which is quite similar to the CD spectrum of the unmodified duplex with ethidium. The spectra of the unmodified and modified duplexes at saturating levels of ethidium are not identical, but the difference may be due to contributions to the CD from the carcinogen or the effect of the carcinogen on the oligonucleotide duplex conformation. The observed transition in the CD spectrum of the modified duplex upon binding ethidium is reminiscent of the changes that accompany the left- to right-handed conversion of poly(dG-dC) in 4.4 M NaCl and is taken as supportive evidence that the modified duplex has a left-handed conformation at low salt and in the absence of ethidium. Proton NMR experiments in progress are expected to provide further details on the conformation of the adduct.

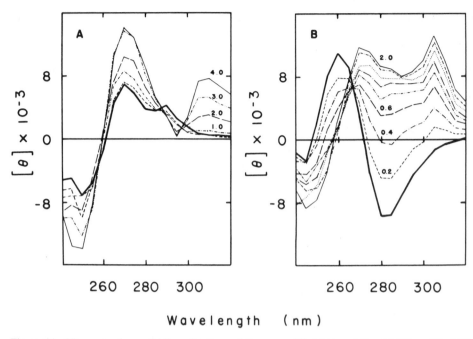

Figure 11. CD spectra from ethidium titrations of the unmodified (A) and the AAAF modified (B) oligonucleotide duplex described in Figure 10. The bold lines indicate the spectra in the absence of ethidium. The numbers above selected spectra indicate the input ratio of ethidium molecules per duplex. Spectra were recorded at 5° C. The molar ellipticity is given in terms of the nucleotide concentration.

Concluding Remarks

The strong preference of ethidium for a right-handed intercalation site is particularly interesting because model building studies do not show stereochemical restrictions for intercalation into a left-handed helix (8). The relative preference for ethidium intercalation into the right- and left-handed forms of poly(dG-dC) in 4.4 M NaCl may be expressed in terms of the ratio of the binding constants as calculated by the allosteric transition model (33); for ethidium, the K_2/K_1 ratio appears to be as large as 300:1 favoring binding to the right-handed form. With actinomycin D, the K_2/K_1 ratio appears to be substantially larger than 300:1, which may arise from the interaction of the cyclic pentapeptides with the outside of the helix (14). The kinetics of ligand binding in the Z-DNA systems are likely to provide new insights into these systems because the activation free energies will contain both electrostatic and hydration contributions, as evidenced for example, by the dependence of the dissociation rate of actinomycin D from poly(dG-dC) on both the NaCl concentration and the solvent composition (39,40). For Z-DNA regions stabilized by supercoiling, the activity of the solvent is presumably the same for both left- and right-handed forms, but the electrostatic terms will be quite different for the two structually different forms, and thus one would expect different rate constants in these systems as well.

B-Z junctions provide a region (albeit, a short region (35)) of unwound DNA which may present an appropriate site for ligand-DNA interactions (41-43). Although the conformation of the B-Z junction, or in our experiments the B'-Z junction, is unknown, it seems reasonable to consider at least a portion of the junction as unwound right-handed DNA which may present a favorable site for the intercalating ligand to bind. This is analogous to the proposal by Sobell and coworkers (41,43) that beta DNA is the target site for intercalation into right-handed DNA. While the Z-DNA region may present one type of target for proteins or ligands, the junction region may present another type of target for a different set of ligands. Indeed, the junctions in these systems may serve as a paradigm for the interactions of ligands with regions of DNA which are structurally unique from the bulk of the polymer. These local structures may arise from the sequence, or they may serve as connecting pieces between two structurally unique regions.

The present experiments have direct pharmacological significance if Z-DNA plays an important role in cellular processes. However, even without this justification, this area will continue to be explored because the ligands have proven to be important probes in studying the conformational properties of poly(dG-dC), as well as carcinogen-modified oligonucleotides.

Acknowledgments

This research was supported by research grant CA-35251 from the National Cancer Institute.

References and Footnotes

1. F. M. Pohl and T. M. Jovin, *J. Mol. Biol. 67*, 375 (1972).
2. A. H.-J. Wang, G. J. Quigley, F. J. Kolpak, J. L. Crawford, J. H. van Boom, G. A. van der Marel and A. Rich, *Nature 282*, 680 (1979).
3. M. Behe and G. Felsenfeld, *Proc. Natl. Acad. Sci. USA 78*, 1619 (1981).
4. F. M. Pohl, T. M. Jovin, W. Baehr and J. J. Holbrook, *Proc. Natl. Acad. Sci. USA 69*, 3805 (1972).
5. J. H. van de Sande and T. M. Jovin, *EMBO 1*, 115 (1982).
6. J. B. Chaires, *Nucleic Acids Res. 11*, 8485 (1983).
7. C. Chen, R. B. Knop and J. S. Cohen, *Biochemistry 22*, 5468 (1983).
8. G. Gupta, M. M. Dhingra and R. H. Sarma, *J. Biomol. Struct. Dyn. 1*, 97 (1983).
9. P. A. Mirau and D. R. Kearns, *Nucleic Acids Res. 11*, 1931 (1983).
10. C. Zimmer, C. Marck and W. Guschlbauer, *FEBS Lett. 154*, 156 (1983).
11. R. H. Shafer, S. C. Brown, A. Delbarre and D. Wade, *Nucleic Acids Res. 12*, 4679 (1984).
12. T. R. Krugh and G. T. Walker, *Studia Biophysica 104*, 133 (1985).
13. G. T. Walker, M. P. Stone and T. R. Krugh, *Biochemistry* in press (1985a).
14. G. T. Walker, M. P. Stone and T. R. Krugh, *Biochemistry* in press (1985b).
15. M. L. Lamos, G. T. Walker, T. R. Krugh and D. H. Turner, *Biochemistry* in press (1986).
16. J. B. Chaires, *Biophys. J. 47*, W-AM-D6 (1985) and personal communication.
17. E. Sage and M. Leng, *Proc. Natl. Acad. Sci. USA 7*, 4597 (1980).
18. E. Sage and M. Leng, *Nucleic Acids Res. 9*, 1241 (1981).
19. R. M. Santella, D. Grunberger, I. B. Weinstein and A. Rich, *Proc. Natl. Acad. Sci. USA 78*, 1451 (1981).

20. R. M. Santella, D. Grunberger, A. Nordheim and A. Rich, *Biochem. Biophys. Res. Comm. 106,* 1226 (1982).
21. A. Moller, A. Nordheim, S. A. Kozlowski, D. J. Patel and A. Rich, *Biochemisty 23,* 54 (1984).
22. P. Rio and M. Leng, *Nucleic Acids Res. 11,* 4947 (1984).
23. D. G. Sanford and T. R. Krugh, *Nucleic Acids Res. 13,* 5907 (1985).
24. B. Malfoy, B. Hartmann and M. Leng, *Nucleic Acids Res. 9,* 5659 (1981).
25. H. M. Ushay, R. M. Santella, J. B. Caradonna, D. Grunberger and S. J. Lippard, *Nucleic Acids Res. 10,* 3573 (1982).
26. R. Durand, C. Job, D. A. Zarling, M. Tissiere, T. M. Jovin and D. Job, *EMBO 2,* (1983).
27. J. J. Butzow, Y. A. Shin and G. L. Eichhorn, *Biochemistry 23,* 4837 (1984).
28. W. Muller and D. M. Crothers, *J. Mol. Biol. 35,* 251 (1968).
29. D. E. Graves and T. R. Krugh, *Biochemistry 22,* 3941 (1983).
30. J.-B. Le Pecq and C. Paoletti, *J. Mol. Biol. 27,* 87 (1967).
31. S. A. Winkle, L. S. Rosenberg and T. R. Krugh, *Nucleic Acids Res. 10,* 8211 (1982).
32. G. Scatchard, *Ann. N. Y. Acad. Sci. 51,* 660 (1949).
33. J. L. Bresloff and D. M. Crothers, *Biochemistry 20,* 3547 (1981).
34. K. S. Dahl, A. Pardi and I. Tinoco, *Biochemistry 21,* 2730 (1982).
35. S. M. Stirdivant, J. Klysik and R. D. Wells, *J. Biol. Chem. 257,* 10159 (1982).
36. D. M. Crothers, *Biopolymers 6,* 575 (1968).
37. J. D. McGhee and P. H. von Hippel, *J. Mol. Biol. 86,* 469 (1974).
38. E. C. Miller, U. Juhl and J. A. Miller, *Science 153,* 1125 (1966).
39. T. R. Krugh, J. W. Hook, S. Lin and F.-M. Chen in *Stereodynamics of Molecular Systems,* Ed., R. H. Sarma, Pergamon Press, New York, p.423 (1979).
40. K. R. Lee, PhD Thesis, University of Rochester, Rochester N. Y. (1981).
41. A. Banerjee and H. Sobell, *J. Biomol. Struct. Dyn. 1,* 253 (1983).
42. L. S. Rosenberg, M. J. Carvlin and T. R. Krugh, *Biochemistry,* in press (1986).
43. H.M. Sobell, *Proc. Natl. Acad. Sci. USA 82,* 5328 (1985).

Biomolecular Stereodynamics III, Proceedings of the Fourth Conversation in the Discipline Biomolecular Stereodynamics, State University of New York, Albany, NY, June 04-09, 1985, Eds., Ramaswamy H. Sarma & Mukti H. Sarma, ISBN 0-940030-14-4, Adenine Press, ©Adenine Press 1986.

Oligodeoxynucleotides Covalently Linked to an Intercalating Agent. Structural and Thermodynamic Studies of a Dodecathymidylate and its Binding to Polynucleotides

C. Hélène*, F. Toulmé[†], M. Delarue*, U. Asseline[†], M. Takasugi*, J.C. Maurizot[†], T. Montenay-Garestier* and N.T. Thuong[†]
*Laboratoire de Biophysique, Muséum National d'Histoire Naturelle
INSERM U.201, CNRS UA. 481
61, Rue Buffon, 75005 Paris

and

[†]Centre de Biophysique Moléculaire,
CNRS, 45045 Orleans Cedex, France

Abstract

A dodecathymidylic acid has been covalently linked to 2-methoxy, 6-chloro, 9-amino acridine. The linker involves a pentamethylene chain attached to the 3′ phosphate of the oligonucleotide and to the 9-amino group of the acridine derivative. Intramolecular interactions between the acridine ring and the oligonucleotide are revealed by absorption, fluorescence and circular dichroism. Binding to poly(rA) and poly(dA) induces large changes in the spectroscopic properties of the acridine ring. Double-stranded complexes are formed with poly(rA) whereas both double-stranded and triple-stranded helices are observed with poly(dA). The double-stranded structures formed with poly(rA) and poly(dA) are characterized by different environments of the acridine ring. Thermodynamic studies show that the intercalating agent provides an additional binding energy which stabilizes the complexes as compared with the unsubstituted oligonucleotide.

Introduction

The regulation of gene expression requires molecules with high specificity and strong affinity for nucleic acid base sequences. These biological processes are usually controlled by specific proteins such as repressors or activating factors (1). More recently the role of small RNAs in the regulation of gene expression has been described (2). In order to build molecules with high affinity for single-stranded nucleic acid base sequences, we have recently synthesized oligodeoxynucleotides covalently linked to an intercalating agent (3). These molecules are expected to bind strongly and selectively to the complementary sequence of the oligonucleotide. The intercalating agent should provide an additional binding energy through stacking interactions with the base pairs of the mini-duplex structure formed by the oligonucleotide with its complementry sequence (figure 1) (4,5). We describe below the results obtained for a dodecathymidylate to which a 9-amino acridine derivative has been covalently attached. The linker involves a pentamethylene chain attached to the 3'-phosphate of the oligonucleotide and to the 9-amino group of 2-methoxy, 6-chloro, 9-aminoacridine [(Tp)$_{12}$(CH$_2$)$_5$Acr]. Structural and thermodynamic studies of its complexes with poly(rA) and poly(dA) are presented.

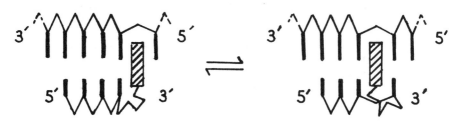

Figure 1. Schematic representation of complexes formed by an oligonucleotide covalently linked to an intercalating agent with its complementary sequence. The hatched box represents the intercalating agent linked to the 3'-phosphate of an oligodT hybridized to a poly(A) matrix.

Intramolecular interactions in (Tp)$_{12}$(CH$_2$)$_5$Acr

The oligonucleotide (Tp)$_{12}$(CH$_2$)$_5$Acr is a dodecathymidylate whose 3'-phosphate has been covalently linked to the 9-amino group of 2-methoxy, 6-chloro, 9-amino acridine *via* a pentamethylene linker. At pH 7 the acridine ring is fully protonated (5). The acridine ring interacts with the thymine bases in the free molecule as shown by an hypochromism and a red-shift of the acridine absorption in the visible.

Complete digestion of the oligonucleotide by nuclease P1 from *Penicillium citrinum* or by calf spleen phosphodiesterase followed by alkaline phosphatase treatment released the acridine derivative with a ω-hydroxy-pentamethylene substituent on its 9-amino group (HO(CH$_2$)$_5$Acr). An hyperchromism and a blue shift of the visible bands were observed with an isosbestic point at 447 nm during the digestion reaction. This allowed us to calculate the extinction coefficient of (Tp)$_{12}$(CH$_2$)$_5$Acr at the maximum wavelength in the visible range $\epsilon_{max} = 8835$ M^{-1} cm^{-1} assuming $\epsilon_{max} = 9750$ M^{-1} cm^{-1} for free HO(CH$_2$)$_5$Acr (7). The λ_{max} is shifted from 421 nm in HO(CH$_2$)$_5$Acr to 425 nm in (Tp)$_{12}$(CH$_2$)$_5$Acr.

Interaction of the acridine ring with thymine bases (or phosphates) in the oligonucleotide was also revealed by an induced circular dichroism (CD) in the visible and near UV bands (figure 2). This CD signal reflects the asymmetric perturbation of the planar acridine ring by the oligonucleotide. The intensity of the induced CD decreased when temperature increased (figure 2). This behavior did not depend on (Tp)$_{12}$(CH$_2$)$_5$Acr concentration indicating that the observed effects were due to intramolecular rather than intermolecular interactions.

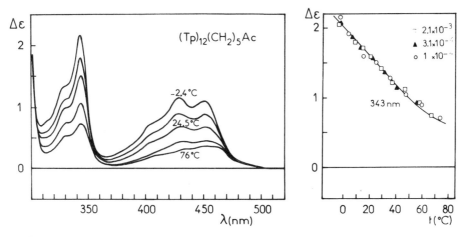

Figure 2. Circular dichroism spectra of 10^{-4} M $(Tp)_{12}(CH_2)_5$Acr at different temperatures in a pH 7 buffer containing 10 mM Na cacodylate and 0.1 M NaCl. The right part shows the decrease of $\Delta\epsilon$ versus temperature at three different concentrations.

Interactions of $(Tp)_{12}(CH_2)_5$Acr with poly(rA) and poly(dA)

a) Absorption studies

Addition of poly(rA) and poly(dA) to $(Tp)_{12}(CH_2)_5$Acr resulted in important changes in the absorption band of the acridine moiety (figure 3). A strong hypochromism was observed in both cases. A new absorption band appeared at long wavelengths in complexes with poly(rA) whereas this band was much weaker with poly(dA). A plot of absorbance changes at 425 nm *versus* polynucleotide concentration at 2°C revealed that $(Tp)_{12}(CH_2)_5$Acr formed only a 1:1 complex with poly(rA) whatever the concentration ratio whereas a 2:1 complex was formed with poly(dA) at low poly(dA) concentrations followed by a 1:1 complex (figure 4). This result is in contrast to what was previously reported for the binding of unsubstituted oligo-thymidylates to poly(rA) and poly(dA). For example oligo(dT)$_{10}$ was shown to form a 1:1 complex with poly(dA) but 2:1 and 1:1 complexes with poly(rA) (6). This difference might originate from the constraint imposed upon the duplex structure by the intercalating agent which might prevent formation of a triple-stranded structure in the case of poly(rA) and allow it in the case of poly(dA). Melting profiles and circular dichroism studies suggest that at a 1:1 ratio a 1:1 complex is formed between $(Tp)_{12}(CH_2)_5$Acr and poly(dA) (see below). In a comparative study we used a dodecathymidylate substituted by an ethyl group on its 3'-phosphate, $(Tp)_{12}$Et, to mimic the effect of the polymethylene linker. This oligonucleotide formed 1:1 complexes at 2°C with both poly(rA) and poly(dA).

The change in absorbance at 425 nm observed upon addition of poly(rA) to $(Tp)_{12}(CH_2)_5$Acr showed a sharp break when the ratio of Adenine to Acridine concentration reached 13.1. This value is slightly higher than what would be expected if each dodecathymidylate covered 12 adenine bases on the polynucleotide lattice.

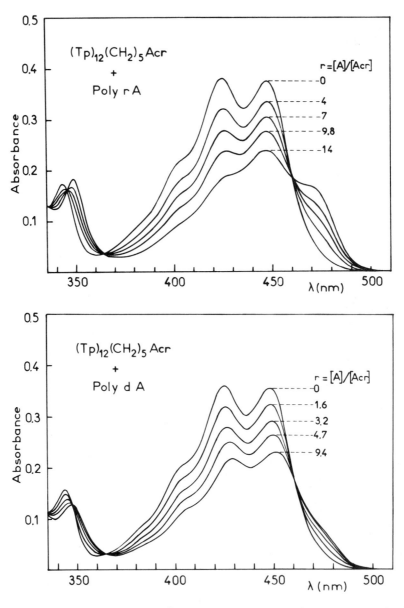

Figure 3. Change in absorption spectra at 2°C observed upon addition of poly(rA) (top) and poly(dA) (bottom) to (Tp)$_{12}$(CH$_2$)$_5$Acr. The oligonucleotide concentrations were 4.3 × 10^{-5} M and 4.1 × 10^{-5} M, respectively. The numbers indicated on the spectra refer to the ratio of adenine to acridine concentrations. The pH 7 buffer used in these experiments contained 10 mM Na cacodylate and 0.1 M NaCl.

This could be due to the exclusion of about one adenine base at the end of each bound oligonucleotide. Alternatively the difference might reflect an erroneous estimation of the extinction coefficient of the acridine derivative attached to the oligonucleotide. The investigation of poly(rA) complexes with a series of oligo(dT)$_n$

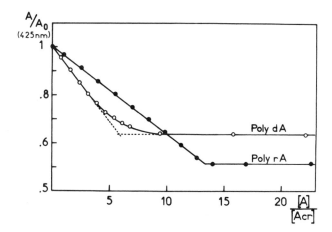

Figure 4. Relative change in absorbance at 425 nm upon addition of poly(rA) or poly(dA) to $(Tp)_{12}(CH)_5Acr$. [A] and [Acr] represent the concentrations of polynucleotide (in nucleotide unit) and of acridine, respectively. Same conditions as in figure 3.

covalently linked to the same acridine derivative revealed a systematic deviation of about 10% in excess over the stoichiometric [A]/[Acr] ratio (rather than a constant value of unity if one adenine was excluded for every bound oligonucleotide). There is considerable variation in the literature about the extinction coefficient of quinacrine which has ring substituents similar to those of the acridine derivative attached to the oligo(dT)s. Values range from 7500 to 9750 $M^{-1} cm^{-1}$ (7). We have chosen this last value to calculate ϵ_{max} (see above). This value might be slightly overestimated.

b) Circular dichroism studies

When $(Tp)_{12}(CH_2)_5Acr$ was mixed in equimolar ratio (base to base) with poly(rA), complex formation led to variations in CD intensity and shifts in the CD bands of the acridine ring. A new band was observed at about 475 nm as already observed in the absorption spectrum of the complex. Its intensity was quite similar to that of the two other bands at about 445 and 425 nm (figure 5).

When a similar experiment was performed using polydA instead of polyrA more drastic changes were observed in the CD spectrum which became negative in the wavelength range 370-460 nm. As observed in the complex with poly(rA) a new band was observed at about 475 nm, but this band was of opposite sign as compared to that of the bands located at shorter wavelengths.

The CD changes observed with poly(rA) and poly(dA) reflect the modification of the environment of the acridine dye occurring upon complex formation. When the oligonucleotide is bound to poly(rA) this environment is different from that obtained when it is bound to poly(dA). It has been proposed that the RNA-DNA hybrid poly(rA)•poly(dT) in aqueous solution adopts a conformation that has similarities

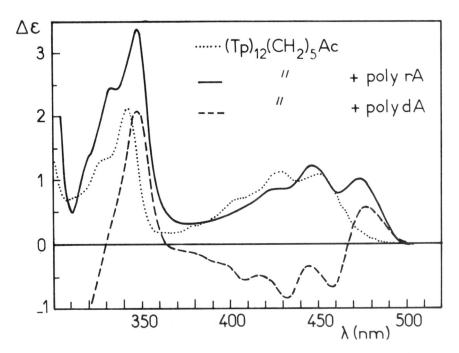

Figure 5. Circular dichroism spectra of 10^{-4} M $(Tp)_{12}(CH_2)_5Acr$ in the absence (dotted line) and in the presence of poly(rA) (full line) and poly(dA) (broken line) at a 1:1 ratio (adenine:thymine). Same buffer as in figure 3.

to B-DNA even though the poly(rA) strand has a RNA-type of sugar pucker (8). Fiber studies have led Arnott *et al.* (9) to propose a heteronomous structure for poly(dA)•poly(dT) with the poly(dA) chain adopting an A-like conformation and poly(dT) a B-like conformation. Solution studies have not yet come to an agreement as to the structure of the poly(dA)•poly(dT) double helix in aqueous solution (10, 11, R. Sarma *et al.* this volume). The helical repeat of poly(dA)•poly(dT) is smaller than that of DNA (10.1 against 10.6 base pairs/turn) which might reflect the property of consecutive dA•dT base pairs to behave like a wedge opened toward the dT side (12). Structural differences might therefore exist in the complexes formed by the oligonucleotide $(Tp)_{12}(CH_2)_5Acr$ with poly(rA) and poly(dA). It is not surprising that differences in nucleic acid structures are reflected in differences in the induced CD spectrum of a bound ligand. For example it has been shown that the induced CD in the visible region of acridine orange bound to double-stranded RNA is quite different in shape from that of the dye bound to DNA (13).

c) Fluorescence studies

The fluorescence quantum yield of the acridine ring of $(Tp)_{12}(CH_2)_5Acr$ excited in the visible absorption band is 2.5-fold higher than that of the parent compound $HO(CH_2)_2Acr$. It should be noted that substituting the 9-NH$_2$ group of 2-methoxy,

6-chloro, 9-amino acridine by an alkyl or an alkoxy group (as in $HO(CH_2)_5Acr$) reduces the fluorescence quantum yield more than 10 times. The intramolecular interactions of the acridine derivative covalently linked to the oligonucleotide (as revealed by CD studies) might partially relieve this quenching of the substituted acridine ring.

Binding of $(Tp)_{12}(CH_2)_5Acr$ to poly(rA) increased the fluorescence quantum yield of the acridine ring as measured upon excitation at the isosbestic wavelength observed at 346 nm in the absorption spectrum upon titration of the oligonucleotide by the polynucleotide (see above). The fluorescence was slightly shifted to longer wavelengths with an isoemmissive point around 482 nm.

The fluorescence decay measured under flash excitation was not exponential neither for the isolated acridine derivative $HO(CH_2)_5Acr$ nor for the acridine ring tethered to the oligonucleotide both in the absence and in the presence of poly(rA). Three components were required to fit the decay curve on the short wavelength side of the fluorescence spectrum (470 nm) whereas two components appeared sufficient in most cases on the long wavelength side (540 nm). A study of fluorescence spectra as a function of temperature and viscosity revealed that this abnormal behavior of the substituted acridine ring was most probably due to solvent relaxation around the excited molecule. A more complete account of these data will be presented elsewhere. Average lifetimes only are given in Table I; these have been used to determine accessibility rate constants (see below).

Table I

Fluorescence parameters for free $HO(CH_2)_5Acr$ and its complex with the double helix poly(rA)·poly(dT) (Acr:A·T = 60) and for $(Tp)_{12}(CH_2)_5Acr$ and its 1:1 complex with poly(rA)

	Fluorescence anisotropy[1]		$\langle \tau_F \rangle$ [2] (ns)		τ_c [3] (ns)	r_∞ [3]	k_Q [4] (M^{-1}·s^{-1})	
$HO(CH_2)_5Acr$	0.03	—	1.8	3.65	< 1	0.004	—	6×10^9
$HO(CH_2)_5Acr$ + poly(rA)·poly(dT)	0.15	0.21	4.6	10.3	15	0.13	1.4×10^8	2.9×10^7
$(Tp)_{12}(CH_2)_5Acr$	0.08	—	3.6	5.7	3.5	0.001	6×10^8	2.5×10^8
$(Tp)_{12}(CH_2)_5Acr$ + poly(rA)	0.18	0.24	7.1	13.5	18	0.12	2.3×10^8	4.5×10^7

(1) Fluorescence anisotropy (r) was measured at 0°C at two excitation wavelengths (420 nm left; 470 nm right). $\lambda_{em} > 520$ nm.

(2) $\langle \tau_F \rangle$ is the average fluorescence lifetime measured at 470 nm (left) and 540 nm (right) upon excitation at 337 nm. Measurements were carried out at 13.5°C in the pH 7 buffer except for $HO(CH_2)_5Acr$ where the values were obtained at 18°C in ethylene glycol.

(3) τ_c is the rotational correlation time and r_∞ the limit anisotropy determined according to equation (1) in the pH 7 buffer at 0°C upon excitation at 420 nm with $\lambda_{em} = 570$ nm.

(4) Rate constant for fluorescence quenching by I$^-$ measured in the pH 7 buffer at 13.5°C at 470 nm (left) and 540 nm (right) according to equation (2) (see text).

The fluorescence anisotropy (r) was measured at $0°C$ in the buffer used for the binding experiments (10 mM Na cacodylate, 0.1 M NaCl, pH 7.0). As shown in Table I, r increased from 0.03 for $HO(CH_2)_5Acr$ to 0.08 for $(Tp)_{12}(CH_2)_5Acr$. This reflects the restriction in mobility of the acridine ring when it is covalently attached to the oligonucleotide. Binding to poly(rA) further increased r as expected for a partial immobilization in the double-stranded structure.

The fluorescence anisotropy decay measured following flash excitation was analyzed according to equation (1)

$$r(t) = r_\infty + (r_o - r_\infty) \exp - t/\theta \tag{1}$$

where r_∞ is the limit anisotropy measured at long times and θ is the rotational correlation time of the fluorescent species. The oligonucleotide $(Tp)_{12}(CH_2)_5Acr$ exhibited a rotational correlation time close to that expected for an oligonucleotide of that length indicating that the acridine ring follows the overall movement of the oligonucleotide. There might be an undetected fast component in the anisotropy decay which could correspond to a rapid movement (< 1 ns) of the acridine with respect to the oligonucleotide.

The anisotropy decay of $(Tp)_{12}(CH_2)_5Acr$ bound to poly(rA) exhibited a limited value r_∞ of 0.12 which reveals a partial immobilization (on the time scale of the fluorescence decay) of the acridine ring in the complex. The measured rotational correlation time (18 ns) indicates that the acridine ring probably follows the local motion of the base pairs at the junction between cooperatively bound oligonucleotides.

The accessibility rate constants (k_Q) of the acridine ring to an external fluorescence quencher (I^- anions) were determined from the reduction of the average fluorescence lifetime $\langle \tau_F \rangle^\circ$ upon addition of potassium iodide (equation (2)). Potassium chloride was used as a control and did not lead to any quenching of the acridine fluorescence.

$$\frac{\langle \tau_F \rangle^\circ}{\langle \tau_F \rangle} = 1 + k_Q \langle \tau_F \rangle^\circ [Q] \tag{2}$$

The values of k_Q were determined at two wavelengths (470 nm and 540 nm) to take into account the wavelength dependence of the fluorescence lifetime and the existence of different excited species. There is a strong reduction in accessibility (10-25 fold) when the acridine ring is tethered to the oligonucleotide. This reflects for the most part the repulsive effect of the negatively charged oligonucleotide on the approach of the negatively charged quencher (I^-). Upon binding to poly(rA) the rate constant k_Q is only reduced 3 to 5 fold indicating that the acridine ring remains accessible probably from the minor groove of the double helical structure. The absolute values of k_Q are not much different from those obtained for $HO(CH_2)_5Acr$ bound to the double helix poly(rA)·poly(dT) at low [dye]/[phosphate] ratio. It should be remembered that the acridine ring undergoes motions of large amplitude in the 15-20 ns time range as revealed by the anisotropy decay. Since these motions occur in the same time range as the fluorescence decay they might make the fluorescent acridine ring accessible to external quenchers before deactivation.

Complex stability

An increase in temperature led to a reversal of the spectroscopic effects observed upon complex formation. Figure 6 shows the melting profiles obtained at 260 nm for $(Tp)_{12}Et$ and at 425 nm for $(Tp)_{12}(CH_2)_5Acr$ at a 1:1 (A:T) concentration ratio. A strong stabilization of the complexes was observed with both poly(rA) and poly(dA) when the acridine ring was covalently attached to the oligonucleotide. The increase in stability was slightly higher for poly(rA) as compared to poly(dA).

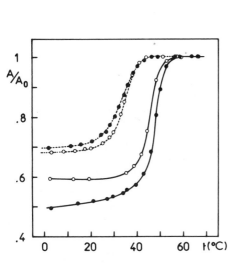

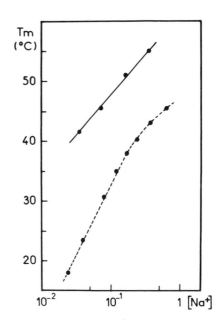

Figure 6. Melting curves for complexes of $(Tp)_{12}Et$ (broken lines) and $(Tp)_{12}(CH_2)_5Acr$ (full lines) with poly(rA) (filled circles) and poly(dA) (open circles). Absorption was measured at 260 nm for $(Tp)_{12}Et$ and 425 nm for $(Tp)_{12}(CH_2)_5Acr$. The thymine: adenine ratio was 1:1 in all cases. The base concentration was 5×10^{-4} M. Same buffer as in figure 3.

Figure 7. Change in melting temperature (T_m) versus the logarithm of NaCl activity for poly(rA) complexes of $(Tp)_{12}Et$ (broken line) and $(Tp)_{12}(CH_2)_5Acr$ (full line). Concentrations: 3.6×10^{-4} M in thymine for $(Tp)_{12}Et$ (T/A = 1) and 5.4×10^{-4} M in thymine for $(Tp)_{12}(CH_2)_5Acr$ (T/A = 0.85). Same conditions as in figure 3.

The melting profiles for $(Tp)_{12}(CH_2)_5Acr$ complexes with poly(rA) and poly(dA) were identical when the absorbance changes were followed at 260 nm and 425 nm. There was no evidence for the formation of triple-stranded structures at a 1:1 ratio neither with poly(rA) nor with poly(dA). A slight increase in absorbance was observed in the visible range for the poly(rA) complex before melting. This effect was even more pronounced in the change in CD intensity (data not shown). This "premelting" phenomenon might have different origins. As shown above the acridine ring experiences different environments in the complexes formed by $(Tp)_{12}(CH_2)_5Acr$ with both poly(rA) and poly(dA). The equilibrium distribution between these different structures is most probably dependent on temperature. Consequently the absorption and CD spectra might vary before dissociation of the double helix takes place. Alternatively the double helix might undergo slight structural perturbations before melting and this might be reflected in the spectroscopic properties of the acridine ring, especially in the CD spectrum which probes the asymmetrical environment of the dye.

The fluorescence polarization ratio decreased only slightly in the premelting region indicating that the mobility of the acridine ring was not appreciably altered before dissociation of the oligonucleotide from the polynucleotide matrix. The fluorescence polarization ratio then dropped abruptly in the melting region until it reached the value of the free oligonucleotide.

The stability of oligonucleotide-polynucleotide complexes depends on ionic concentration (14). As expected, an increase in NaCl concentration led to an increase in melting temperature for poly(rA) complexes of both $(Tp)_{12}Et$ and $(Tp)_{12}(CH_2)_5Acr$ (figure 7). When T_m was plotted *versus* log [NaCl] a straight line was obtained for $(Tp)_{12}(CH_2)_5Acr$. A curvature was observed for $(Tp)_{12}Et$ at high NaCl concentration. The slopes of the linear parts were 13.4°C and 20.4°C for $(Tp)_{12}(CH_2)_5Acr$ and $(Tp)_{12}Et$, respectively. The last value corresponds to that expected for the melting of a double helical structure. The lower value obtained with $(Tp)_{12}(CH_2)_5Acr$ might reflect the contribution of the positively charged acridine ring to the electrostatic potential of the double helix and of the free oligonucleotide. However, the same value of $\approx 13.4°C$ was obtained for shorter oligo(dT) covalently linked to the same acridine derivative, namely $(Tp)_8(CH_2)_6Acr$ and $(Tp)_4(CH_2)_5Acr$. Since changes in melting temperature with salt concentration reflect the difference in cation binding between the separated and the associated molecules (14) this result might mean that cation binding to the free oligonucleotides and to their complexes with poly(rA) are such that the difference in the apparent number of cations bound per phosphate remains constant. When the oligonucleotide length decreases the relative contribution of the positively charged acridine increases in both the free oligonuleotide and its complex with poly(rA).

Thermodynamic parameters for the binding of $(Tp)_{12}(CH_2)_5Acr$ to poly(rA) and poly(dA) were calculated from the dependence of melting temperature (T_m) *versus* oligonucleotide concentration (C_m) according to equation (3)

$$\frac{1}{T_m} = \frac{\Delta S}{\Delta H} + \frac{2.3R}{\Delta H} \log C_m \qquad (3)$$

ΔH values of -88.9 kcal·mole^{-1} and -100.9 kcal·mole^{-1} were obtained for binding to poly(rA) and poly(dA), respectively. For the binding of oligo(dT)$_8$ to poly(dA) a ΔH value of -6.8 kcal/mole of base pair was previously reported (15). For a dodecathymidylate this would give a total ΔH of -81.6 kcal·mole^{-1} for binding to poly(dA). The presence of the acridine ring has therefore stabilized the complex by about 19.3 kcal/mole of oligonucleotide. Using the published value of $\Delta S = -19.5$ e.u. per mole of A·T base pair, the total ΔS for the binding of (dT)$_{12}$ to poly(dA) should be -234 e.u. The value obtained for (Tp)$_{12}$(CH$_2$)$_5$Acr is lower by ≈ 59 e.u. This leads to a partial compensation of the stabilization afforded by the change in ΔH due to the interaction of the acridine ring.

The above ΔH and ΔS values were determined for cooperative binding of the oligonucleotide along the polynucleotide lattice. Experiments using a shorter oligonucleotide, (Tp)$_8$(CH$_2$)$_6$Acr, have shown that a large part of the stabilization afforded by the intercalating agent corresponds to isolated site binding and not to cooperative interactions between contiguously bound oligonucleotides (5 and unpublished data).

Under identical concentration conditions, the complexes formed by (Tp)$_{12}$(CH$_2$)$_5$Acr with poly(rA) and poly(dA) exhibit quite similar melting temperatures (figure 6). We used the difference in shape between the CD spectrum of the acridine dye in the complexes with poly(rA) and poly(dA) to perform competition experiments between these complexes at low temperature (4°C) where we expected that their lifetimes are long enough compared to the time required to run a CD spectrum. Two experiments were performed. To a preformed complex between poly(dA) and (Tp)$_{12}$(CH$_2$)$_5$Acr was added an equimolar quantity of poly(rA). The CD signal did not change for a period of time as long as 72 hours. In another experiment poly(dA) was added to a preformed complex of poly(rA) and (Tp)$_{12}$(CH$_2$)$_5$Acr. The CD signal slowly evolved to reach after 72 hours that obtained with poly(dA). These experiments demonstrate that poly(dA) is able to displace poly(rA) from its complex with (Tp)$_{12}$(CH$_2$)$_5$Acr whereas poly(rA) cannot displace poly(dA). Therefore at low temperature the complex formed between the modified oligonucleotide and poly(dA) is more stable than that formed with poly(rA). In addition the kinetics of the exchange reaction reveal a long lifetime for these complexes (several hours for the poly(rA) complex at 4°C).

Conclusions

The experiments described above for (Tp)$_{12}$(CH$_2$)$_5$Acr binding to poly(rA) and poly(dA) and those previously published for other oligothymidylates (4,5) demonstrate that oligonucleotides covalently linked to an intercalating agent bind strongly and selectively to their complementary sequence. The spectroscopic effects induced upon binding to poly(rA) and poly(dA) were not observed with polynucleotides which did not contain repeated adenine sequences. Complex stability is increased when compared to the same oligonucleotide without acridine. This is due to the additional binding energy provided by the interaction of the acridine ring with the base pairs of the duplex structure formed by the oligonucleotide with its complementary sequence. Proton and phosphorus nuclear magnetic resonance experiments carried out with a tetrathymidylate covalently linked to the intercalating agent have shown that the acridine ring intercalates between A·T base pairs (16 and G. Lancelot, unpublished data). The acridine derivative used in our experiments is similar to that of quinacrine which has been proposed to intercalate between DNA base pairs (7). Fluorescence studies using external quenchers have shown, however, that the acridine ring remains accessible in these complexes which probably reflects the dynamic state of the local dye environment.

Oligonucleotides covalently linked to an intercalating agent represent a new family of molecules which can be targeted to any single-stranded nucleic acid sequence. The strength and specificity of the complexes make them suitable for control of gene expression at different levels. We have recently shown that transcription initiation can be inhibited by such molecules directed against the open complex formed by RNA polymerase with its promoter (17). Binding to a messenger RNA can also block gene expression at the translation level (17).

Acknowledgements

We wish to thank M. Chassignol and V. Roig for expert technical assistance in chemical synthesis. We are indebted to other colleagues from both laboratories for helpful discussions.

References and Footnotes

1. C. Hélène and G. Lancelot, *Prog. Biophys. Mol. Biol. 39,* 1 (1982).
2. J. Coleman, P.J. Green and M. Inouye, *Cell 37,* 429 (1984).
3. U. Asseline, N.T. Thuong and C. Hélène, C.R. *Acad. Sci. Paris 297,* 369 (1983).
4. U. Asseline, F. Toulmé, N.T. Thuong, M. Delarue, T. Montenay-Garestier and C. Hélène, *EMBO J. 3,* 795 (1984).
5. U. Asseline, M. Delarue, G. Lancelot, F. Toulmé, N.T. Thuong, T. Montenay-Garestier and C. Hélène, *Proc. Nat. Acad. Sci. USA 81,* 3297 (1984).
6. G.R. Cassani and F.J. Bollum, *Biochemistry 8,* 3928 (1969).
7. W.D. Wilson and I.G. Loop, *Biopolymers 18,* 3025 (1979).
8. S.B. Zimmerman and B.H. Pheiffer, *Proc. Nat. Acad. Sci. USA 78,* 78 (1981).
9. S.Arnott, R. Chandrasekaran, I.H. Hall and L.C. Puigjaner, *Nucl. Acids Res. 11,* 4141 (1983).
10. G.A. Thomas and W.L. Peticolas, *J. Am. Chem. Soc. 105,* 993 (1983).
11. B. Jollès, A. Laigle, L. Chinsky and P.Y. Turpin, *Nucl. Acids. Res. 13,* 2075 (1985).
12. A. Prunell, I. Goulet, Y. Jacob and F. Goutorbe, *Eur. J. Biochem. 138,* 253 (1984).
13. M. Zama and S. Ichimura, *Biopolymers 15,* 1693 (1976).
14. M.T. Record, C.P. Woodbury and T.M. Lohman, *Biopolymers 15,* 893 (1976).
15. R.C. Pless and P.O.P. T'so, *Biochemistry 16,* 1239 (1977).
16. G. Lancelot, U. Asseline, N.T. Thuong and C. Hélène, *Biochemistry 24,* 2521 (1985).
17. C. Hélène, T. Montenay-Garestier, J.J. Toulmé, T. Saison-Behmoaras, N.T. Thuong, U. Asseline, G. Lancelot, J.C. Maurizot and F. Toulmé, *Biochimie,* in press (1985).

Biomolecular Stereodynamics III, Proceedings of the Fourth Conversation in the Discipline Biomolecular Stereodynamics, State University of New York, Albany, NY, June 04-09, 1985, Eds., Ramaswamy H. Sarma & Mukti H. Sarma, ISBN 0-940030-14-4, Adenine Press, ©Adenine Press 1986.

Crystalline Hydrates of a Drug-Nucleic Acid Complex

Helen M. Berman and Stephan L. Ginell

The Institute for Cancer Research
Fox Chase Cancer Center
7701 Burholme Avenue, Philadelphia, PA 19111

Abstract

A new form of the crystalline hydrate of 2:2 complex of proflavine and deoxycytidylyl 3'5' guanosine (d(CpG)-proflavine) was studied using x-ray crystallographic methods. The results of this analysis demonstrate conclusively that relatively small differences in the crystalline environment can affect the conformation of the molecular species. This work also gives us additional insights about the nature of the geometry of water in biological systems.

Introduction

The crystal structure of the 2:2 complex of proflavine and deoxycytidylyl 3'5' guanosine (d(CpG)-proflavine) demonstrated for the first time the existence of a semi-clathrate like water network involving a biologically significant molecular system (1,2). The network consists of several edge-linked pentagons and a polygon disk. Because of this hydration pattern and also because the structure is relatively small, well determined and biologically interesting, this d(CpG)-proflavine crystal structure has proven itself to be an attractive model system for theoretical studies of intercalation (3,4) and crystal hydration (5,6). These latter studies have also pointed up some inadequacies of the theoretical methods as well as the experimental data. It has become apparent to us that at least some of the problems would best be approached by a careful experimental reanalysis of this crystalline system under a variety of environmental conditions. Among the variables we sought to clarify were the effects of temperature and humidity. We report here our preliminary findings.

Methods

Fresh crystals of average dimensions, 0.5 × 0.4 × .15 mm, were grown at 4°C. As shown in Table I, the unit cell volume of this crystal is significantly larger than those reported for the two crystal forms previously analyzed. At first we attributed this effect to the lowering of temperature. However, when the liquid around the crystal was allowed to distill away by creating a temperature gradient in the capillary tube, it was possible to reproduce the unit cell previously obtained. It was also possible to obtain unit cell volumes of intermediate value in the process of rehydrating

131

Table I

Characteristics of three crystals of the 2:2 complex of d(CpG)-proflavine

	Crystal A	Crystal B	Crystal C
Unit cell	a = 32.991 Å	a = 32.871 Å	a = 32.867 Å
	b = 21.995 Å	b = 22.187 Å	b = 22.356 Å
	c = 13.509 Å	c = 13.506 Å	c = 13.461 Å
	Volume = 9,802 Å^3	Volume = 9,850 Å^3	Volume = 9891 Å^3
		Space group P2$_1$2$_1$2	
Resolution limit	0.83 Å	1.00 Å	.89 Å
No. of water molecules	26	27.5	28.5

the crystal. Two sets of three dimensional intensity data were collected on the highly hydrated crystals (form C) at $-2°C$ to .89Å resolution. There were 2863 independent reflections with intensities above the 2σ threshold for one crystal and 2758 for the other. After the data were corrected for Lorentz-polarization, absorption effects and crystal decay, the structure was reanalyzed by systematic Fourier methods. The water molecule positions and sugar conformations in both crystals yielded the same results. Both block diagonal and Konnert-Hendrickson refinement (7) on the crystal that had the larger number of observed intensities led to the present R value of 15.6%.

Results and Discussion

The original description of this structure was based on an analysis of two crystals designated A and B. One crystal, labelled A in Table I, diffracted well and yielded a very reliable structure. The asymmetric unit consists of an intercalated duplex with another proflavine stacked above the base plane. One DNA chain (strand 1) contains two C3′ endo sugars and the other (strand 2) a C3′ and a C2′ endo sugar. The intercalated duplexes are stacked along the $\vec{c}$ axis. The packing of the duplexes is such that there are large cavities formed by the major grooves and flat narrow cavities formed by the minor grooves. The major groove cavities are filled with first and second hydration shell water organized similar to a semi-clathrate. The minor groove water molecules form a flat polygon disk. The unit cell determination of crystal B indicates clearly that there are more waters in that crystal. These water molecules are arranged in a similar way to what was found in crystal A with the additional waters located in the major groove cavity. Interestingly, one of the guanosine 3′ deoxyribose sugars appears to have C3′ endo and C2′ endo conformational disorder.

Several theoretical studies have sought to simulate the water structure within the crystalline complex (5,6). Among the difficulties and ambiguities encountered in these simulations were uncertainties about the exact temperature of the data collection, questions about the amount and nature of the disordered water, and the

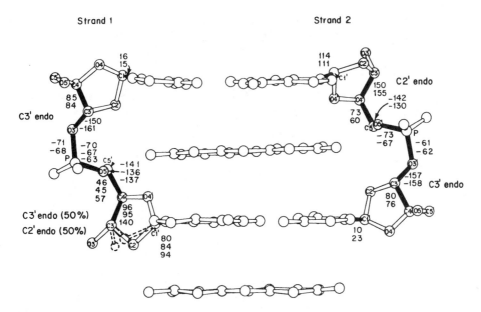

Figure 1. The conformations of two forms of d(CpG)-proflavine. The first number given is the torsion angle found in the crystal A. The second number is the value found in crystal C and the third number is the value of the C2′ endo form of the disordered sugar.

possibility of the presence of ionic species in the solvent. The present analysis resolves some of the questions, and as is often the case, raises a few more.

The structure of the molecular complex in crystal C (Figure 1) is similar to what was found for crystal B in that the 3′ sugar of strand 1 is disordered between C2′ endo and C3′ endo. In this case, however, that disordering is clearly resolved in the Fourier maps with discrete density found for each half occupancy O3′ hydroxyl atom. The disorder does not extend to the guanosine base. This clarity is probably attributable to the lower temperature of the data collection leading to lower thermal motion and generally better quality data. Although the major conformational differences between the dinucleoside phosphates in this crystal and those in crystal A are found around the disordered deoxyribose ring, there are significant differences observed among other torsion angles. For example, the glycosidic torsion angles of the cytosine moiety in strand 2 is different in crystal A and crystal C as are the torsion angles around C4′-C5′ in that strand.

As expected, more molecules of water were located in the electron density maps than had been located for the previous crystals. This is due partially to the fact that the crystal contains more water and partially to the fact that more of the water is ordered. The new waters are all located in the major groove cavity and serve to substantially narrow the gap previously observed in that area (Figure 2). One of these new water molecules is hydrogen bonded to the amino group of the cytosine in strand 2. It is noteworthy that this is the one whose glycosidic torsion angle

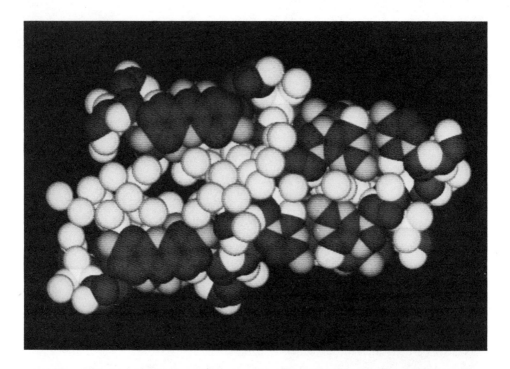

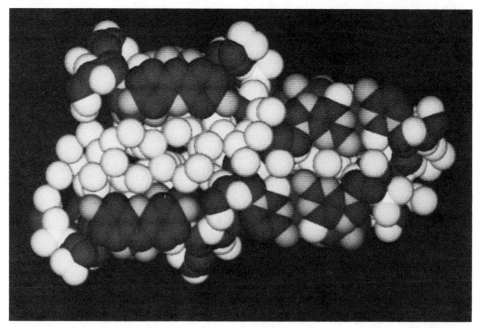

Figure 2. a) A space filling figure showing the packing of four d(CpG)-proflavine duplexes in crystal A. The lightest colored spheres are water molecules. The major groove cavity is on the left. The minor groove cavity is on the right. Note the black channels which probably contain disordered water molecules. b) The same view as in (a) of crystal C. Note that the black channels are smaller in size.

differs significantly in crystals A and C. Another new water molecule is hydrogen bonded to the O3′ hydroxyl of the 3′ sugar in strand 1 when it is in the C2′ endo conformation. This water has half occupancy and does not coexist with the sugar when it has the C3′ endo pucker.

All the water molecules that were involved in the pentagonal network in crystal A are present in crystal C. However, because their positions are as much as .6Å different, some of the presumptive hydrogen bonds are broken and new ones have formed (Figures 3,4). For example, the distance between water molecules 23 and 9 is too long to form a hydrogen bond in crystal C but 17 and 23 are close enough to hydrogen bond. The overall effect of the shifts in the positions of these molecules is to create a hexagonal network from a pentagonal one. In crystal C water 13 and 20 are too far apart to hydrogen bond resulting in the breakup of another pentagon. On the other hand, waters 11 and 20 are in close proximity thus serving to form a new pentagon between stacked duplexes. Only one pentagon bounded by water molecules 22,9,17,14, and 2 is in common between the two structures. The polygon disk is disrupted by the absence of one water molecule (26) that is present in crystal A.

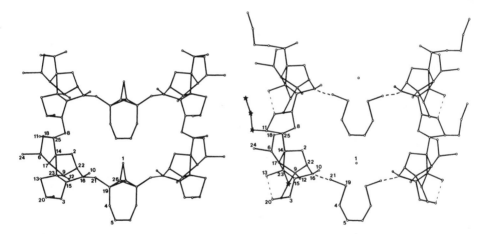

Figure 3. a) The water network in crystal A. b) The water network in crystal C. Solid lines indicate water-water distances less than 3.3Å. The dotted lines show places in crystal C where the distances have become greater than 3.3Å. The stars are newly found water molecules.

In summary, this structure analysis confirms and extends the previous observation (1) that relatively small differences in the crystalline environment can perceptibly affect the conformation of the molecular species contained in the crystal. It also demonstrates that it is possible to change both solvent-solute and solute-solute interactions by changing the humidity and temperature around the crystal. Further experiments are now in progress to explore the full range of these variables.

Acknowledgments

This research was supported by the National Institutes of Health through grants

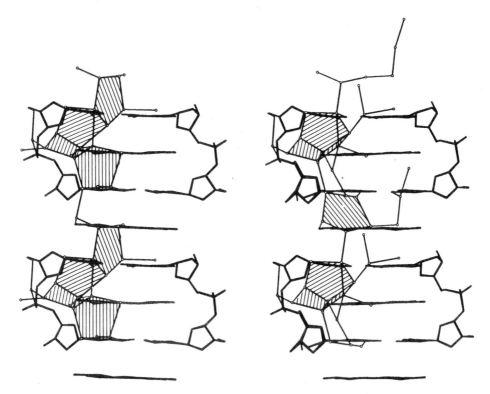

Figure 4. a) The interaction between the water network with solute in crystal A. b) The interaction of the water network with solute in crystal C. Note the hexagonal patterns as well as the formation of the new pentagon.

GM21589, CA06927, RR05539 and through an appropriation from the Commonwealth of Pennsylvania.

We thank Carol Afshar for growing the crystals and David Beveridge and Eric Westhof for many stimulating discussions on this subject.

References and Footnotes

1. S. Neidle, H.M. Berman and H.-S. Shieh, *Nature 288,* 129 (1980).
2. H.-S. Shieh, H.M. Berman, M. Dabrow and S. Neidle, *Nucleic Acids Research 8,* 85 (1980).
3. S.A. Islam and S. Neidle, *Acta Cryst 40,* 424 (1984).
4. A. Dearing, P. Weiner and P.A. Kollman, *Nucleic Acids Research 9,* 1483 (1981).
5. M. Mezei, D.L. Beveridge, H.M. Berman, J.M. Goodfellow, J.L. Finney and S. Neidle, *J. Biomol. Struc. & Dyn. 1,* 287 (1983).
6. K.S. Kim and E. Clementi, *J. Am. Chem. Soc. 107,* 227 (1985).
7. W.A. Hendrickson and J. Konnert, *Biomolecular Structure, Conformation, Function and Evolution,* ed. R. Srinivasan (Pergamon, Oxford), 43-57 (1979).

Biomolecular Stereodynamics III, Proceedings of the Fourth Conversation in the Discipline Biomolecular Stereodynamics, State University of New York, Albany, NY, June 04-09, 1985, Eds., Ramaswamy H. Sarma & Mukti H. Sarma, ISBN 0-940030-14-4, Adenine Press, ©Adenine Press 1986.

Mechanistic Studies of DNA Gyrase

Anthony Maxwell,[1] Donald C. Rau[2] and Martin Gellert[1]
Laboratory of Molecular Biology[1] and Laboratory of Chemical Biology,[2]
NIDAK, National Institutes of Health
Bethesda, Maryland 20205

Abstract

We have studied the interaction of DNA gyrase with DNA by examining the dependence of the ATPase activity of this enzyme on DNA length and by analyzing the electric dichroism of gyrase-DNA complexes. Using linear DNA molecules of 46-171 bp (base pairs), we have found that under conditions where those of 100 bp or more show maximal stimulation of the gyrase ATPase, those of 70 bp or less show little or none. However, short DNA molecules (<70 bp) at much higher concentrations will stimulate the ATPase. The behavior of long and short DNA molecules with respect to ATPase stimulation is also reflected in the binding of these DNA molecules to DNA gyrase. We suggest a model for the interaction of DNA and ATP with DNA gyrase.

We have also examined complexes between DNA gyrase and DNA by the method of transient electric dichroism. Both the DNA dichroism extrapolated to infinite field and the rotational diffusion coefficients of these complexes suggest that the DNA is wrapped around the enzyme in a single turn with the entry and exit points of the DNA located close together. Studies of complexes of DNA gyrase, DNA and an ATP analog indicate that a significant conformational change is induced upon binding of the nucleotide.

Introduction

Bacterial DNA gyrase is a member of a group of enzymes known as topoisomerases which catalyze linking number changes in DNA (for reviews, see 1, 2 and 3). All DNA topoisomerases can relax negatively supercoiled DNA but only DNA gyrase has been shown to also introduce negative supercoils into DNA (4). The free energy required for negative supercoiling by gyrase is derived from the hydrolysis of ATP. Mechanistic studies of DNA gyrase and other topoisomerases have suggested that the essential steps of topoisomerase action are DNA binding, DNA cleavage, translocation of a segment of DNA through the break and resealing of the broken phosphodiester bonds. DNA gyrase is perhaps the best studied of the DNA topoisomerases and much experimental evidence has accumulated concerning its mechanism of action.

DNA gyrase is composed of two proteins, A and B, with two copies of each protein making up the active enzyme (5,6,7). Gyrase is specifically inhibited by two classes of antibiotics. One class, the quinolones (e.g., oxolinic acid and norfloxacin), interrupts the breakage and reunion of DNA by the enzyme and acts via the A protein (8,9). The other class, the coumarins (e.g., novobiocin), inhibits ATP hydrolysis and acts via the B protein (5,10). This and other evidence identifies the A protein as being principally involved in the breakage and reunion of DNA and the B protein as being responsible for ATP hydrolysis.

DNA gyrase and DNA form stable complexes whose properties have been examined by several methods. Digestion of such complexes with staphylococcal nuclease showed that gyrase protects about 140 bp (base pairs) of DNA from nuclease digestion (11). When probed with pancreatic DNase I, the protected region was found to contain sites of enhanced nuclease sensitivity spaced 10-11 bp apart (11,12,13,14), consistent with wrapping of the DNA around the enzyme. Other experiments have shown that the bound DNA is wrapped with a positive superhelical sense; this handedness probably determines the directionality of the supercoiling reaction (15).

When DNA gyrase and DNA are incubated in the presence of quinolone drugs and sodium dodecyl sulfate is subsequently added, DNA cleavage results (8,9). Both strands are cleaved and the A protein of gyrase is found covalently bound to the newly formed 5'-phosphoryl termini (16,17). These cleavage sites are thought to represent sites at which breakage and reunion of DNA occur during the supercoiling reaction of gyrase. Several of these cleavage sites have been found to be roughly centrally located in regions of DNA protected by gyrase from DNase I digestion (12,13,14). It is likely, therefore, that the breakage and reunion of DNA by gyrase occurs approximately in the center of the wrapped DNA segment.

ATP hydrolysis by DNA gyrase is not required for DNA binding or DNA cleavage but is essential for catalytic supercoiling. However, the nonhydrolyzable ATP analog, $(\beta,\gamma$-imido) ATP, will support limited supercoiling by gyrase (10). It is proposed that nucleotide binding promotes one round of supercoiling but that hydrolysis is required for the enzyme to turn over.

Recent work in our laboratory has involved studies of the ATPase reaction of DNA gyrase and analysis of the structure of gyrase-DNA complexes using the method of transient electric dichroism.

The ATPase Reaction

Although covalently closed circular DNA is the substrate for the DNA gyrase super-coiling reaction, other forms of DNA (e.g., linear DNA) will also interact with the enzyme and stimulate the ATPase activity (5,18). Using linear double-stranded DNA fragments 46-171 bp in length, we have determined the minimum size of DNA required to activate the gyrase ATPase reaction (19). Under conditions where DNA molecules of 100 bp or greater in length stimulated the gyrase ATPase reaction, those of 70 bp or less did not (Fig. 1). No sequence dependence was apparent from these data and

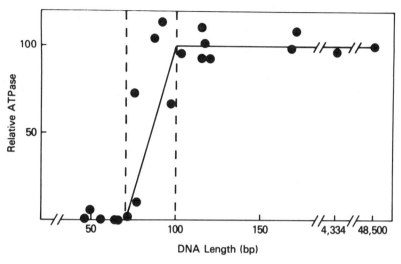

Figure 1. Dependence of ATPase on DNA length. Reaction mixtures containing ATP (0.1 mM), gyrase A protein (28 nM dimer), gyrase B protein (14 nM dimer), and DNA (10 μg/ml) were incubated for 2 hr at 25°C. The amount of ATP hydrolyzed in the presence of each DNA was determined and is expressed relative to that for λ DNA (48,500 bp). Points lying above the dashed line at 100 bp show little variation ($\pm$10%) in ATPase compared with λ DNA; those lying below the dashed line at 70 bp show very low levels (<5%) of ATPase activity. The bold line connecting the points is arbitrary. (From Ref. 19).

we concluded that the discrimination shown in Fig. 1 is based on DNA length and that gyrase requires about 85 bp of DNA to stimulate the ATPase reaction.

Although short DNA molecules (<70 bp) were ineffective in stimulating the gyrase ATPase reaction under the conditions of Fig. 1, at much higher DNA concentrations (e.g., 100-200 μg/ml) significant stimulation of the gyrase ATPase reaction was found. Moreover, the dependence of the ATPase activity on DNA concentration was found to be sigmoidal, suggesting that two or more DNA molecules interact with each gyrase molecule. By contrast, long DNA molecules (>100 bp) did not show a sigmoidal dependence and gave maximal stimulation of the gyrase ATPase reaction at much lower DNA concentrations (2-3 μg/ml).

A similar discrimination with respect to DNA length was also found in DNA binding. Both filter binding and a gel method showed that gyrase binds tightly to long DNA molecules (>100 bp), forming a biomolecular complex. In contrast, the binding of gyrase to short DNA molecules (<70 bp) required higher DNA concentrations and gave sigmoidal binding isotherms, again suggesting the interaction of two or more short DNA fragments with each gyrase molecule. We conclude from the ATPase and binding data that gyrase requires the binding of DNA to two or more sites on the enzyme before ATP hydrolysis can occur. These sites are so disposed on the enzyme that only DNA molecules of 100 bp or greater can bridge them. A single short DNA molecule (<70 bp) cannot contact both sites, so that binding of two or more short DNA molecules is then required. These conclusions are incorporated

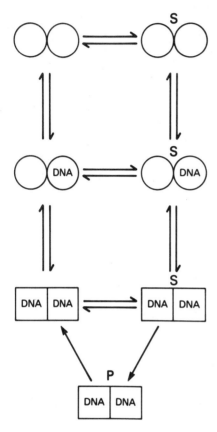

Figure 2. The interaction of DNA gyrase with DNA and ATP. The model shows the proposed cycle of DNA binding and ATP hydrolysis by gyrase. The gyrase molecule is represented as a dimer with each half containing a single binding site for DNA. The enzyme is represented in two states, the circles being inactive in ATP hydrolysis and the squares active. Binding of DNA to both halves of the enzyme mediates the transition between the two states. S is ATP and P is ADP and phosphate. (From Ref. 19).

into a model for the interaction of gyrase with DNA and ATP (Fig. 2) in which the cooperative binding of DNA to two sites on the enzyme promotes a conformational change whereby it becomes an active ATPase. In relation to the proposed structure of the gyrase-DNA complex (see below), it is likely that the protein-DNA contacts necessary to turn on the gyrase ATPase are interactions between the enzyme and the wrapped segment of DNA.

Although the model in Fig. 2 satisfies the ATPase and binding data, it is not clear from these data alone how many DNA binding sites on gyrase are involved in the mediation of the ATPase reaction. To clarify this point, we have now examined the binding of a short DNA fragment (55 bp) to DNA gyrase and determined the number of DNA molecules bound per enzyme molecule (Fig. 3). In this experiment, reaction mixtures containing DNA gyrase and ^{3}H-labelled DNA were applied to a polyacrylamide gel and electrophoresed in the presence of Mg^{2+}. Under these conditions, complexes between gyrase and DNA are stable during electrophoresis (19). After electrophoresis, bands containing free DNA and gyrase-DNA complexes are excised and the amount of radioactivity in each determined. At high concentrations of the 55 bp fragment, we found that two molecules of this DNA were bound

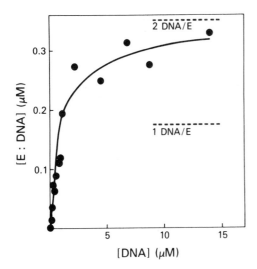

Figure 3. The binding of DNA gyrase to a 55-bp DNA fragment. Reaction mixtures containing gyrase A protein (176 nM dimer), gyrase B protein (176 nM dimer) and a ^{3}H-labelled 55-bp DNA fragment (0-14.4 μM) were incubated for 1 hr at 25°C, then applied to a polyacrylamide gel and electrophoresed in the presence of Mg^{2+}. Bands containing free DNA or gyrase-DNA complex were excised from the gel and the amount of DNA determined by scintillation counting. The concentration of gyrase-DNA complex (E:DNA) was determined at each DNA concentration. The dashed lines indicate the concentrations at saturation for complexes containing 1 and 2 DNA molecules per enzyme tetramer.

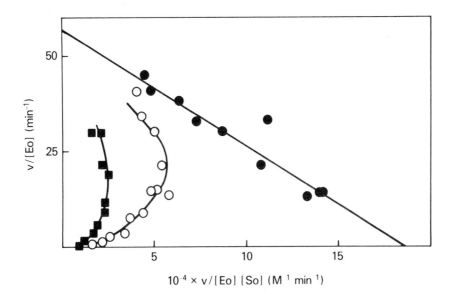

Figure 4. The kinetics of ATP hydrolysis. Initial rates of ATP hydrolysis by DNA gyrase in the presence of a saturating concentration of λ DNA were determined at a range of ATP concentrations in the absence and presence of ADP. Reaction mixtures containing ATP (0.03-1 mM), gyrase A protein (20-120 nM dimer), gyrase B protein (10-60 nM dimer) and λ DNA (10 μg/ml) were incubated at 25°C and the amount of ATP hydrolyzed was determined at various times. Initial rates were calculated from tangents at time zero to the curve of product concentration versus time. Results are displayed as Eadie-Hofstee plots, where v is initial velocity, [Eo] is total enzyme concentration and [So] is initial substrate concentration. ●, no ADP; ○, 0.1 mM ADP; ■, 0.25 mM ADP.

per enzyme tetramer, supporting the notion that there are two DNA binding sites on gyrase which must be occupied in order to stimulate the ATPase reaction.

In the model in Fig. 2, the hydrolysis of a single ATP molecule per catalytic cycle is shown. However, in view of the proposed A_2B_2 structure of DNA gyrase, it is likely that the enzyme has two ATP binding sites. This possibility has also been raised by analysis of the steady-state kinetics of ATP hydrolysis by gyrase both in the absence and the presence of the inhibitor ADP (Fig. 4). In the absence of ADP, reaction rates show the expected hyperbolic dependence on ATP concentration and give a linear Eadie-Hofstee plot. However, in the presence of ADP, rates of ATP hydrolysis show a sigmoidal dependence on ATP concentration and the Eadie-Hofstee plots deviate markedly from linearity (Fig. 4), suggesting a departure from Michaelis-Menten kinetics. Using computer-assisted modelling, we have found that the data in Fig. 4 are consistent with a scheme involving the binding of two ATP molecules to each gyrase molecule prior to hydrolysis.

Electric Dichroism

Transient electric dichroism has proven to be a particularly informative technique for determining the solution structures of DNA and complexes of DNA with various drugs and proteins (20,21,22; for review see 23). This technique entails orienting macromolecules in solution by applying strong electric fields and measuring the change in absorbance of light polarized either parallel or perpendicular to the field. The reduced linear dichroism for either polarization direction, $\rho_\parallel$ or $\rho_\perp$, is defined as:

$$\rho_\parallel = \frac{A_\parallel(E) - A_\parallel(0)}{A_\parallel(0)}$$

$$\rho_\perp = \frac{A_\perp(E) - A_\perp(0)}{A_\perp(0)}$$

where A(E) is the steady-state absorbance in a field of strength E and A(0) is the sample absorbance with no applied field. The total dichroism, ρ, is usually defined as:

$$\rho = \rho_\parallel - \rho_\perp.$$

For well-behaved samples in which the change in absorbance is due solely to orientation and not to a change in the isotropic absorbance, these dichroisms are simply related by:

$$\rho_\perp = -1/2\,\rho_\parallel$$

and

$$\rho = 3/2\,\rho_\parallel.$$

Structural information is extracted by measuring the dichroism as a function of field strength and extrapolating to infinite field strength where the orientation is perfect. The dichroism at infinite field ($\rho(E \Rightarrow \infty)$) is determined by the distribution of angles that the absorbing chromophores make with the orienting dipole axis (see for example, 24). DNA is especially well suited for this technique since it absorbs light strongly at 260 nm (where protein absorbance at the same weight concentration is negligible) and the chromophores (the base pairs) are oriented almost perpendicularly to the helix axis. The orientation of base pairs at complete alignment in the field determines the orientation of the helix axis with respect to the dipole axis. Additionally, DNA, like other polyelectrolytes, has very strong induced dipole moments due to the polarization of the surrounding counterions with respect to the phosphate changes of the DNA backbone (25). For this type of dipole, the orienting axis is simply the longest DNA axis.

Another valuable piece of structural information that can be extracted from transient electric dichroism experiments is the decay or relaxation of the dichroism after the field is turned off. For well defined structures with little or no internal motion, the dichroism at time t, $\rho(t)$, relative to the steady-state dichroism in the field, $\rho(o)$, is given by:

$$\rho(t)/\rho(o) = \exp(-6D_o t),$$

where D_o is the diffusion coefficient for rotation about the orienting axis. This parameter is very sensitive to molecular size and shape. For spheres of radius R or cylinders of length L, the relaxation times ($\tau = 1/6 D_o$) vary as R^3 or approximately as L^3. Numerical calculations for the rotational diffusion coefficients of a number of other structures are given by Garcia de la Torre and Bloomfield (26).

We have examined the electric dichroism of complexes between DNA gyrase and three DNA fragments of 172, 207 and 256 bp, derived from the sea urchin 5s rRNA gene (27). Digestion with pancreatic DNase I shows that gyrase protects a well-defined region of approximately 105 bp centrally located in the 172 and 207 bp fragments. Quinolone-induced DNA cleavage by gyrase identifies two strong gyrase cleavage sites spaced 11 bp apart, located approximately in the middle of the protected region. The unprotected regions of DNA indicate that the complexes between DNA gyrase and these fragments are expected to have two free DNA tails of approximately equal lengths ($\sim$ 30 bp for the 172 bp fragment, $\sim$ 50 bp for the 207 bp fragment).

In contrast, no clear pattern of protection from DNase I digestion by gyrase could be found with the 256-bp fragment. Instead protection extended along the entire length of the DNA, suggesting that there was no single preferred gyrase binding site. Quinolone-induced gyrase cleavage revealed that in addition to the strong cleavage sites observed with the 172-bp and 207-bp fragments, there were several other cleavage sites. It is likely, therefore, that complexes between DNA gyrase and the 256-bp fragment are heterogeneous, having DNA tails of varying lengths.

For complexes of DNA gyrase with each of these DNA fragments, the dependence of the dichroism on field strength and also the dichroism decay curves were measured. Both the ionic strength dependence of the dipole moment and the symmetry of the dichroism build-up and decay at low fields indicate that orientation is due to the counterion polarization dipole characteristic of polyelectrolytes. More importantly, the dichroism curves of the gyrase complexes show a good fit to the expected behavior of molecules with pure induced dipole moments (23). These results indicate that any permanent dipole arising from an asymmetric distribution of protein and DNA charges is small by comparison with the induced dipole from counterion polarization. Values of $\rho_{||}$ (E $\Rightarrow \infty$) can be determined either by fitting $\rho_{||}$ versus E data to theoretical curves or, if the dipole moment is of sufficient magnitude, by direct extrapolation at $\rho_{||}$ versus $1/E^2$ at very high field strengths.

The intrinsic dichroism for each of the three complexes can be partitioned into two components; that due to DNA bound to DNA gyrase (the "core"), and that due to DNA not bound to the enzyme (the "tails"). If $\rho_{||,c}$ and $\rho_{||,T}$ are the intrinsic dichroisms of core and tail DNA and if fc is the fraction of DNA in the core, then:

$$\rho_{||} \ (E \Rightarrow \infty) = fc\cdot\rho_{||,c} + (1-fc)\cdot\rho_{||,T}.$$

For each DNA fragment of N bp, fc = 105/N, assuming that the core DNA is the 105 bp protected from nuclease digestion. Using the extrapolated dichroism of the three fragments, we have sufficient information to assign $\rho_{||,c}$ and $\rho_{||,T}$, unambiguously. For a 105-bp core we find $\rho_{||,c} = -0.24 \pm .01$. This value is consistent with a simple, single turn circle of DNA around gyrase as shown in Fig. 5. If this DNA is

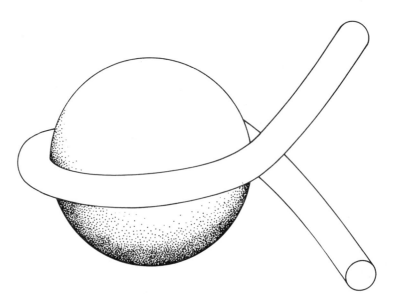

Figure 5. Complex between DNA gyrase and DNA. The figure gives a schematic representation of the simplest structure between gyrase and DNA consistent with the electric dichroism data presented in this paper. DNA gyrase is represented as a sphere around which the DNA makes a single turn.

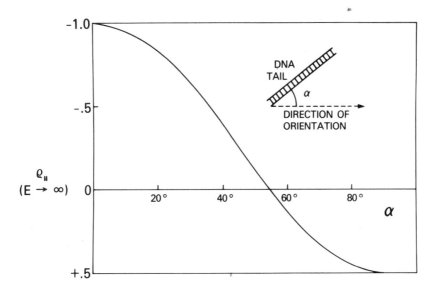

Figure 6. The calculated dependence of the intrinsic dichroism of a DNA tail on the angle the tail makes to the direction of orientation. The orientation axis bisects the angle between the two tails. Experimentally, we determine $\rho_{\parallel}$ for tails to be $+0.13$, such that the angle that each tail makes to the direction of orientation is about $60°$ and the angle between two tails is approximately $120°$.

wound as a supercoil, then the superhelical pitch must be less than 40 Å. The value of $\rho_{\parallel,T}$ is found to be $+0.13$ for these three complexes. For tails of equal length we can interpret this dichroism in terms of an angle between a tail and the orientation axis. Fig. 6 shows the expected dichroism as a function of this angle, assuming the tails are rigid rods. From this graph we estimate the angle between the tails to be $120°$.

The dichroism decay of the gyrase-DNA complexes shows single exponential kinetics consistent with a stable, well defined structure. For each complex we have determined the rotational diffusion coefficient and have found that these are in excellent agreement with the structures derived from the dichroism measurements (Fig. 5).

We have also made preliminary measurements of the dichroism of gyrase-DNA complexes in the presence of the quinolone drug norfloxacin and the non-hydrolyzable ATP analog (β,γ-imido) ATP. As norfloxacin can lead to DNA cleavage by gyrase and as (β,γ-imido) ATP supports limited supercoiling, we expected to detect dichroism changes in the presence of these ligands. Complexes between DNA gyrase, DNA and norfloxacin show no difference in either the dichroism or the relaxation time when compared to complexes without the drug. This suggests that drug binding induces little or no change in the configuration of the complexed DNA. However, in the presence of (β,γ-imido) ATP, both the dichroism and relaxation time change significantly; the extrapolated dichroism is more positive and the relaxation time is shorter. These differences are consistent with a substantial conformational transition of the gyrase-DNA complex. As (β,γ-imido) ATP supports limited supercoiling by gyrase, it is likely that these changes in dichroism are associated with the supercoiling process.

References and Footnotes

1. Gellert, M., *Ann. Rev. Biochem. 50,* 879-910 (1981).
2. Wang, J.C., *Sci. Am. 247,* 94-109 (1982).
3. Cozzarelli, N.R., *Cell 22,* 327-328 (1980).
4. Gellert, M., Mizuuchi, K., O'Dea, M.H. and Nash, H.A., *Proc. Natl. Acad. Sci. USA 73,* 3872-3876 (1976).
5. Mizuucchi, M., O'Dea, M.H. and Gellert, M., *Proc. Natl. Acad. Sci. USA 75,* 5960-5963 (1978).
6. Higgins, N.P., Peebles, C.L., Sugino, A. and Cozzarelli, N.R. *Proc. Natl. Acad. Sci. USA 75,* 1773-1777 (1978).
7. Klevan, L. and Wang, J.C. *Biochemistry 19,* 5229-5234 (1980).
8. Gellert, M., Mizuuchi, K., O'Dea, M.H., Itoh, T. and Tomizawa, J. *Proc. Natl. Acad. Sci. USA 74,* 4772-4776 (1977).
9. Sugino, A., Peebles, C.L., Kreuzer, K.N. and Cozzarelli, N.R. *Proc. Natl. Acad. Sci. USA 74,* 4767-4771 (1977).
10. Sugino, A., Higgins, N.P., Brown, P.O., Peebles, C.L. and Cozzarelli, N.R. *Proc. Natl. Acad. Sci. USA 75,* 4838-4842 (1978).
11. Liu, L.F. and Wang, J.C., *Cell 15,* 984-979 (1984).
12. Fisher, L.M., Mizuuchi, K., O'Dea, M.H., Ohmori, H. and Gellert, M. *Proc. Natl. Acad. Sci. USA 78,* 4165-4169 (1981).
13. Kirkegaard, K. and Wang, J.C., *Cell 23,* 721-729 (1981).
14. Morrison, A. and Cozzarelli, N.R., *Proc. Natl. Acad. Sci. USA 78,* 1416-1420 (1981).
15. Liu, L.F. and Wang, J.C., *Proc. Natl. Acad. Sci. USA 75,* 2098-2102 (1978).
16. Morrison, A. and Cozzarelli, N.R., *Cell 17,* 175-184 (1979).
17. Morrison, A., Brown, P.O., Kreuzer, K.N., Otter, R., Gerrard, S.P. and Cozzarelli, N.R. in *Mechanistic Studies of DNA Replication and Genetic Recombination,* Eds., B.M. Alberts and C.F. Fox, Academic Press, New York, pp. 785-806 (1981).
18. Sugino, A. and Cozzarelli, N.R. *J. Biol. Chem. 255,* 6299-6306 (1980).
19. Maxwell, A. and Gellert, M., *J. Biol. Chem. 259,* 14472-14480 (1984).
20. Dickmann, S., Hillen, W., Jung, M., Wells, R.D. and Porschke, D. *Biophys. Chem. 15,* 157-167 (1982).
21. Fritzche, H., Triebel, H., Chaires, J.B., Dattagupta, N. and Crothers, D.M., *Biochemistry 21,* 3940-3946 (1982).
22. McGhee, J.D., Nickol, J.M., Felsenfeld, G. and Rau, D.C., *Cell 33,* 831-841 (1983).
23. Fredericq, E. and Houssier, C., *Electric Dichroism and Electric Birefringence.* Oxford, Clarendon Press (1973).
24. Rill, R.L., *Biopolymers 11,* 1929-1941 (1972).
25. Rau, D.C. and Charney, E., *Biophys. Chem. 14,* 1-9 (1981).
26. Garcia de la Torre, J. and Bloomfield, V.A., *Quart. Rev. Biophys. 14,* 81-139 (1981).
27. Simpson, R.T. and Stafford, D.W., *Proc. Natl. Acad. Sci. USA 80,* 51-55 (1983).

Biomolecular Stereodynamics III, Proceedings of the Fourth Conversation in the Discipline Biomolecular Stereodynamics, State University of New York, Albany, NY, June 04-09, 1985, Eds., Ramaswamy H. Sarma & Mukti H. Sarma, ISBN 0-940030-14-4, Adenine Press, ©Adenine Press 1986.

Direct Comparison of the Membrane Bound and Structural Forms of the Coat Protein of the Filamentous Bacteriophage fd

L.A. Colnago, G.C. Leo, K.G. Valentine, and S.J. Opella

Department of Chemistry
University of Pennsylvania
Philadelphia, Pennsylvania 19104

Abstract

The coat protein of the filamentous bacteriophage fd exists in a membrane bound form as well as in the structural form of the virus during its lifecycle. The changes in the dynamics of the coat protein that accompany the infection of *E. coli* and the assembly of new virus particles are described with NMR experiments. Solid state NMR is used to describe the dynamics of isotopically labelled peptide backbone and amino acid sidechain sites. Motional averaging of powder pattern lineshapes gives qualitative information about the amplitudes and rates of motions.

Introduction

The filamentous bacteriophages are large nucleoprotein complexes (1). fd consists of 90% by weight coat protein and 10% DNA. fd infects *E. coli* and reproduces in a non-lytic infection. At infection the DNA enters the bacterial cell cytoplasm and the coat protein becomes associated with the inner cell membrane. Many copies of viral DNA and coat protein are synthesized within the infected cells (2). These new coat protein molecules are stored along with the salvaged ones from the infecting virus particles in the cell membrane. New virus particles are assembled at the cell membrane from this total pool of new and old coat proteins and DNA. In the assembly process the coat protein changes from its membrane bound form to the form it adopts as the major structural element of the virus particles (3). The changes in the coat protein that occur during infection and assembly processes must be reversible. The changes in the dynamics of the coat protein are the subject of this paper. This is an initial account of the findings derived from comparisons of the solid state NMR studies of the coat protein in the virus (4) and in reconstituted membrane bilayers (5).

Each coat protein molecule of fd has 50 residues. The amino acid sequence shows the protein to have a central hydrophobic midsection of about 20 residues, sur-

Sequence of fd Coat Protein

(hydrophilic)

$NH_3{}^+$-ALA-GLU-GLY-ASP-ASP-PRO-ALA-LYS-ALA-ALA-PHE-ASP-SER-LEU-GLN-ALA-SER-ALA-THR-GLU-...

(hydrophobic)

...-TYR-ILE-GLY-TYR-ALA-TRP-ALA-MET-VAL-VAL-VAL-ILE-VAL-GLY-ALA-THR-ILE-GLY-ILE-...

(hydrophilic)

...-LYS-LEU-PHE-LYS-LYS-PHE-THR-SER-LYS-ALA-SER-COO-

Figure 1. Amino acid sequence of fd coat protein (6a,b).

rounded by relatively hydrophilic C- and N- terminal regions. The sequence of the protein is given in Figure 1 (6a,b).

The two major techniques employed in these studies are biosynthetic isotopic labelling of the coat protein in the virus and the analysis of powder pattern lineshapes from the labelled sites obtained in solid state NMR experiments. Since *E. coli* can be grown on chemically defined media when infected with fd, labels can be readily incorporated into specific sites of amino acids in the viral proteins. Since the major coat protein of fd is small there are only one or several of each amino acid type, enabling a high degree of selectivity through labelling of one amino acid at a time.

Solid state NMR methods, rather than solution NMR methods, are essential for these studies, since neither the virus nor the coat protein-membrane complex reorient rapidly in solution. The lineshape analysis of powder patterns provides a direct way to describe molecular dynamics in molecules that are not undergoing overall rapid reorientation. These methods have been used successfully to describe the dynamics of several proteins, including bacteriorhodopsin in membrane bilayers (7-9), collagen in fibrils (10), and the coat proteins of the filamentous bacteriophages (4). Figure 2 shows a collection of calculated lineshapes for powder patterns with and without motional averaging from chemical shift and quadrupole interactions, specifically for ^{15}N in peptide linkages and ^2H in sidechains (11). The shape and breadth of the rigid lattice powder patterns are determined by the fundamental properties of the nuclear spin interactions. Sites that do not have large amplitude motions that are rapid compared to the frequency breadth of the spin interaction have powder patterns that are very similar to those of the rigid lattice. Large amplitude, rapid motions have substantial effects in averaging the rigid lattice patterns. Given the rigid lattice lineshapes and specific models for motion, then motionally averaged lineshapes can be calculated using standard procedures. Lineshapes can be calculated for other nuclei, spin interactions, and types of motion. Peptide backbone and sidechain sites have been shown to have large amplitude, rapid motions in some protein sites and to be immobile in others by these methods.

Materials and Methods

Sample Preparation

Isotopically labelled fd was grown and purified as described previously(4). The isotopically labelled amino acids were obtained from commercial sources with the

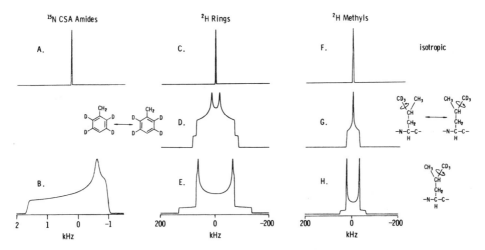

Figure 2. Calculated lineshapes based on static values from model compounds. (A) ^{15}N chemical shift anisotropy of amide group averaged by isotropic motion. (B) ^{15}N chemical shift anisotropy of static amide group. (C) ^{2}H quadrupole averaged by isotropic motion. (D) ^{2}H quadrupole averaged by two-fold ring flips. (E) ^{2}H quadrupole of static C-D bond. (F) ^{2}H quadrupole averaged by isotropic motion. (G) ^{2}H quadrupole averaged by two-site tetrahedral jump plus three-fold methyl reorientation. (H) ^{2}H quadrupole averaged by three-fold methyl reorientation.

exception of d_4-Thr which was synthesized using published procedures (12). The spectra of the intact virus were obtained from solutions with a concentration of 200 mg/ml in 40 mM borate buffer at pH 8. The bilayer samples were prepared by sonicating the virus in the presence of phospholipids followed by lyophilization and rehydration (13). The membrane samples contained about 100 mg protein and slightly greater amount of lipid in excess water.

Spectroscopy

The NMR experiments were performed on home-built spectrometers with 3.5T and 5.7T fields and a modified JEOL GX-400WB spectrometer with a 9.4T field. The ^{2}H NMR spectra were obtained using the quadrupole echo pulse sequence (14) using 2-4 μsec pulses and 20-50 μsec interpulse delays. The ^{15}N NMR spectra were obtained using the spin-lock version of cross-polarization (15) with a 1 msec mixing interval and proton decoupling during data acquisition.

Results

The experimental results described in this section are in the form of powder pattern lineshapes. Direct comparisons are made in the Figures between the membrane bound form of the coat protein in phospholipid bilayers and the structural form of the same protein in the virus. Specific models for motion are examined by comparing the experimental lineshapes to the calculated motionally averaged lineshapes in Figure 2 of the Introduction.

Backbone Sites

^{15}N NMR is an effective way of studying backbone dynamics of proteins even when all nitrogen sites are uniformly labelled. Nearly all of the nitrogens in proteins are in the peptide groups of the backbone. The ^{15}N chemical shift anisotropy is a relatively small interaction at the magnetic fields used in these experiments, resulting in a timescale for motional averaging of 1-10 kHz. A static powder pattern is observed for a labelled site when all orientations are present and large amplitude, rapid motions are absent. Large amplitude motions that occur more frequently than the breadth of the powder pattern result in an averaging of the lineshape. The lineshapes for the two extreme cases of a rigidly held and an isotropically reorienting amide group are shown in Figure 2. Labelling of individual amino acids adds greatly to the specificity of these experiments by allowing the lineshapes for one or a few sites to be examined.

Figure 3 compares the ^{15}N chemical shift spectra for the coat protein in membrane bilayers (top row) to the coat protein in its structural form in the virus (bottom row). Spectra from protein samples that are uniformly labelled with ^{15}N in all sites, labelled only in the four glycine residues, and labelled only in the two leucine residues are presented in Figure 3. These spectra are distinctive in showing a heterogeneity of protein backbone dynamics in both the membrane bound and structural forms of the protein. This is seen most directly in the spectra of the

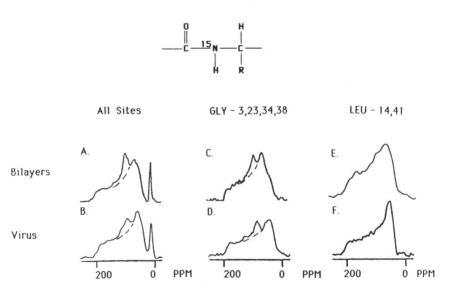

Figure 3. ^{15}N NMR spectra of labelled fd coat protein. (A) All sites uniformly labelled in the coat protein in the bilayers. (B) All sites uniformly labelled in the coat protein in the virus. (C) Glycines labelled in the coat protein in bilayers. (D) Glycines labelled in the coat protein in the virus. (E) Leucines labelled in the coat protein in bilayers. (F) Leucines labelled in the coat protein in the virus.

uniformly labelled and of the glycine labelled protein samples which have a narrow resonance near the isotropic frequency superimposed on the powder pattern lineshapes. Many of the amino acids in the protein have been individually labelled in order to identify which are mobile and which are immobile in both the membrane bound and structural forms. For example, Gly-3 is mobile while Gly-23, 34, 38 are all immobile in both forms of the protein based on the spectra in Figures 3c and 3d and the assignment of the single isotropic resonance to Gly-3 through proteolytic cleavage experiments. Leu-14 and 41 are immobile in both forms of the protein as seen in Figures 3e and 3f.

The sidechain of alanine is a methyl group bonded to the alpha carbon; therefore, the dynamics of the methyl group directly reflect on those of the backbone for these residues. The ^{2}H NMR spectrum of a CD_3 group is a narrowed powder pattern due to the reorientation of the methyl group about its C_α-C_β bond axis as shown in Figure 2h. This occurs whether the backbone is mobile or immobile at the residue. Additional motions can only occur in the methyl sidechain as a result of backbone motions. The spectra in Figure 4 show that the ten alanine residues in the coat protein show heterogeneous dynamics in both the membrane and in the virus, since a narrow central resonance is superimposed on the powder pattern from a reorienting methyl group. There is an increase in the magnitude of the isotropic

ALA − 1,7,9,10,16,18,25,27,35,49

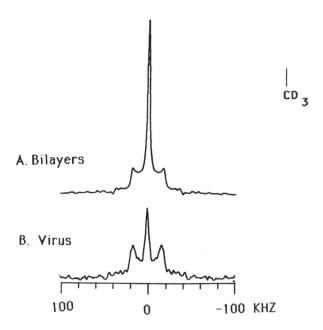

Figure 4. ^{2}H NMR spectra of d_3-Ala labelled fd coat protein. (A) Coat protein in bilayers. (B) Coat protein in the virus.

resonance in the membrane bound form over the structural form of the coat protein. This indicates that a larger percentage of the backbone sites are mobile in the membrane bound form of the protein.

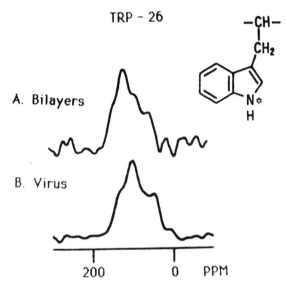

Figure 5. ^{15}N NMR spectra of ^{15}N-Trp labelled fd coat protein. (A) Coat protein in bilayers. (B) Coat protein in the virus.

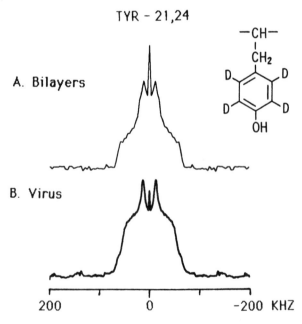

Figure 6. ^{2}H NMR spectra in d[4]-Tyr labelled fd coat protein. (A) Coat protein in bilayers. (B) Coat protein in the virus.

Aromatic Sidechains

fd coat protein has five aromatic residues. The single Trp (26) residue and the two Tyr (21,24) residues are in the hydrophobic midsection of the protein, while the three Phe (11,42,45) residues are in the terminal regions. Figure 5 compares the ^{15}N chemical shift anisotropy powder patterns for the Trp sidechain in the membrane bound and structural forms of the coat protein. In both cases the sidechain is immobile, since the powder patterns are not averaged compared to the rigid lattice model of polycrystalline tryptophan. Many of the initial results on protein dynamics by solid state NMR were obtained on aromatic sites of proteins, showing some Tyr and Phe residues to undergo two fold jumps about their C_β-C_γ bond axis (4,7,8). Figure 6 contains the ^{2}H NMR spectra from d_4-Tyr labelled coat protein samples. These experimental powder pattern lineshapes show that both of the tyrosine sidechains are undergoing rapid two-fold jumps in both forms of the protein by comparison with the lineshapes calculated for two-fold jump motions of this sidechain in Figure 2d. Phenylalanine sidechains have been generally found to behave like tyrosine sidechains in proteins, either being immobile or undergoing two-fold jumps. The ^{2}H NMR spectra in Figure 7 show the phenylalanine sidechains in the coat protein in membrane bilayers to have a great deal of additional motion as compared to the same sites in the structural form of the coat protein and other proteins.

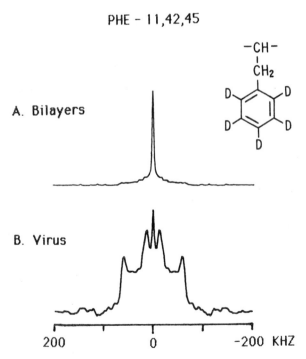

Figure 7. ^{2}H NMR spectra of d_5-Phe labelled fd coat protein. (A) Coat protein in bilayers. (B) Coat protein in the virus.

Aliphatic Sidechains

The alanine sidechains discussed above were used to monitor the backbone sites of those residues. The CD_3 labelled methyl sites of the leucine residues are capable of additional modes of motion, even with immobile backbone sites as observed from the ^{15}N labels in Figure 3. Leucine sidechains have been shown to have a two-site hop motion superimposed on the methyl rotation in several systems (16,17); the calculated lineshape for this motion is in Figure 2g. Figure 8 shows that this is the case for both Leu-14 and -41 in the membrane bound and structural forms of the coat protein.

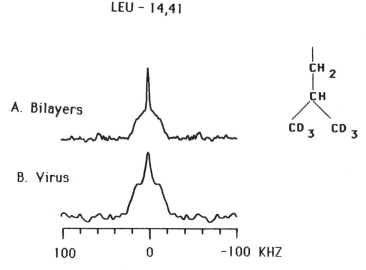

LEU - 14,41

Figure 8. ^2H NMR spectra of d_{10}-Leu labelled fd coat protein. (A) Coat protein in bilayers. (B) Coat protein in the virus.

The methyl group in threonine also provides an interesting monitor of protein dynamics (17). The threonine sidechain can clearly undergo additional modes of motion, even when the backbone at that residue is immobile. The methyl rotation powder lineshape is additionally averaged by a flip about the C_α-C_β bond. The membrane bound coat protein demonstrates that all three threonine residues undergo methyl rotation as well as the C_α-C_β jump motion. In the virus, there appears to be only one threonine sidechain which exhibits both types of motion. The remaining two threonine sidechains appear to have only the methyl rotation. The data in Figure 9 for the labelled coat protein in the virus and in the membrane bound form show differences for the two forms. These well defined lineshapes also serve to demonstrate the absence of large amplitude backbone motions for the threonine residues in both forms of the protein.

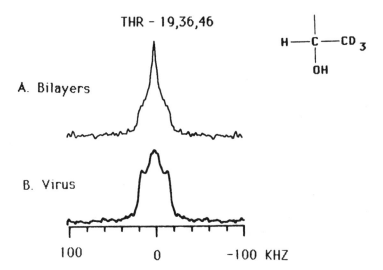

Figure 9. ^{2}H NMR spectra of d_4-Thr labelled fd coat protein. (A) Coat protein in bilayers. (B) Coat protein in the virus.

Methionine sidechains are capable of many motions (18). The ^{2}H NMR spectra in Figure 10 are for CD_3 labelled methionine in both forms of the coat protein. This is a particularly favorable case, because there is only a single Met residue in the protein. Methyl rotation combined with one flip motion gives a calculated spectrum that is similar to the experimental lineshape. This interpretation indicates that the protein backbone is immobile at Met-28 in both forms of the coat protein.

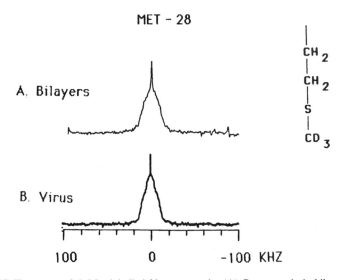

Figure 10. ^{2}H NMR spectra of d_3-Met labelled fd coat protein. (A) Coat protein in bilayers. (B) Coat protein in the virus.

Discussion

The major coat protein of fd goes from the structural form of the virus to the membrane bound form upon infection of the bacterial cell. The reverse occurs in the assembly process. A number of experiments indicate that the protein changes substantially in going between these two forms; for example circular dichroism (CD) spectra have been interpreted to show that the protein has about 90% alpha helix in the virus and about 50% alpha helix in the membrane (3). This suggests that substantial rearrangements of the protein occur. This contrasts to the conformational changes in allosteric proteins which typically have barely observable changes in secondary structure or even coat proteins of other viruses which apparently change very little during assembly.

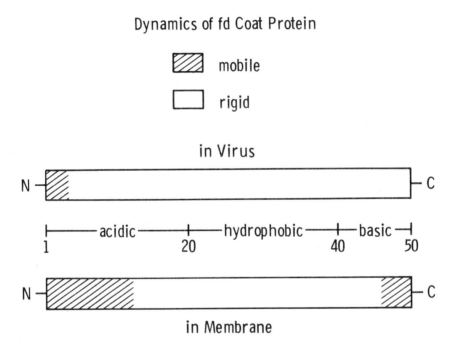

Figure 11. Representation of the distribution in backbone dynamics in fd coat protein.

The results presented here indicate that the changes in the coat protein that accompany infection and assembly must be explained in terms of dynamics as well as structure. Table I qualitatively summarizes the dynamics of the protein sites discussed in this paper. A clear trend emerges for increased mobility in the membrane bound form of the protein. Approximately 20-30% of the protein backbone is isotropically reorienting in the membrane bound form, in contrast to only 5-10% in the virus. The change of about 10 residues in the amino terminal region from immobile to isotropic motion on the kHz timescale is the largest influence of assembly on the protein. Changes in sidechain dynamics are also observed. Figure

Table I
Coat Protein Dynamics

Residue Type	Labeled Position	Structural Form Coat Protein	Membrane Bound Coat Protein
All sites	^{15}N amide backbone	4 mobile 46 static	~12 mobile ~38 static
GLY 3, 23, 34, 38	^{15}N amide backbone	1 mobile 3 static	1 mobile 3 static
LEU 14, 41	^{15}N amide backbone	2 static	2 static
	^{2}H methyl sidechain	2 methyl rotation with tetrahedral flip	2 methyl rotation with tetrahedral flip
ALA 1, 7, 9, 10, 16, 18, 25, 27, 35, 49	^{2}H methyl sidechain	1 mobile 9 static	~4-6 mobile ~4-6 static
TRP 26	^{15}N aromatic sidechain	1 static	1 static
TYR 21, 24	^{2}H aromatic sidechain	25°C 2-180° ring flip 40°C 2-180° ring flip	25°C 2-180° ring flip 40°C 2-180° ring flip with additional motion
PHE 11, 42, 45	^{2}H aromatic sidechain	3-180° ring flip	2-180° ring flip 1-180° ring flip with additional motion
THR 19, 36, 46	^{2}H methyl sidechain	2-methyl rotation 1-methyl rotation with tetrahedral flip	3-methyl rotation with tetrahedral flip
MET 28	^{2}H methyl sidechain	1-methyl rotation with twofold flip	1-methyl rotation with twofold flip

11 illustrates the distribution of the mobile and rigid regions of the backbone of the coat protein in the virus and membrane forms. If the protein spans the membrane bilayers with the C-terminal region outside the cell, then the mobile N-terminal region on the inside of the cell may be involved with displacing the gene 5 protein and wrapping about the DNA during the assembly process (19).

Acknowledgments

This research is being supported by grants GM-24266, AI-20770, and GM-34343 from the N.I.H. and by Smith, Kline, and French Laboratories. L.A.C. is supported by a fellowship from Conselho Nacional de Desenolvimento Cientifico e Tecnolgico (CNPq) Brazil.

References and Footnotes

1. L. Makowski in *Biological Macromolecules and Assemblies: Vol. 1,* Ed. A. McPherson, J. Wiley and Sons, p. 203 (1984).
2. H. Smilowitz, J. Carson and P. Robbins, *J. Supramol. Struct. 1,* 1 (1972).
3. Y. Nozaki, B.K. Chamberlain, R.E. Webster and C. Tanford Nature (Lond). 259, 35 (1976).
4. C.M. Gall, T.A. Cross, J.A. DiVerdi and S.J. Opella, *Proc. Natl. Acad. Sci. U.S.A. 79,* 101 (1982).
5. M.J. Bogusky, G.C. Leo and S.J. Opella in *Magnetic Resonance in Biology and Medicine,* Ed., G. Govil, C. Khetrapal and A. Saran, Tata McGraw Hill, New Delhi, 375 (1985).
6. a. F. Asbeck, K. Beyreuther, H. Kohler, G. vonWettstein and G. Braunitzer *Hoppe-Seyler's Z. Physiol. Chemi. 350,* 1047 (1969); b. Y. Nakashima and W. Konigsberg, *J. Mol. Biol. 73,* 598 (1974).
7. S. Schramm, R.A. Kinsey, A. Kintanar, T.M. Rothgeb and E. Oldfield in *Proceedings of the Second SUNYA Conversation in the Discipline of Biomolecular Stereodynamics Vol. II,* Ed. R.H. Sarma, Adenine Press, New York, p. 271 (1981).
8. D.M. Rice, A. Blume, J. Herzfeld, R.J. Wittebort, T.H. Huang, S.K. DasGupta and R.G. Griffin in *Proceedings of the Second SUNYA Conversation in the Discipline of Biomolecular Stereodynamics Vol. II,* Ed. R.H. Sarma, Adenine Press, New York p. 255 (1981).
9. M.A. Keniry, H.S. Gutowsky and E. Oldfield, *Nature (Lond.) 307,* 383 (1984).
10. D.A. Torchia, *Meth. Enzymol. 82,* 174 (1982).
11. S.J. Opella, *Meth. Enzymol.* in press (1985).
12. M. Sato, K. Okawa and S. Akabori, *S. Bull. Chem. Soc. Japan 30,* 937 (1957).
13. S.P.A. Fodor, A.K. Dunker, Y.C. Ng, D. Carsten and R.W. Williams in *Bacteriophage Assembly,* Ed. M.S. Dubow p. 441 (1981).
14. J.H. Davis, K.R. Jeffrey, M. Bloom, M.I. Valic and T.P. Higgs, *Chem. Phys. Lett. 42,* 390 (1976).
15. A. Pines, M.G. Gibby and J.S. Waugh, *J. Chem. Phys. 59,* 569 (1973).
16. L.S. Batchelder, C.E. Sullivan, L.W. Jelinski and D.A. Torchia, *Proc. Natl. Acad. Sci. U.S.A. 79,* 386 (1982).
17. M.A. Keniry, A. Kintanar, R.L. Smith, H.S. Gutowsky and E. Oldfield, *Biochemistry 23,* 288 (1984).
18. T.M. Rothgeb and E. Oldfield, *J. Biol. Chem. 256,* 1432 (1981).
19. A. Kornberg in *DNA Synthesis,* W.H. Freeman, San Francisco, p. 473 (1974).

Biomolecular Stereodynamics III, Proceedings of the Fourth Conversation in the Discipline Biomolecular Stereodynamics, State University of New York, Albany, NY, June 04-09, 1985, Eds., Ramaswamy H. Sarma & Mukti H. Sarma, ISBN 0-940030-14-4, Adenine Press, ©Adenine Press 1986.

Solid-State ^{13}C- and ^{15}N-NMR Investigations of the Mechanism of Proton-Pumping by Bacteriorhodopsin's Retinal Chromophore

S.O. Smith[a], G.S. Harbison[c], D.P. Raleigh[a,b], J.E. Roberts[a], J.A. Pardoen[d], S.K. Das Gupta[a], C. Mulliken[c], R.A. Mathies[e], J. Lugtenburg[d], J. Herzfeld[c] and R.G. Griffin[a]

[a]Francis Bitter National Magnet Laboratory and
[b]Department of Chemistry
Massachusetts Institute of Technology
Cambridge, MA 02139
[c]Department of Physiology and Biophysics
Harvard Medical School
Boston, MA 02115
[d]Department of Chemistry
University of Leiden
2300 RA Leiden, The Netherlands
[e]Department of Chemistry
University of California
Berkeley, CA 94720

Abstract

Magic angle sample spinning NMR spectra together with specific isotopic labeling have been used to study several features of the retinal chromophore conformation in dark adapted bacteriorhodopsin (bR^{DA}). ^{13}C spectra indicate that bR contains a mixture of *all-trans* and 13-*cis* retinals in agreement with other experiments. In addition, the spectra show that the C=N bond is *anti* and *syn* in all-*trans* and 13-*cis* forms, respectively, and in both cases the retinals are in the 6-s-*trans* rather the 6-s-*cis* forms such as are found in solution. ^{15}N spectra of ϵ-^{15}N-lys-bR indicate that bR^{DA} contains protonated Schiff base which are weakly hydrogen bonded.

Introduction

Solid-state NMR techniques provide a powerful approach for studying the structure and function of membrane proteins. When combined with specific isotopic substitution, solid state NMR can be used to probe specific sites in macromolecules which are inaccessible by other techniques.

We have been studying the light-activated proton-pumping mechanism of bacteriorhodopsin (bR), a 26000 dalton protein in the "purple membrane" of *Halobacterium halobium* (1). The light-sensitive prosthetic group in bR$_{568}$ is the all-*trans* protonated Schiff base (PSB) of retinal. As shown in Figure 1, photochemical 13-*trans* to 13-*cis* isomerization of the retinal chromophore (bR$_{568}$ $\Rightarrow$ K$_{625}$) translates the Schiff base

159

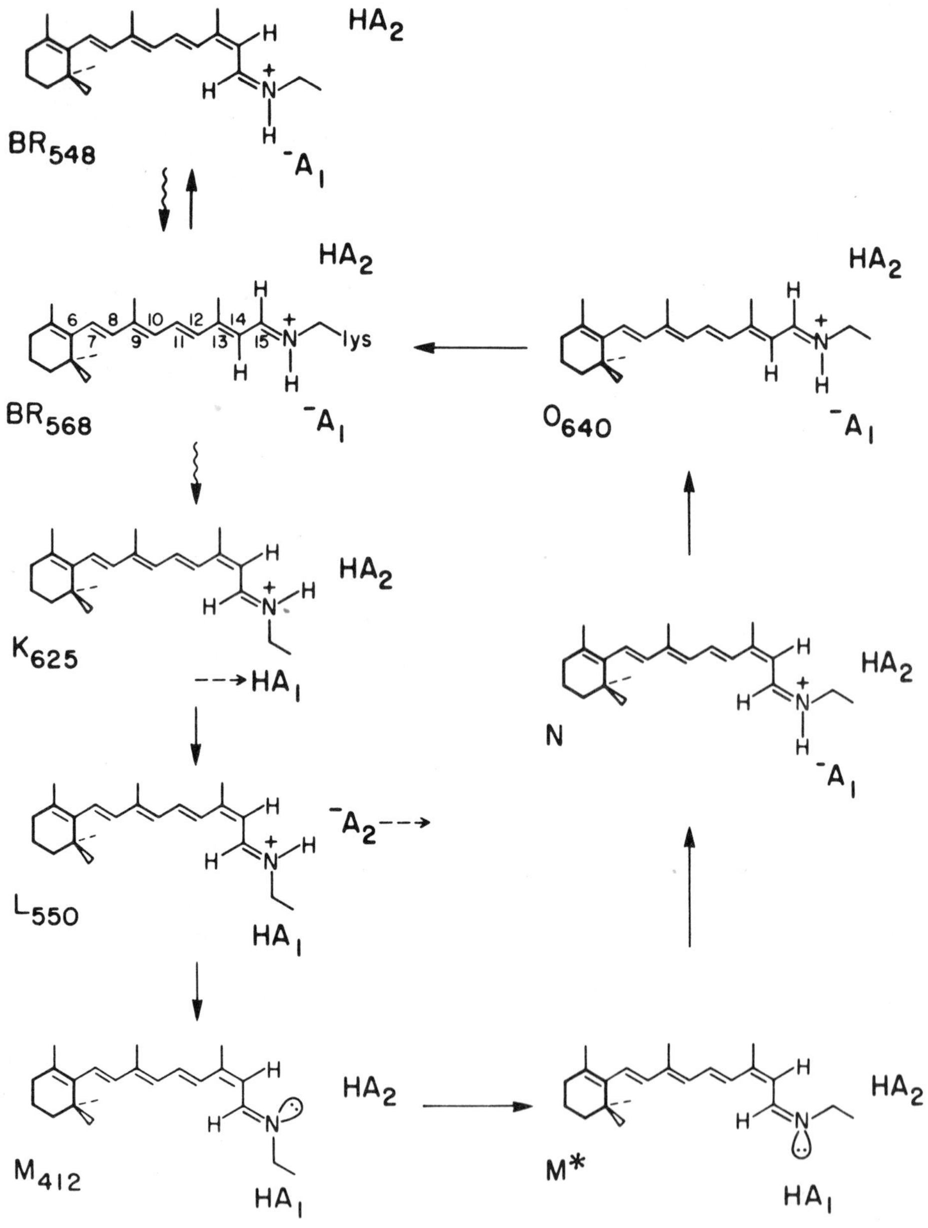

Figure 1. Photochemical reaction scheme of bacteriorhodopsin showing the structural changes involved in the proton-pumping photocycle and in dark-adaptation (3). The model illustrates how isomerization of the retinal provides a "switch" for proton transfer between protein residues A_1 and A_2. Photoisomerization about the C_{13}=C_{14} bond in the $bR_{568} \Rightarrow K_{625}$ transition translates the Schiff base proton from A_1 to A_2. Subsequent nitrogen inversion is proposed as a mechanism for allowing the Schiff base to reprotonate from A_1 (3). The direction of proton transfer (cell interior $\Rightarrow$ exterior) is indicated by dashed arrows. The structure of the retinal chromophore and the protonation changes of protein residues in bR are key elements of the proton-pumping mechanism, and can be probed with ^{13}C and ^{15}N solid-state NMR. Subscripts refer to room temperature absorption maxima.

proton across the retinal binding site where it is transferred to the protein ($K_{625} \Rightarrow$ $L_{550} \Rightarrow M_{412}$). The retinal then reisomerizes and reprotonates ($M_{412} \Rightarrow O_{640} \Rightarrow bR_{568}$), resetting the proton-pumping mechanism (2,3). In the dark, bacteriorhodopsin converts to dark-adapted bR which contains a 40:60 mixture of bR_{568} and bR_{548} (4). Recent solid-state NMR (5) and resonance Raman (6) studies have shown that bR_{548} contains a 13,15-di*cis* protonated Schiff base chromophore. The retinal chromophore in bR functions both as a "gate" to insure the flow of protons in only one direction across the membrane, and also as an "impeller", providing the driving force for proton transport. However, the pathway for protons through the protein has not been established. Fourier transform infrared (FTIR) difference spectroscopy (7,8) and kinetic UV absorption (9) studies have indicated changes in protonation of aspartate, glutamate and tyrosine residues occur during the proton-pumping photocycle. These studies are consistent with the proposal that protons are transported along a proton wire connecting the cytoplasmic and exterior surfaces of the protein through a chain of hydrogen-bonded amino acids (10). In this paper we discuss recent ^{13}C and ^{15}N solid-state NMR studies on two important aspects of the proton-pumping mechanism of bR. First, the structure and chemical environment of the retinal chromophore is investigated using bR containing ^{13}C-labeled retinal and ^{15}N-labeled lysine. Second, we discuss how the protonation state of amino acid side chains in bR can be probed using specific ^{13}C-labeling.

Solid-state NMR can be used to investigate the structure of both the retinal chromophore and proton-transporting amino acids in bR. High resolution ^{13}C and ^{15}N spectra can be obtained using magic angle sample spinning (MASS) and cross-polarization (CP) techniques (11). MASS increases resolution by breaking up broad ^{13}C and ^{15}N shift anisotropy powder patterns into sharp centerbands and rotational sidebands. Typically, a single chromophore or amino acid position is isotopically labeled to increase the sensitivity of the resonance over natural abundance levels. An additional increase in sensitivity is attained by transfer of magnetization between ^{1}H and ^{13}C or ^{15}N spins by cross-polarization. In contrast, in solution NMR the low rotational correlation time of the large purple membrane fragments produces incomplete averaging of the dipolar interactions and broad NMR resonances. Further, the dilute solutions required to prevent aggregation of the protein in these experiments result in a reduction in signal-to-noise ratios.

Since molecular reorientation is stopped or restricted in the solid, the anisotropic components of the chemical shift are retained in the MASS NMR experiment. This information can be "extracted" from the relative intensities of the rotational sidebands which are observed in spectra of slow-spinning samples (12). As will be seen below the principal values of the chemical shift tensor are invaluable in determining the *origin* of the differences in isotopic chemical shift observed in the protein.

^{13}C-MASS NMR of all-trans Retinal

The interpretation of ^{13}C-NMR spectra of the retinal prosthetic group in bR is based in part on studies of retinal and its unprotonated and protonated Schiff bases

(13). Figure 2 presents the MASS NMR spectrum of all-*trans* retinal spinning at 2.8 kHz. The lines are numbered according to their assignments and the centerband portion of the spectrum is expanded and shown in the inset. The assignments of the ^{13}C resonances were made using several approaches. (1) The ^{13}C chemical shifts in the crystalline and solution spectra of all-*trans* retinal are similar, and many of the assignments can be made by comparison. However, differences in the isotropic chemical shift of up to 4 ppm can result from differences in steric interactions, local charge distributions, dielectric and local diamagnetic shielding, and in the case of closely spaced lines positive assignments are not possible. (2) Specific isotropic labelling of the retinal chromophore with ^{13}C permits assignments for selected ^{13}C sites. Isotropic enrichment not only provides a means for spectral simplification, but also allows a much clearer measurement of the chemical shift tensor of a particular ^{13}C resonance. (3) Protonated carbons can be distinguished from nonprotonated carbons using a variable delay between proton decoupling and sample acquisition which takes advantage of the longer T_2 values observed in nonprotonated carbons (14).

In addition to the isotropic chemical shifts, the principle values of the chemical shielding tensors can be obtained from the all-*trans* retinal MASS NMR spectrum

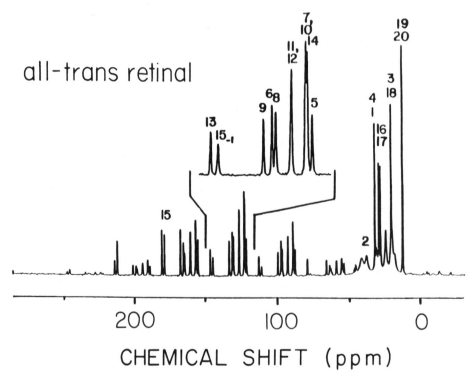

Figure 2. ^{13}C-MASS NMR spectrum of all-*trans* retinal obtained at $\nu_R = 2.8$ kHz. $\nu_{13_C} = 79.9$ MHz. The numbers correspond to the centerbands for particular carbons. The subscript -1 indicates a sideband for C-15.

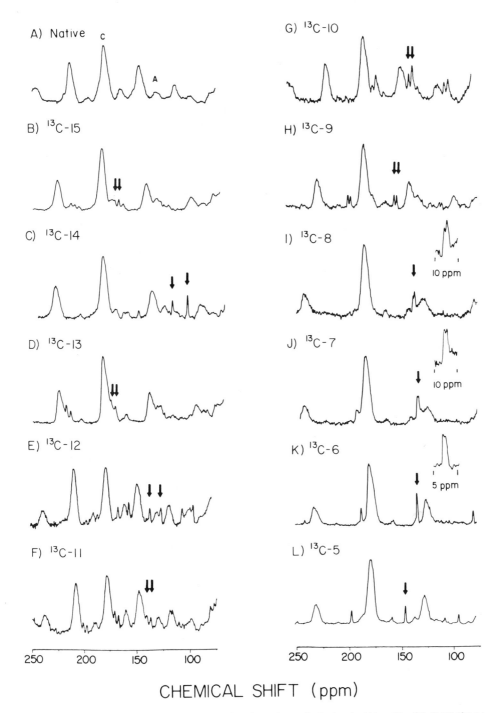

Figure 3. ^{13}C-MASS NMR spectra of natural abundance bacteriorhodopsin (A) and its ^{13}C-15 (B) ^{13}C-14 (C), ^{13}C-13 (D), ^{13}C-12 (E), ^{13}C-11 (F), ^{13}C-10 (G), ^{13}C-9 (H), ^{13}C-8 (I), ^{13}C-7 (J), ^{13}C-6 (K) and ^{13}C-5 (L) derivatives. In (A), C indicates the carbonyl centerband and A the aromatic centerband.

by analyzing the intensities of the rotational sidebands using the method of Herzfeld and Berger (12). These data often provide a basis for distinguishing between specific structural or environmental changes which influence the position of the isotropic chemical shift. For instance, *trans* $\Rightarrow$ *cis* isomerization of a C$=$C bond in the retinal chain causes an upfield shift of the ^{13}C resonance one bond away. This shift, termed the γ-effect, results from steric interaction between the hydrogen atoms across the *cis* double bond. Comparison of the shielding tensors of ^{13}C-12 all-*trans* and ^{13}C-12 13-*cis* retinal show that the γ-effect is localized in the σ_{11} element (13).

^{13}C-MASS NMR of Retinal in bR

Differences in the ^{13}C-NMR spectrum of the retinal PSB after incorporation into bacterio-opsin can result from both changes in chromophore structure and from interactions between the chromophore and the protein-binding site. To investigate the structure of the retinal chromophore and its protein environment in both bR$_{568}$ and bR$_{548}$, we have obtained spectra of dark-adapted bR using purple membrane regenerated with retinal labeled with ^{13}C at positions 5 through 15. These spectra are presented in Figure 3 and illustrate that high resolution spectra can be obtained of an intrinsic membrane protein isotopically enriched at a single site. The chemical shifts observed for the two isomers in these dark-adapted samples in addition to those of the all-*trans* protonated Schiff base are presented in Table I.

Table I

Chemical shifts of the 13-*cis* and all-*trans* isomers of bR, compared with those of all-*trans*-retinylidene butylimmonium chloride, in the solid-state and solution

	bR$_{568}$	bR$_{548}$	NRBH$^+$Cl$^-$ Solid[d]	NRBH$^+$Cl$^-$ Solution[e]
C-5	144.8	144.8[a]	128.7	131.8
C-6	135.4	134.9[a,f]	138.8	137.4
C-7	129.5	130.7[a]	128.8	132.0
C-8	132.7	131.6[a]	140.8	136.9
C-9	146.4	148.4[a]	142.1	145.3
C-10	133.0	129.7[b]	135.0	129.5
C-11	139.1	135.4[b]	138.9	137.4
C-12	134.3	124.2[b]	135.0	133.6
C-13	169.0	165.3[a,f]	161.8	162.3
C-14	122.0	110.5[c]	122.5	120.1
C-15	163.2	160.4[a,f]	167.0	163.6
C-18	22.0	22.0[a]	23.3	21.9
C-19	11.3	11.3[b]	14.0	13.2
C-20	13.3	13.3[b]	14.0	14.3

[a]ref. 15
[b]ref. 16
[c]ref. 5
[d]ref. 13
[e]ref. 17
[f] assignment to bR$_{548}$ or bR$_{568}$ is not known.

$C_{13} \colon C_{14}$ and C$\colon$N Configuration in bR. The most significant differences between the isotropic chemical shifts of bR$_{568}$ and bR$_{548}$ are at C-14 and C-12. The chemical shifts of C-14 and C-12 in bR$_{568}$ are very similar to those of the all-*trans* PSB model compound. However, upfield shifts of 11.5 (C-14) and 10.1 ppm (C-12) are observed for the bR$_{548}$ isomer. These differences result from a γ-effect between the C-15 and C-12 protons, and between the C-14 proton and the CH$_2$ protons of the ϵ-carbon of lysine. In both cases, the shift is localized in the σ_{11} element of the shift tensor. The γ-effect on the C-12 chemical shift is in agreement with previous chemical extraction (4) and resonance Raman (6) experiments demonstrating a 13-*cis* chromophore in bR$_{548}$, while the γ-effect at C-14 was unexpected and indicated that the C$\colon$N bond was in the *syn* (or *cis*) configuration in this pigment.

The Schiff Base Environment in bR. Comparison of the bR chemical shifts with those of the all-*trans* PSB shows several differences. First, a 3.8-6.6 ppm difference is observed in the ^{13}C-15 chemical shift. This difference results from a change in hydrogen-bonding of the Schiff base proton with its counterion, which appears to be weak in bR (see below). This difference vanishes when the counterion for the PSB model compound is more weakly hydrogen-bonded as in the trichloroacetate salt.

C_6-C_7 Conformation and the Opsin Shift. The most distinctive differences between the bR chemical shifts and those of the PSB occur at the C-8 and C-5 positions. Figure 4 present the MASS spectrum of hydrated ^{13}C-5-bR. The ^{13}C-5 resonance is

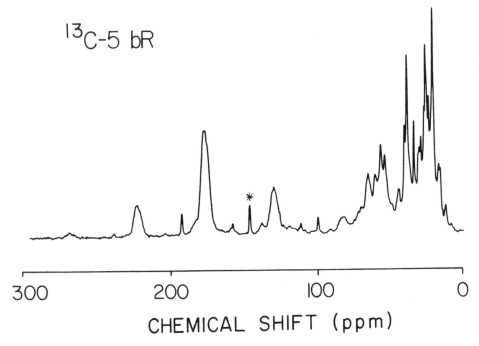

Figure 4. ^{13}C-MASS NMR spectrum of ^{13}C-5-bR. $\nu_R = 3.7$ kHz. $\nu_{13_C} = 79.9$ MHz.

~16 ppm downfield from its position in the all-*trans* protonated Schiff base. This difference has been shown to arise from two very different effects—isomerization about the C_6-C_7 bond and a protein charge perturbation centered near C-5 (15).

The effects of C_6-C_7 isomerization can be gauged by comparing the chemical shifts between the 6-s-*cis* and 6-s-*trans* forms of retinoic acid (13). The most pronounced differences (6-s-*cis* $\Rightarrow$ 6-s-*trans*) are observed at C-5 (+7.1 ppm), C-8 (−8.0 ppm), and C-9 (−6.7 ppm). An analysis of the shift anisotropies from the sideband patterns shows that the shift of C-5 is localized in the σ_{33} element, which moves 20 ppm, with essentially no change occurring in the σ_{11} and σ_{22} elements. In bR, a similar shift of the σ_{33} element is observed in comparison with the 6-s-*cis* protonated Schiff base arguing that the C_6-C_7 bond is *trans* in the pigment. This conclusion is supported by measurements of the T_1's of the ^{13}C-18 methyl group. In 6-s-*cis* retinal model compounds, the T_1's for the C-18 methyl group are relatively short (0.4-4.0 sec), while for the 6-s-*trans* retinals and bR, they are much longer, ~25 sec and 17 sec, respectively (15).

In contrast to the model compounds, we observe an additional 27 ppm downfield shift in the σ_{22} element which could be caused by a negative bacterio-opsin charge close to the C-5 position of the ionone ring (15). The 16 ppm chemical shift difference between bR and the all-*trans* PSB can therefore be decomposed into an ~7 ppm shift due to C_6-C_7 isomerization and an ~9 ppm shift due to protein perturbation.

A protein charge localized near C-5 might be expected to exhibit effects on carbon chemical shifts beyond the site of perturbation. However, if we take into account the changes in isotropic chemical shift expected for 6-s-*cis* to 6-s-*trans* isomerization, there are only very small differences (< 3 ppm) in the chemical shifts of C-6, C-7 and C-8. One possible explanation for the absence of chemical shift differences in the vicinity of C-7 is that the downfield shift induced by a negative charge at C-5 is counteracted by a positive bacterio-opsin charge near C-7. There are several attractive features to this idea. First in the hydrophobic interior of the protein, an ion pair would be more stable than an isolated negative charge. The expected ion-counterion distance of 2.5-3 Å corresponds well with the distance between C-5 and C-7. Secondly, we have previously noted (16) an upfield shift at C-19 which might be due to the same counterion, particularly if it is a guanidinium group of an arginine residue whose positive charge may be distributed over three nitrogens. Finally, recent studies on bacteriorhodopsin containing dihydroderivatives of retinal indicate that there is a larger opsin shift in the 7,8-dihydroderivative (3500 cm^{-1}) than in 5,6-dihydro-bR (2300 cm^{-1}) supporting the placement of a positive charge near C-7 (18,19).

^{15}N-MASS NMR of the Retinal Schiff Base in bR

Isotopic labeling of the Schiff base requires incorporating ^{15}N-labeled lysine into the protein by growing *H. halobium* on enriched media (20). Figure 5 presents the ^{15}N spectrum of hydrated bR along with spectra of the ^{15}N unprotonated and

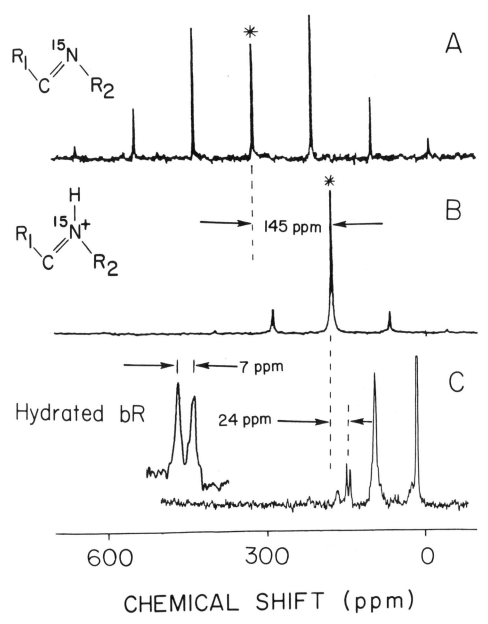

Figure 5. ^{15}N-MASS NMR spectra of the unprotonated (A) and protonated (B) retinal ^{15}N-butylamine Schiff bases, and ϵ-^{15}N-lysine-bR (C). ν_R (spectrum C) = 2 kHz. ν_{15_N} = 32.2 MHz. Note the large difference between the isotropic and anisotropic chemical shifts of the protonated and unprotonated Schiff bases.

protonated retinal Schiff base model compounds. In the model compound spectra (Figures 5A and 5B) the centerband is marked with an asterisk and is flanked by rotational sidebands. In ϵ-^{15}N-lysine bR (Figure 5C), the six free lysine residues give rise to the intense resonance on the right of the spectrum. The strong line downfield from the lysine resonance is due to the natural-abundance amide backbone, and the small doublet to the left of the amide line arises from the retinal-lysine Schiff base nitrogen. The splitting in the Schiff base resonance, as in the ^{13}C-bR spectra, is due to the presence of bR_{568} and bR_{548} in the dark-adapted sample.

The ^{15}N chemical shift of Schiff bases is extremely sensitive to protonation. Protonation of the retinal N-butylamine Schiff base results in an 145 ppm movement of the isotropic chemical shift from 317 to 172 ppm. Further, the shift anisotropy changes from an $\eta \simeq 1$ powder pattern with a breath of 600 ppm to an $\eta \simeq 0.5$ pattern with a breath of 270 ppm. Thus, a simple measurement of the isotropic or anisotropic shift is sufficient to distinguish between protonated and nonprotonated Schiff bases. The average chemical shift shown in Figure 5C is 148 ppm and is clearly much closer to the protonated Schiff base than to the unprotonated Schiff base. The 24 ppm difference between the Schiff base Cl$^-$ salt and bR is probably due to the presence of a weakly hydrogen-bonded counterion in the protein (20).

^{13}C-NMR of Tyrosine

Ionizable amino acid sidechains, particularly those of anionic residues—aspartate, glutamate and tyrosine—are expected to play a central role in the proton transfer process. The presence or absence of charged amino acids in bR can be easily tested by ^{13}C-NMR. For example, it has been known for many years that the ^{13}C chemical shifts of aromatic ring carbons of phenols depend strongly on whether the phenol is protonated or deprotonated. In particular, the carbon proximal to the hydroxyl group shifts downfield by 12 ppm in simple phenols upon deprotonation (21). A similar 10-12 ppm shift change was observed for a tyrosyl residue in a simple peptide on going from pH 7 to 11, as a result of ionization of the phenolic OH (22,23).

Studies with model compounds in this laboratory (24) have shown that the same effect occurs in the solid-state. For example, in Figure 6 we compare the chemical shift of the ^{13}C-OH carbon in cresol and sodium cresolate. The isotropic shift changes from 153 to 166 ppm (relative to TMS) upon deprotonation. The strong downfield shift is due primarily to a 35 ppm shift in the σ_{22} element of the shielding tensor. The result is a decrease in the tensor asymmetry parameter from 0.9 to 0.5. Such a dramatic change is easily detected in a spinning spectrum and provides another means of distinguishing between a protonated and an unprotonated phenol.

By incorporating tyrosine labeled with ^{13}C at the 4' position (proximal to the -OH group) into bR in the same manner as was done with ^{15}N-lysine (see above) we can use these characteristic changes in chemical shift to probe the protonation state of tyrosine residues in the protein. For example an isotropic shift of 157 ppm is the expected value for the protonated tyrosine in the solid state spectrum (22,23) and is identical to the solution chemical shifts measured for protonated tyrosine residues in peptides and other proteins. The same chemical shift has also been measured for crystalline tyrosine and for simple dipeptide tyrosylphenylalanine in the solid state

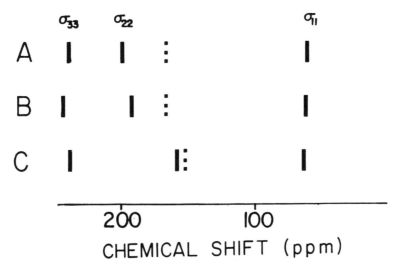

Figure 6. Isotropic shifts (dashed lines) and principal values (solid lines) of the shielding tensors for protonated and deprotonated cresol. (A) Potassium cresolate. (B) Sodium cresolate. (C) Cresol. Clearly, the 13 ppm change in isotropic shift upon deprotonation is due to a large shift in the σ_{22} element of the shielding tensor. All shifts are referenced to external TMS.

(25). Furthermore, pH titration of bR should reveal if any of the tyrosine's are exposed to the surface of the protein. NMR was first used to follow the titration of tyrosine resonances within a protein by Maurer et al (26), but complete titration (from fully protonated to fully ionized) was first accomplished by Wilbur and Allerhand (22). They showed that in several myoglobins, "surface" tyrosines gave pK's close to those of free tyrosine and of small tyrosine-containing peptides (pK $\approx$ 10-11). In contrast, residues known from the crystal structure to be "buried" were essentially untitratable (pK$_a$ > 12.5). Most structural maps of bR, however, place several tyrosines in the "link" regions between the helices and several others near the surface. The titration data may suggest that these tyrosines are, in fact, "buried" in bR or perhaps hydrogen-bonded in such a way that they are not freely exposed to the solvent. Experiments with 4'-^{13}C tyrosine-labeled bR are currently in progress.

Conclusions

The results presented in this paper illustrate that high-resolution ^{13}C- and ^{15}N NMR spectra can be obtained of an intrinsic membrane protein isotopically enriched at a single site. Furthermore, the isotropic chemical shift and chemical shift tensor values provide detailed information on both molecular structure and environment. We have been interested in understanding the mechanism of proton-pumping in bacteriorhodopsin. The key elements of this light-driven pump are the retinal chromophore, and protein residues which form proton conduction paths between the chromophore and the surfaces of the protein. Figure 7 presents space filling models of the retinal protonated Schiff base and shows the structural changes involved in dark-adaptation (bR$_{568}$ $\Rightarrow$ bR$_{548}$) and in the primary photochemistry

$(bR_{568} \Rightarrow K_{625})$. Our results have shown that bR_{568} and bR_{548} have 13-*trans* and 13-*cis* retinal protonated Schiff base chromophores, respectively, in agreement with previous chemical extraction (4) and resonance Raman (27,28) studies. More importantly, the conformation about the C_6-C_7 and C=N bonds have been determined in bR_{568} and bR_{548}, information which is lost upon chromophore extraction. The C_6-C_7 bond is s-*trans* in both pigments, in contrast to the 6-s-*cis* structures of the retinal model compounds. In the s-*trans* conformation, the π-electrons of the retinal chain are more fully conjugated with the ionone ring which produces a red-shift in the absorption band of the retinal. The C=N configuration is shown to be *anti* (or *trans*) in bR_{568} and *syn* (or *cis*) in bR_{548}. This is based on an upfield shift of the ^{13}C-14 resonance of bR_{548}, the γ-effect, which results from steric interaction between the C-14 proton and the ϵ-CH_2 protons of lysine across the *cis* C=N bond. The close approach of these protons, as well as the C-12 and C-15 protons across the *cis* C_{13}=C_{14} bond in bR_{548}, is clearly seen in Figure 7. Isomerization about both the C_{13}=C_{14} and C=N bonds in the $bR_{568} \Rightarrow bR_{548}$ conversion does not change the orientation of the Schiff base proton and leaves the chromophore in a roughly linear geometry. In contrast, isomerization about the C_{13}=C_{14} bond alone in the $bR_{568} \Rightarrow K_{625}$ conversion (6) translates the Schiff base proton into a new protein environment favorable for deprotonation, and results in an ~60° bend in the conjugated retinal chain.

The ^{13}C-retinal NMR spectra of bR also provide evidence for a negative bacterio-opsin charge near C_5 and a positive charge near C_7 of the retinal. These results are supported by recent studies on bR containing dihydroderivatives of retinal (18,19). The protein charges stabilize a more delocalized electronic structure for the conjugated retinal chain which exhibits a shift in the retinal absorption band from 440 nm (λ_{max}) for the protonated Schiff base in solution to 568 nm in bR_{568}. The lower C=C bond orders found in bR_{568} facilitate the ground-state isomerization of the retinal during the thermal reactions of the proton-pumping cycle.

Finally, our spectra of ϵ-^{15}N-lysine-bR demonstrate the ability to examine specific amino acid residues in the protein. Detailed studies of the protonation state of key amino acid groups, such as tyrosine, in bR_{568} and the photocycle intermediates will allow us to identify those residues involved in the pump mechanism.

Acknowledgements

This research was supported by the U.S. National Institutes of Health (GM-23316, GM-23289, EY-02051, and RR-00995), the U.S. National Science Foundation (CHE-8116042 and DMR-8211416), the Netherlands Foundation for Chemical Research (SON) and the Netherlands Organization for the Advancement of Pure Research (ZWO). R.A.M. is a recipient of an NIH Research Career Development Award. D.P.R. is an NSF Predoctoral Fellow and S.O.S. is an NIH Postdoctoral Fellow. We thank V. Copie for experimental assistance.

References and Footnotes

1. Stoeckenius, W., and Bogomolni, R.A., *Annu. Rev. Biochem. 51*, 587-616 (1982).

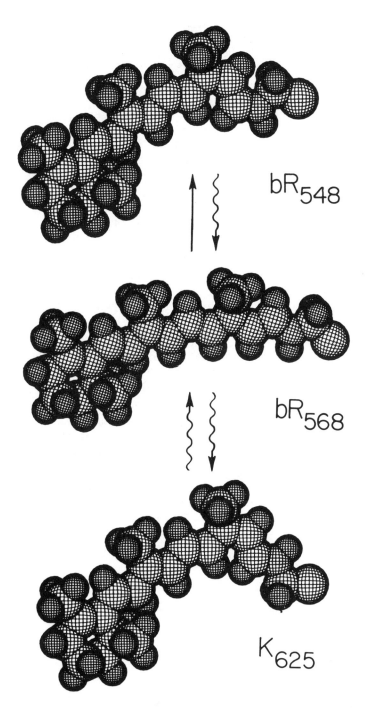

Figure 7. Space-filling models of the retinal chromophore in the bR_{548}, bR_{568} and K_{625} intermediates of bacteriorhodopsin. The structural changes involved in dark adaptation ($bR_{568} \Rightarrow bR_{548}$) have been studied using solid state NMR, and differ from the changes involved in the primary photochemistry ($bR_{568} \Rightarrow K_{625}$) primarily in the configuration about the C=N bond (see text).

2. Lozier, R.H., Bogomolni, R.A., and Stoeckenius, W., *Biophys. J. 15,* 955-962 (1975).

3. Smith, S.O., Horung, I., van den Steen, R., Pardoen, J.A., Braiman, M.S., Lugtenburg, J., and Mathies, R., *Proc. Natl. Acad. Sci. USA,* in press (1986).

4. Pettei, M.J., Yudd, A.P., Nakanishi, K., Henselman, R., and Stoeckenius, W., *Biochemistry 16,* 1955-1959 (1977).

5. Harbison, G.S., Smith, S.O., Pardoen, J.A., Winkel, C., Lugtenburg, J., Herzfeld, J., Mathies, R., and Griffin, R.G., *Proc. Natl. Acad. Sci. USA 81,* 1706-1709 (1984).

6. Smith, S.O., Myers, A.B., Pardoen, J.A., Winkel, C., Mulder, P.P.J., Lugtenburg, J., and Mathies, R., *Proc. Natl. Acad. Sci. USA 81,* 2055-2059 (1984).

7. Rothschild, K., and Marrero, H., *Proc. Natl. Acad. Sci. USA 79,* 4045-4049 (1982).

8. Engelhard, M., Gerwert, K., Hess, B., and Siebert, F., *Biochemistry 24,* 400-407 (1985).

9. Bogomolni, R.A., Stubbs, L., and Lanyi, J.K., *Biochemistry 17,* 1037-1041 (1978).

10. Nagle, J.F., and Morowitz, H.J., *Proc. Natl. Acad. Sci. USA 75,* 298-302 (1978).

11. Griffin, R.G., *Methods in Enzymology 72,* 108-174 (1981).

12. Herzfeld, J., and Berger, A.G., *J. Chem. Phys. 74,* 6021-6030 (1980).

13. Harbison, G.S., Mulder, P.P.J., Pardoen, J.A., Lugtenburg, J., Herzfeld, J., and Griffin, R.G., *J. Am. Chem. Soc. 107,* 4809-4816 (1985).

14. Munowitz, M.G., Griffin, R.G., Bodenhausen, G., and Huang, T.H., *J. Am. Chem. Soc. 103,* 2529-2533 (1981).

15. Harbison, G.S., Smith, S.O., Pardoen, J.A., Courtin, J.M.L., Lugtenburg, J., Herzfeld, J., Mathies, R.A., and Griffin, R.G., *Biochemistry 24,* 6955-6962 (1985).

16. Harbison, G.S., Smith, S.O., Pardoen, J.A., Mulder, P.P.J., Lugtenburg, J., Herzfeld, J., Mathies, R., and Griffin, R.G., *Biochemistry 23,* 2662-2667 (1984).

17. Shriver, J., Abrahamson, E.W., and Mateescu, G.D., *J. Am. Chem. Soc. 98,* 2407-2409 (1976).

18. Okabe, M., Balogh-Nair, V., and Nakanishi, K., *Biophys. J. 45,* 272a (1984).

19. Lugtenburg, J., Muradin-Szweykowska, M., Harbison, G.S., Smith, S.O., Heeremans, C., Pardoen, J.A., Herzfeld, J., Griffin, R.G., and Mathies, R.A., *J. Am. Chem. Soc.* submitted (1986).

20. Harbison, G.S., Herzfeld, J., and Griffin, R.G., *Biochemistry 22,* 1-5 (1983).

21. Levy, G.C., Litchter, R.C., and Nelson, G.L. (1980) in: "Carbon-13 Nuclear Magnetic Resonance Spectroscopy", John Wiley and Sons, New York, NY.

22. Wilbur, D.J., and Allerhand, A., *J. Biol. Chem. 251,* 5187-5194 (1976).

23. Norton, R.S., and Bradbury, J.H., *J. Chem. Soc. Chem. Commun. 102,* 870-871 (1974).

24. Harbison, G.S., Raleigh, D.P., Das Gupta, S.K., C.M. Mulliken, Herzfeld, J., and Griffin, R.G., to be submitted.

25. Harbison, G.S., Raleigh, D.P., Herzfeld, J., and Griffin, R.G., *Jour. Magn. Reson. 64,* 284-295 (1985).

26. Maurer, W., Haar, W., and Rüterjans, H., *Z. Phys. Chem. 93,* 119-129 (1974).

27. Braiman, M., and Mathies, R., *Biochemistry 19,* 5421-5428 (1980).

28. Smith, S.O., Lugtenburg, J., and Mathies, R.A., *J. Memb. Biol. 85,* 95-109 (1985).

Biomolecular Stereodynamics III, Proceedings of the Fourth Conversation in the Discipline Biomolecular Stereodynamics, State University of New York, Albany, NY, June 04-09, 1985, Eds., Ramaswamy H. Sarma & Mukti H. Sarma, ISBN 0-940030-14-4, Adenine Press, ©Adenine Press 1986.

Elasticity of the Polypentapeptide of Elastin Due to a Regular, Non-random, Dynamic Conformation: Review of Temperature Studies

Dan W. Urry

Laboratory of Molecular Biophysics, School of Medicine
University of Alabama at Birmingham
P.O. Box 311/University Station
Birmingham, Alabama 35294

This manuscript is dedicated to the memory of Paul Flory who contributed so extensively to the analyses of elastomers and to whom the manuscript was sent for comments but whose death occurred a few months later before a response was obtained.

Abstract

The elasticity and structure of the polypentapeptide of elastin, $(L \cdot Val^1-L \cdot Pro^2-Gly^3-L \cdot Val^4-Gly^5)_n$ where $n \simeq 200$, are reviewed as a function of temperature. It is noted that the γ-irradiation cross-linked polypentapeptide appears to be dominantly an entropic elastomer at temperatures greater than 45°C. On the basis of the classical theory of rubber elasticity this would be taken to mean that the undeformed state of the polypentatpeptide would be characterized in this temperature range as a network of random chains. At a fixed elongation the elastomeric force of the cross-linked polypentapeptide, however, increases dramatically on raising the temperature from 20° to 40°C. As a means of understanding this dramatic increase in elastomeric force, four independent physical methods are used to characterize the structure of the polypentapeptide as a function of temperature. Dielectric relaxation studies on a state of 62% water-38% peptide by weight demonstrate the development of an intense, localized 25 MHz peptide librational mode as the temperature is raised from 25 to 40°C. Nuclear magnetic resonance relaxation studies demonstrate the occurrence in the same state of an inverse temperature transition where carbonyl mobility decreases on raising the temperature from 25° to 37°C. These transitions occur without significant changes in composition. Circular dichroism studies demonstrate a change from a state of limited order at 20°C to a regular structure containing a repeating Type II β-turn at 40°C. Temperature dependence of aggregation (coacervation) in the 25° to 40°C is shown in combination with light microscopy and scanning and transmission electron microscopies to be an intermolecular ordering process resulting in self assembly to form anistropic fibers. Each of these four physical characterizations show the order of the polypentapeptide in water to increase as the temperature is increased from 25° to 40°C and the increase in order correlates with the dramatic increase in elastomeric force. In contrast to the expectations of the classical theory of rubber elasticity there is little elasticity when there is limited order and high elasticity develops in proportion to the increase in polypentapeptide order.

173

A new mechanism of elasticity, called the librational entropy mechanism, is discussed as a basis for understanding the polypentatpeptide elasticity. It is based on a loose helical structure called a dynamic β-spiral that the polypentapeptide in water is considered to form on raising the temperature. It is a structure with non-restricting interturn hydrophobic contacts and with suspended peptide segments wherein peptide moieties can undergo large librational motions. It is the entropy of these librational motions and their becoming damped on stretching that is proposed as the basis for the intrachain entropic component of the elastomeric force.

Introduction

The elastomeric force, f, exhibited by a stretched elastomer is considered in terms of internal energy, f_e, and entropy, f_s, components, i.e.,

$$f = f_e + f_s \tag{1}$$

Temperature dependences of elastomeric force at fixed length (thermoelasticity studies) have been used to estimate the relative magnitudes of f_e and f_s usually by means of determining the f_e/f ratio (1). When this ratio for an elastomer is small, less than a few tenths, the elastomeric force is comfortably said to be dominantly entropic in origin. In terms of the classical theory of rubber elasticity such small f_e/f ratios have been interpreted to indicate that the relaxed state of the elastomer is a state characterized by a network of random chains (2-4). In the present review this perspective is examined subsequent to consideration of a number of temperature studies on the polypentapeptide (L·Val[1]-L·Pro[2]-Gly[3]-L·Val[4]-Gly[5])$_n$ where n is of the order of 200. This polypentapeptide is the most striking primary structural feature of the precursor protein of the biological elastic fiber (5-7). The biological elastic fiber, on isolation and purification including removal of a fine microfibrillar glycoprotein coat, becomes the cross-linked protein elastin. This elastomeric matrix, comprised of a single precursor protein, has been shown to exhibit a small f_e/f ratio of the order of 0.1 (8-10). The structural interpretation has been that elastin consists of a random network of cross-linked polypeptide chains (8,10). The magnitude of the f_e/f ratio has also been estimated for the γ-irradiation cross-linked polypentapeptide, as seen in Figure 1, to be about -0.1 with a standard deviation of ±0.25 in the 45° to 60°C temperature range (11). On the basis of the classical theory of rubber elasticity, therefore, the γ-irradiation cross-linked polypentapeptide would be describable as a network of random chains.

The temperature studies reviewed here utilize four independent physical methods to address the question of the structure of the polypentapeptide under conditions of composition and temperature that are equivalent to those of the thermoelasticity experiment. After first considering the phase diagram which describes composition of the polypentapeptide-water system as a function of the temperature, the independent physical characterizations are applied to a state that is within $\pm2\%$ of 62% water-38% peptide by weight. The physical characterizations are (1) dielectric relaxation, (2) nuclear magnetic resonance relaxation, (3) circular dichroism, and

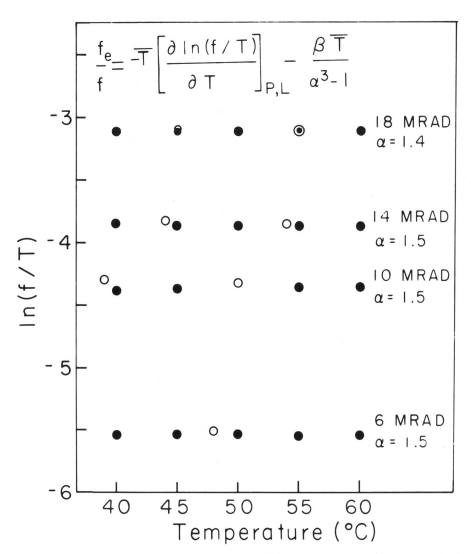

Figure 1. Thermoelasticity studies on the γ-irradiation cross-linked polypentapeptide (62% water, 38% peptide by weight) immersed in water for several mega radiation absorbed doses (MRAD). The solid circles are the values on heating, the open circles on cooling. The degree of elongation, α, is defined as L, the stretched length, divided by L_i, the initial length of the water equilibrated elastomer. As seen by the equation in the upper part of the figure the f_e/f ratio is proportional to the slope. Estimates of this value for the 45° to 60°C temperature range were -0.1 ± 0.25 when setting β, the thermal expansion coefficient, to zero. Based on the data giving rise to Figure 2, present estimates of β are 2.4×10^{-4}/deg such that the correction term would be -0.03. Adapted with permission from reference 11.

(4) temperature dependence of aggregation coupled with microscopic examination of the aggregated state using light microscopy, scanning electron microscopy and transmission electron microscopy with the negative staining technique and optical diffraction of the resulting electron micrograph. The consistent conclusion of each

of these four physical methods is at odds with the interpretations based on the classical theory of rubber elasticity. Each set of data argue that the elastomeric state of the polypentapeptide is not a network of random chains. Quite the inverse, elastomeric force increases as intermolecular and intramolecular order of the polypentapeptide increases; the state of the polypentapeptide of least order is the state exhibiting the least elastomeric force. Consistent with this, it is then argued that a regular structure can give rise to the decrease in configurational entropy on stretching. The regular structure is a class of dynamic β-spirals and the mechanism is called the librational entropy mechanism of elasticity.

Temperature-Composition Phase Diagram of the Polypentapeptide-Water System

The polypentapeptide elastomer is a two-component system (polypentapeptide + water) such that a phase diagram can be constructed plotting percent by weight of polypeptide (or water) as a function of temperature (12). This is done in Figure 2. Polypentapeptide (PPP) and water are seen to be miscible in all proportions below 25°C (region A). When there is less than 37% PPP by weight, at temperatures above 25°C, two phases coexist (region C), an equilibrium solution and a viscoelastic coacervate. The composition of the coacervate is only very slightly temperature dependent with a small amount of water extrusion (less than 3% by weight) on raising the temperature from 30° to 60°C. Above 60°C however, there is a very slow extrusion of large amounts of water from the coacervate. This is designated by the boundary between regions B and D. The process of coacervation, i.e., the development of two phases when raising the temperature above 25°C of solutions containing more than 63% water by weight, is entirely reversible. It is the effects of raising the temperature above 20°C on elastomeric force of γ-irradiation cross-linked polypentapeptide coacervate and on the structure of the polypentapeptide in the coacervate phase that is to be reviewed here.

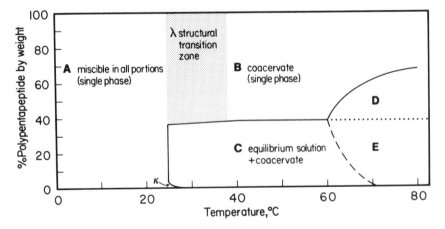

Figure 2. Phase diagram of the polypentapeptide-water system as a function of temperature. See text for discussion.

Temperature Dependence of Elastomeric Force

Coacervates of polypentapeptide with n of the order 200 can be cross-linked into an insoluble matrix suitable for characterization of elasticity by exposure to as little as 6 Mrad of γ-irradiation (11,13). The trace amounts of altered amino acids resulting from even 30 Mrad are so small as to make their quantitation difficult (e.g. less than 1 per 1000). The elastomeric matrix formed on γ-irradiation can be studied for its mechanical properties just as elastin itself has been studied (8-10). As in Figure 1, the thermoelasticity study is carried out by stretching the elastomer to an extension of about 50%. Holding the elastomeric band at a fixed extension, the elastomeric force is determined as a function of temperature. This data is plotted in Figure 3 for a 20 Mrad cross-linked coacervate that was extended to 60% (14). The measured elastomeric force has been divided by temperature ($^{\circ}$K) in order to remove a linear dependence of elastomeric force on temperature. A dramatic rise in elastomeric force is seen on increasing the temperature from 20° to 40°C. The shaded region is from 25° to 37°C for reasons that will be seen in the NMR studies given below. The nature of the process that gives rise to this large change in elastomeric force is

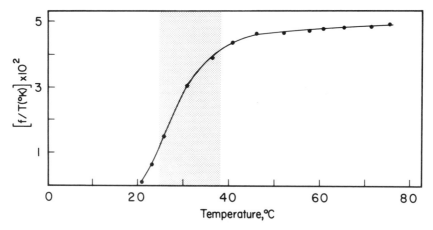

Figure 3. Temperature dependence of elastomeric force of γ-irradiation cross-linked (20 MRAD) polypentapeptide (62% water-38% peptide by weight). Note the dramatic increase in elastomeric force as the temperature is increased from 20° to 40°C. The shaded region delimits an inverse temperature transition as defined by the data in Figure 6. Adapted with permission from reference 14.

examined below using a number of independent physical methods. At this stage the issue of composition changes as a possible source for the dramatic rise in elastomeric force with temperature can be considered based on the data giving rise to Figure 2. The change in moles of water with increase in temperature, $(\partial n_{H_2O}/\partial T)_{P,L_i}$ at constant pressure and for one cubic centimeter of undeformed coacervate, has been determined in the 30° to 40°C range to be of a magnitude less than or equal to -1.2×10^{-4} mole/deg and in the 50 to 60°C range to be -0.5×10^{-4} mole/deg. For the 40° to 50°C range $(\partial n_{H_2O}/\partial T)_{P,L_i} = 0 \pm 1.2 \times 10^{-5}$ mole/deg (12). This suggests that change in composition itself cannot be responsible for the change in force with temperature but rather that one should look to polypeptide structural changes.

Temperature Dependence of Dielectric Permittivity

When dielectric relaxation studies are carried out on the polypentapeptide-water system that had been equilibrated at 40°C for a day, it is found at 20°C that the real part of the dielectric permittivity is a monotonic increasing function on scanning the frequency from 1000 MHz to 1 MHz. This is shown in Figure 4 (15). Beginning at about 25°C and increasing in intensity as the temperature is increased, however, is a localized relaxation with a characteristic frequency of 25 MHz. The composition of the coacervate, containing water and peptide moieties as the only dipolar moieties; the viscosity of the coacervate being between 10^3 and 10^4 centipoise; the intensity of the relaxation; the characteristic frequency of the relaxation; and the low temperature dependence of the characteristic frequency have led to the assignment of the relaxation to the rocking motion or libration of peptide moieties (15). The localized nature of the relaxation requires that each pentamer contributing to the relaxation be in the same conformation. This is because the energetics and hence

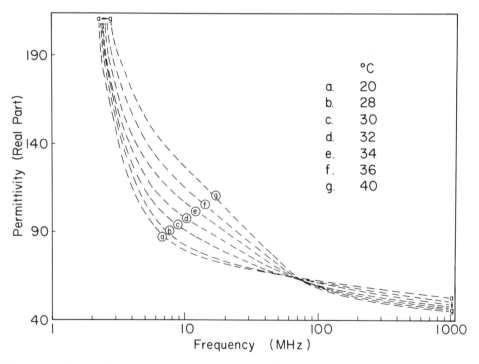

Figure 4. Real part of the dielectric permittivity of, ϵ', for 38% polypentapeptide, 62% water by weight. At 20° there is a monotonic increase in ϵ' as the frequency is decreased from 1000 MHz toward 1 MHz, but beginning about 25°C there is the development of an intense relaxation that is well-fit by a single Debye expression with a frequency of 25 MHz. Reproduced with permission from reference 15.

the frequency of peptide libration would necessarily be different for pentamers in different conformations. The decrease in magnitude of the real part of the dielectric permittivity in the 100 to 1000 MHz range, as the temperature is raised, is greater than expected for water alone and is considered to be due in part to a loss of relaxation frequencies occurring across the observation window and arising from pentamers in different conformations. As the temperature is raised and a regular conformation develops, relaxation frequencies, which were spread over a wide range, shift to 25 MHz and become a localized, Debye-type relaxation. Since the rocking of a single peptide moiety involves some motion of adjacent atoms, this has been called a peptide librational mode of the repeating pentamer unit (15).

When the real part of dielectric permittivity at a given frequency, e.g. 3.9 MHz, is divided by the absolute temperature and plotted as a function of temperature, as in Figure 5, the transition is found to be essentially localized within the shaded 25° to 37°C temperature range. This is the same temperature range where the elastomeric force increases so dramatically. Thus, there is a correlation of the increase in elastomeric force with the insensity of the relaxation, that is, with the development of a regular conformation wherein each pentamer is in a common conformationl state.

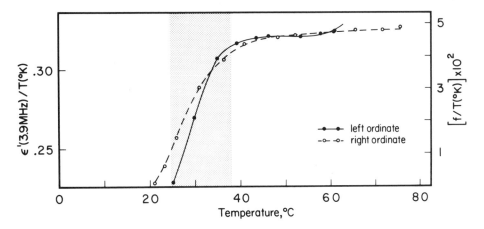

Figure 5. Real part of the dielectric permittivity, ϵ', at 3.9 MHz, plotted as a function of temperature as the solid line with respect to the left-hand ordinate. The development of the relaxation of Figure 4 is seen to correlate with the development of elastomeric force plotted as the dashed line on the right-hand ordinate. Adapted with permission from reference 14.

Temperature Dependence of NMR-derived Correlation Times: Chacterization of an Inverse Temperature Transition (16)

It is well-appreciated from the Second Law of Thermodyamics that the entropy of the total system, e.g. polypentapeptide + water, will increase as the temperature increases. Yet it was just argued from the dielectric relaxation studies above that the order of the polypentapeptide increased on going from 25° to 40°C, i.e. that the entropy of the polypentapeptide decreased over this temperature range. This would require, as has been well-stated by Kauzman (17), by Tanford (18) and earlier by Frank and Evans (19), that there should occur over the same temperature range an increase in entropy of the water that would be greater than the decrease in entropy of the polypentapeptide. This is considered to occur as clathrate-like water (relatively ordered water) surrounding the valyl and prolyl side chains converts to relatively less ordered bulk water during association of the hydrophobic side chains. With respect to the polypentapeptide this is an inverse temperature transition. Since in the fundamental sense temperature is a measure of molecular motion, an inverse temperature transition for the polypentapeptide might be expressible in terms of a decrease in mobility of the polypentapeptide as the temperature is raised through this transition region from 25° to 40°C. It is particularly relevant to test this perspective directly in relation to the peptide librational process. The dielectric relaxation studies demonstrate an increase in the number of peptide moieties that oscillate at the 25 MHz frequency, that is with a 6 to 7 nsec correlation time. This increase in peptide motion with a 6 to 7 nsec correlation time would then seem necessarily to occur with a simultaneous decrease in the mean mobility of the peptide moieties of the polypentapeptide. Accordingly, nuclear magnetic resonance relaxation studies were undertaken on carbonyl carbon labelled polypentapeptides. Noted here will be results on $(1-^{13}C)Pro^2$-polypentapeptide. Equivalent studies with the same result have been carried out on $(1-^{13}C)Gly^5$-PPP, and work is in progress on each of the other singly labelled carbonyls of the polypentapeptide, i.e. $(1-^{13}C)Val^1$-PPP, $(1-^{13}C)Gly^3$-PPP and $(1-^{13}C)Val^4$-PPP. In this way the relative mobilities of the five peptide moieties may also be assessed. This may prove useful in identifying the peptide moieties primarily responsible for the 25 MHz relaxation frequency of Figure 4.

Carbon-13 nuclear magnetic resonance longitudinal relaxation studies were carried out at 25 MHz using the inversion recovery method, and the rotational correlation times, τ_c, were calculated assuming the dipole-dipole relaxation mechanism for isotropically reorienting moieties and utilizing the hydrogens within three Angstoms of the carbonyl carbon. The correct solution of τ_c over the temperature range, of the two possible values from the T_1 study, was determined by first calculating the correlation times as a function of temperature from the carbonyl carbon half linewidths. From the linewidth study the rotational correlation times were seen to decrease (i.e. mobility increase) up to 25°C, to increase from 25° to 40°C and then to resume the decrease as the temperatures were increased above 40°C (20). This allowed choice of the correct solution of τ_c from the T_1 values which were considered to provide more realistic correlation times, uncompromised by conformational and hence chemical shift heterogeneity contributing to line width within this viscous medium of long chained polymer. A plot of log τ_c as a function of temperature is shown in Figure 6 where the mobility is seen to increase and peak at 25°C, to decrease to a second extremum at 37°C and then to increase above 37°C (16). These extrema provide good delineation of the temperature range for the transition and are the basis for the width of the shaded region. As had been anticipated, an inverse temperature transition is observed for the calculated mean rotational

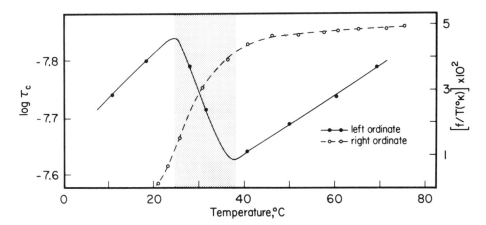

Figure 6. Plotted on the left-hand ordinate is the temperature dependence of the mean rotational correlation time, τ_c, of the Pro^2 carbonyl carbon of (1-^{13}C) Pro^2-polypentapeptide for 63% water-37% polypentapeptide by weight. The mobility is seen to increase from 10° to 25°C, to decrease sharply from 25° to 37°C and to resume the increase above 37°C. This provides the narrowest limits of an inverse temperature transition which is seen to correlate with the dramatic rise in elastomeric force of the γ-irradiation cross-linked polypentapeptide, plotted with respect to the right-hand ordinate. Adapted with permission from reference 16.

correlation times of the peptide carbonyl. The calculated mean correlation times range from 15 to 23 nsec; these values are satisfyingly close to the 6 to 7 nsec correlation time of the dielectric relaxation studies which can be obtained by fitting the data in Figure 4 to a simple Debye function (15). Plotted also in Figure 6 with respect to the right-hand ordinate is the elastomeric force/T(°K). The inverse temperature transition correlates strikingly with the dramatic increase in elastomeric force of the same polypentapeptide preparation that had been γ-irradiation cross-linked in the viscoelastic coacervate state.

The temperature dependence of the mean correlation time above 40°C also provides an estimate of the energy of activation for peptide motion. In spite of just having completed an inverse temperature transition resulting in greater polypeptide order, the activation energy is only 1.2 kcal/mole which is the order of magnitude of a motional barrier required for polymer elasticity.

Temperature Dependence of Ellipticity of the Polypentapeptide

The circular dichroism spectrum of the polypentapeptide dispersed in water is shown by the solid curves in Figure 7 for a polymer with n = 10 to 15 (21). The spectrum is characterized by a negative shoulder near 220 nm and a negative extremum at 197 nm. As a large negative extremum at 195 nm of the order of -4×10^4 is characteristic of disordered or random polypeptides, this may be considered the least ordered state of the polypentapeptide. It is probably not correct to consider this spectrum as indicating that the polypentapeptide is entirely random because of the lower magnitude of the negative extremum and the presence

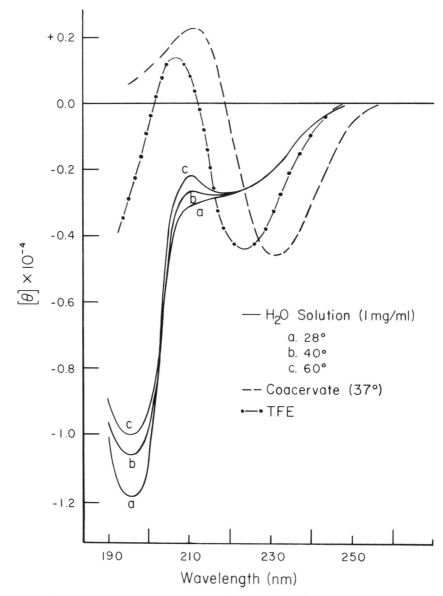

Figure 7. Circular dichroism of the polypentapeptide with n = 10 to 15 at a concentration of 1 mg/ml in water (solid curves) and in trifluoroethanal (TFE) and as a coacervate film at 37°C (the dashed curve). At this low molecular weight there is only a small shift of the curves for water as the temperature is raised but it is in the direction of the TFE and coacervate curves. These latter curves are characteristic of polypeptides containing Type II β-turns whereas the curves in water approach a curve characteristic of disordered polypeptide. For a higher molecular weight, e.g. n ≃ 200, the temperature dependence of the transition from the state of limited order to the Type II β-turn structure can be followed at 197 nm as shown in Figure 8. Reproduced with permission from reference 20.

of the negative shoulder and also because proton and carbon-13 nuclear magnetic resonance studies indicate the presence of some weak secondary structure in water with probabilities of occurrence in the lower temperature range of about 30% (22).

When the molecular weight is low, raising the temperature has a small effect on the CD spectrum, shifting it toward that obtained for coacervate films (see the dashed curve in Figure 7). The CD spectrum of the coacervate film is characteristic of a regular structure comprised of Type II β-turns (21,23). This is, of course, not surprising as the Pro-Gly sequence was subsequently shown to be the most probable sequence for inserting β-turns by the Chou and Fasman (24) survey of 29 protein crystal structures. In that effort the β-turn conformational feature was found to be more probable than the β-pleated sheet conformations and almost as probable as the α-helix. The regularly repeating β-turn of the polypentapeptide was first shown by NMR to involve a hydrogen bond between the Val[4] NH and Val[1] C-O moieties placing Pro[2] and Gly[3] at the corners of the β-turn (25) and then shown by means of the nuclear Overhauser effect to be a Type II β-turn (26). Also the Type II β-turn of the repeating pentamer has been observed in a water-peptide crystalline state as shown in Figure 11 A below (27). Thus the evidence is quite clear that the polypentapeptide in the coacervate state occurs as a regular structure containing a regularly repeating β-turn. However at low temperature in water, the polypentapeptide is in its least ordered state. The issue now is to use circular dichroism to characterize the temperature dependence of the conversion from the state of limited order to the state characterized by the regularly recurring Type II β-turns.

When the number of repeating pentamers in the polypentapeptide is much greater than 10 to 15, e.g. when n is the order of 200 or more, it is possible to determine by circular dichroism the temperature dependence for the conversion of the poly-pentapeptide from its state of least order below 20°C to the regular structure of the coacervate state. This is shown with the solid curve in Figure 8 which is a plot of the mean residue ellipticity at 197 nm for 2.3 mg/ml polypentapeptide in water (28).

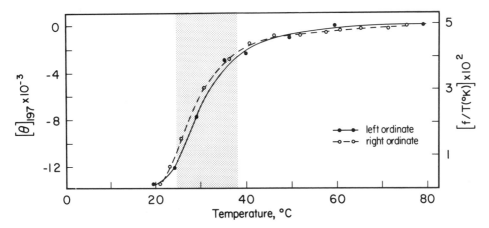

Figure 8. Temperature dependence of the ellipticity at 197 nm of polypentapeptide in water (2.3 mg/ml) plotted as the solid curve with respect to the left-hand ordinate. The conversion from a polypentapeptide structure of limited order at 20°C to a structure with a regularly repeating Type II β-turn as the temperature increases shows a striking parallel to the increase in elastomeric force of the γ-irradiation cross-linked polypentapeptide (dashed curve plotted with respect to the right-hand ordinate). Adapted with permission from reference 28.

The polypentapeptide is seen to undergo its structural transition from a state of limited order at 20°C to the state of regularly recurring β-turns by 40°C and to do so in parallel with the increase in elastomeric force of the γ-irradiation cross-linked polypentapeptide coacervate. It would appear that not only is the elastomeric state non-random, the more nearly random state exhibits little elasticity. This demonstration appears to be exactly contrary to the perspective of the classical theory of rubber elasticity wherein elasticity derives from a network of random chains.

Temperature Dependence of Coacervation and Microscopy of the Coacervate State

Temperature dependence of coacervation and the associated microscopy has been previously reviewed in a number of contexts (29-31). Results are included here for completeness of this review, principally because microscopy of structures formed during coacervation presents what appears to be an unequivocal demonstration that coacervation is an intermolecular ordering process.

When there is more than 63% water by weight, the temperature profile for coacervation can be determined by monitoring the turbidity of the polypentapeptide-water system. This is shown in Figure 9 for different concentrations. The concentration dependence for the onset of coacervation of high molecular weight polypentapeptide, n>200, is shown by the curved boundary marked κ in Figure 2 between regions A and C. The boundary between regions A and C of the phase

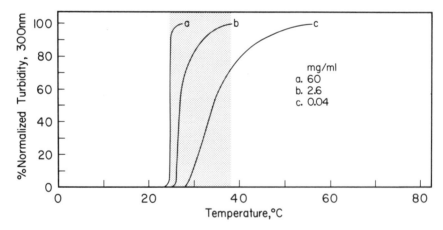

Figure 9. Temperature dependence of turbidity due to the aggregation (coacervation) of polypentapeptide with n ≃ 200 in water at different concentrations. The 60 mg/ml sample is at or above the high concentration limit, i.e. a concentration for which further increases do not shift the temperature for coacervation to lower values. The midpoint temperature (i.e. the temperature at half height) for the high concentration limits, however, depend on molecular weight in that the midpoint temperature shifts to higher values as the molecular weight decreases. The shaded region indicates the width of the structural transition when there is 63% water-37% peptide by weight. When a droplet of the cloudy solution is negatively stained and examined in the transmission electron microscope, the aggregates are seen to be comprised of parallel aligned filaments. When chemical cross-linking is carried out just above the onset temperature for coacervation the polypentapeptide self-assembles into fibers seen in the light microscope as shown in Figure 10.

diagram in Figure 2 is defined as the temperature for the onset of coacervation for the high concentration limit of the polymer, that is, the onset temperature for coacervation for that concentration above which the onset of coacervation no longer shifts to lower temperatures. The concentration dependence is an obvious intermolecular effect; there is also an intramolecular aspect to coacervation. The intramolecular element to coacervation, reflecting intramolecular hydrophobic association, is demonstrated by the observation that the midpoint temperature for the coacervation profile for the high concentration limit of different molecular weight preparations can be used to estimate molecular weight. The relationship appears to be quite simple; a plot of log (molecular weight) against the midpoint temperature, for the temperature profile of coacervation for the high concentration limit of each preparation, is a straight line (11,12). Care must be taken in the synthesis to avoid random racemization in order for the relationship to hold. While this is an interesting intramolecular effect of coacervation, primary interest here is in the intermolecular aspect of coacervation.

Coacervation can be carried out under conditions where cross-links are formed that, in effect, fix the developing structural relationships. Two different polypenta-peptide syntheses are carried out; one where one out of every half dozen Val[4] residues is replaced by a lysine residue and a second where the Val[4] residue is similarly replaced by a glutamic acid residue. The two syntheses are combined at the appropriate ratio, and under coacervation conditions the two polypentapeptides are cross-linked by means of a water soluble carbodiimide (32). When the product is observed in a light microscope with no fixative or stain, as shown in Figure 10 A, the polypentapeptides are seen to have self-assembled into fibers (30). When a several micron diameter fiber is examined in the scanning electron microscope, it is found to be comprised of parallel aligned fibrils (see Figures 10 B and C). When the fibrils are examined in the transmission electron microscope, as in Figure 10 D, using a negative staining technique (oxalic acid, uranyl acetate, pH 6.2), they are seen to be comprised of parallel aligned filaments (33). This is the state generally seen for the initial small aggregates of coacervation where the optical diffraction of the electron micrograph shows diffraction spots with the primary spot indicating an approximate 5.5 nm spacing between filaments (34). Coacervation, therefore, can be said to be the intermolecular ordering to form filaments by polypentapeptide molecules which were initially molecularly dispersed in solution. The filaments then align to form fibrils which further associate to form fibers. It would seem that the chemically cross-linked polypentapeptide is clearly an anisotropic, fibrillar elastomer (32).

Structural Consequences of the Temperature Studies and the Dynamic β-spiral Conformation of the Polypentapeptide

Conformational details of the polypentapeptide have been derived from proton and carbon-13 nuclear magnetic resonance studies using two independent approaches; they are the most ordered state approach and an approach that identifies and characterizes in great detail a cyclic oligopentapeptide with a conformation nearly

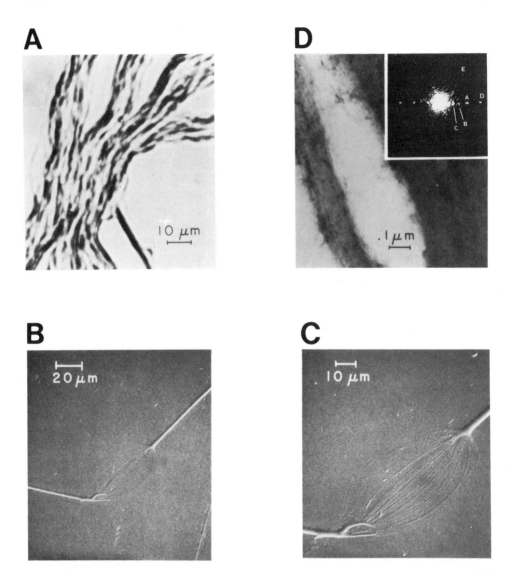

Figure 10. A. Light micrograph of self-assembled fibers of polypentapeptide formed on chemical cross-linking just above the onset temperature for coacervation. Reproduced with permission from reference 30. B and C. Scanning electron micrograph of self-assembled fiber seen to be comprised of fine fibrils that splay-out and recoalesce back into the same sized fiber. Reproduced with permission from reference 32. D. Self-assembled fibers, negatively stained and examined in the transmission electron micrograph, seen to be comprised of parallel aligned filaments of about 5 nm diameter. Adapted with permission from reference 33. Insert: optical diffraction pattern obtained from electron micrograph of negatively stained polypentapeptide aggregates. Reproduced with permission from reference 34.

identical to the linear polymer (22,29). Both approaches give closely, appropriately related results. These structural studies, which involve identification of specific hydrogen bonds, torsion angles ranges and proximity of specific hydrogens as well as molecular mechanics calculations, have been verified in detail by the particularly relevant crystal structure of the cyclopentadecapeptide (27). In solution the cyclopentadecapeptide exhibits very nearly the same conformation as the linear polypentapeptide and exhibits the same response to temperature and solvent perturbations (35). Also the solution and crystal conformations have been shown to be very similar (36). In the crystal the cyclopentadecapeptide maintains its three-fold symmetry; the molecules stack exactly on top of each other; the 30% solvent (water) by weight in the crystal occurs in triple helical fashion within the stack of cyclic molecules; the intermolecular interactions, both between molecules within stacks and between stacks, are hydrophobic; the Type II Pro^2-Gly^3 β-turn is present as seen in Figure 11 A; and the Val^4-Gly^5-Val^1 peptide segment is suspended between the β-turns (27). It is conceptually a relatively simple process to convert a stack of cyclic molecules into a helical structure. When this is done for the stack of cyclopentadecapeptide molecules (37), a loose right-handed helical structure is obtained in which the β-turns function as spacers between the turns of helix by means of intramolecular hydrophobic contacts as depicted in Figure 11 C and E. There is space for water within the helix (see Figure 11 D); and the Val^4-Gly^5-Val^1 peptide segment is suspended in a manner that allows for large amplitude rocking or libration of the included peptide moieties. The resulting loose helical structure is called a dynamic β-spiral because the dominant repeating secondary structural feature is the β-turn.

The dynamic β-spiral of Figure 11 was derived entirely independently of the temperature studies reviewed here. These temperature studies even though their information is of a more general nature, completely confirm the detailed studies and in addition they identify the conditions under which the transition to the dynamic β-spiral occurs. The circular dichroism data of Figures 7 and 8 demonstrate the development of a regularly recurring Type II β-turn as the temperature is raised from 20° to 40°C. The nuclear magnetic resonance relaxation studies, shown in Figure 6, confirm this inverse temperature transition and demonstrate that a decrease in mobility does occur on ordering but that the energy barrier to motion in the relatively ordered repeating β-turn structure is low of the order of one or a few kT. The dielectric relaxation studies (see Figures 4 and 5) show the development of a regular structure where the pentamers are each in a common conformation, as in a helix, where peptide moieties can exhibit large amplitude librations, as in the suspended segment, and where the barrier to this motion as measured by the low temperature dependence of the correlation time (15) is low, as in the NMR results. The interturn hydrophobic contacts of Figure 11 E are as expected for this increase in intramolecular order with increasing temperature and they are consistent with the molecular weight dependence of the coacervation temperature for the high concentration limits of different molecular weight polypentapeptides.

Thus the class of structures, represented in Figure 11 and referred to as dynamic β-spirals, was derived by detailed NMR studies and molecular mechanics calculations and was strongly supported by a most relevant crystal structure, and the class of structures was found to be consistent with the structural inferences of the temperature

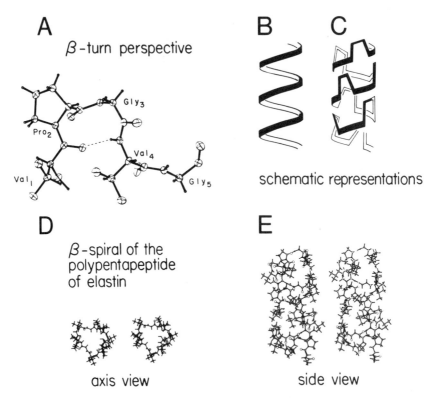

A. β-turn perspective

B   C

schematic representations

D   β-spiral of the polypentapeptide of elastin

E

axis view   side view

Figure 11. A. β-turn perspective of the repeat pentamer in the crystal structure of the cyclopenta-decapeptide. The 10 atom hydrogen bonded ring of the β-turn is indicated by the dashed line. The orientation of the end peptide moiety, i.e. the peptide moiety between the Pro and Gly residues, is such that the Pro² α hydrogen and the Gly³ NH hydrogen are on the same side of the β-turn. This defines a Type II β-turn and this proximity of the Pro² αCH and the Gly³ NH protons has been observed in solution using the Nuclear Overhauser effect (26). Structure adapted with permission from reference (27). B. Schematic representation of an open helix. C. Representation of an open helix in which β-turns are schematically shown functioning as spacers between the turns of the helix. This by definition of a β-turn recurring in a helical array is called a β-spiral. This also schematically shows the segment connecting the β-turns. Reproduced with permission from reference 30. D. Stereo pair of the axis view of the right-handed β-spiral of the polypentapeptide. Here the suspended segment connecting β-turns is easily seen. The open helix contains water that can exchange with extra spiral water through spaces in the surface adjacent to the suspended segments. E. Stereo pair of the side view of the dynamic β-spiral of the polypentapeptide. The interturn contacts are seen to involve the hydrophobic Pro and Val side chains. As in C the β-turns are seen to function as spacers between the turns of the spiral. Openings are apparent in the surface of the β-spiral adjacent to the Val⁴-Gly⁵-Val¹ suspended segment. This allows for intraspiral water to exchange with extraspiral water and allows for the freedom of the peptide moieties of the suspended segment to librate. Reproduced with permission from reference 37.

studies. In turn the temperature studies show dramatic increase in elastomeric force with increase in temperature to correlate with the increase in order. In the next section, it will be shown how a new mechanism of elasticity can be based on the class of dynamic β-spiral conformations.

A Librational Entropy Mechanism of Elasticity as an Alternative to the Classical Theory of High Elasticity of Polymeric Networks

The undeformed polypentapeptide chains in water have been shown by four independent physical methods to undergo an inverse temperature transition to increased order on raising the temperature from 20° to 40°C. The ordered state is described as a dynamic β-spiral conformation defined as the recurrence of a β-turn on a helical axis. Strikingly it is shown by the same methods that the development of this order closely correlates with the development of elastomeric force of the γ-irradiation cross-linked polypentapeptide coacervate (see Figures 5, 6, 8 and 9). Elasticity is defined as the property whereby a material in response to an applied force resists and recovers from deformation. For the polypentapeptide of elastin when there is limited order in the undeformed state there is little elasticity; and where there is more order in the undeformed state there is greater elasticity. This is not as expected from the now classical theory for high elasticity of polymeric networks (2). The first basic postulate of that theory has been stated (2): "1. Undeformed polymeric chains adopt random spatial configurations in the amorphous state even when packed to bulk density in the absence of diluents. The same holds for elastic networks of chains." In short, elastic networks of undeformed polymeric chains adopt random spatial configurations. In view of the data reviewed here it is more difficult to see how this can be correct for the polypentapeptide of elastin.

The second basic postulate is (2), "2. The stress resulting from deformation of such a network originates within the chains and not (to an appreciable extent) from interactions between them." With deletion of the phrase "of such a network" this postulate could well be consistent with the physical characterizations of the polypentapeptide of elastin. The intermolecular interactions are dominantly hydrophobic so that the stress resulting from deformation of the polypentapeptide chains could reasonably be borne primarily within the individual polypentapeptide chains and not by interactions between chains. The dynamic β-spiral of the polypentapeptide could be viewed as analogous to greased helices which would readily slip past each other. This could possibly alleviate the problems of chain entanglements and the assumptions of phantom chains common to the classical theories of rubber elasticity based on random networks (38,39). The low values for the f_e/f ratio obtained from the data in Figure 1 (for the 45° to 60°C temperature range, i.e. once the structural transition is essentially complete) suggest that the dominant component of the elastomeric force is entropic. It is then the configurational entropy of the dynamic β-spiral conformation that is of interest. It is, therefore, in the context of the second postulate and the configurational entropy of the β-spiral of the polypentapeptide that the elasticity will now be discussed.

The experimental data argue that the entropic elasticity of the polypentapeptide of elastin can not be based on a network of random chains in the undeformed state. It is natural, therefore, to look for the presence of large configurational entropy and the potential of large configurational entropy changes within the proposed dynamic β-spiral of the polypentapeptide. In this regard it should be appreciated that polypeptide states with describable order, e.g. α-helix, β-pleated sheet, β-helix, β-spiral, etc., are nonetheless states of widely different configurational entropy. Thus it has been shown by Lotz et al. (40) for poly-γ-benzyl glutamate that films cast at room temperature result in the α-helical conformation of the polypeptide, that holding the temperature at 130°C for a number of hours and cooling results in formation of the single stranded β-helical conformation (41), π_{LD}-helices following our previous nomenclature (42), and that further heating at 220° to 230°C and cooling results in formation of the double stranded β-helical conformation (41), referred to as $\pi\pi_{DL}$ in the Lotz et al. nomenclature (40). It has also been shown in solution for the pentadecapeptide, hydrogenated gramicidin A (41) that heating relatively high concentrations converts the conformation from α-helical to β-helical with an entropy change per residue of approximately 1.6 cal/mole-deg (41,43). In these cases the primary source of increased entropy is considered to arise from the greater librational freedom of peptide moieties within β-helices (43). Thus there is evidence that describable regular structures, even those characterizable by X-ray diffraction, occur with widely different amounts of configurational entropy. To get significant entropy changes it is not necessary that one of the states involve a random or disordered state. It would seem to be quite sufficient if there be greater librational freedom in one state. This provides experimental and initial conceptual insight for the proposed librational entropy mechanism of elasticity (44).

The Librational Entropy Mechanism of Elasticity

The presence of two valyl residues and one prolyl residue in the repeating pentamer a priori restrict the possible ψ and ϕ torsion angles accessible to the pentamer. While flexible, the Type II Pro²-Gly³ β-turn, with abundant evidence for its presence in the elastomeric state, further limits the accessible ranges of torsion angles contained within the β-turn. In this regard DeBolt and Mark within the formalism of the Flory classical theory of rubber elasticity found introduction of the β-turn into the polypentapeptide to be useful in obtaining correlation with the experimental f_e/f value for elastin (45). Thus the Val and Pro side chains and the β-turn restrictions on available configuration (conformation) space would seem to be a necessary aspect of a theory treating elasticity of the polypentapeptide. The additional restrictions in configurational entropy due to the β-spiral conformation would arise from the intramolecular (interturn) hydrophobic interactions considered important in developing the regular helical structure. The addition of the loose helical constraints of the β-spiral would then only add restrictions to the torsion angles of the Val⁴-Gly⁵-Val¹ suspended segment. But just how limiting is this? And is this limiting of available configuration space considerable when compared to the limitations imposed by the realities of chain entanglements present in the random networks?

The question of the restrictions imposed by the loose helical structure can be addressed by fixing in the helical constraint the two Val¹ α-carbons delimiting a single pentamer repeat and by determining the number of conformational states that the pentamer with a flexible β-turn could exhibit and yet fill the helical gap. The answer is that there are many conformational states within a low energy cutoff above the lowest energy conformation of the β-spiral. Using a 5° step in torsion angles, using the potential functions and helical constraint approach of Scheraga and coworkers (46-49), and using a cutoff energy of 2 kcal/mole-residue, there are some 1853 states (44). On stretching the helix 130%, the number of states, identically counted, decreases to 162. The number of states drop in a systematic way as lower cutoffs are used, e.g. 1.5, 1.0 and 0.6 kcal/mole-residue (43,44). With this counting of states, an estimate of the entropy change, ΔS, can be obtained from the Boltzmann relation, i.e.

$$\Delta S = S^r - S^e = R \ln W^r/W^e \qquad (2)$$

where the superscripts r and e stand for relaxed and extended states; R is the gas constant (1.987 cal/mole-deg.), and W is the number of a priori equally probable states accessible to the pentamer. The change in entropy per residue is close to 1 cal/mole-deg whichever cutoff is used. The Boltzmann summation over states, $\sum_i e^{-\epsilon_i/kT}$, can also be used as part of the statistical mechanical definition of entropy, i.e.

$$S = R \ln \sum_i e^{-\epsilon_i/kT} + \frac{E}{T} \qquad (3)$$

where E is the total internal energy of the system, ϵ_i is the energy of each conformational state, k is the Boltzmann's constant, and T is the absolute temperature (°K). For the change in entropy this gives

$$\Delta S = R \ln\left(\sum_i^r e^{-\epsilon_i^r/kT} / \sum_i^e e^{-\epsilon_i^e/kT} \right) + (E^r - E^e)/T \qquad (4)$$

Taking the elastomeric force to be entirely entropic on extension, the change in internal energy would be zero, i.e. $(E^r - E^e) = 0$, and using each conformational state in the summation up to a 2 kcal/mole-residue cutoff, again 1 entropy unit (cal/mole-deg) per residue is obtained. It would seem that there is more than sufficient entropy change to account for an elastic modulus of 1×10^6 dynes/cm^2.

Of particular interest is the relationship among the allowed states. If ψ_i is plotted against ϕ_{i+1} for the residue 4-5 and 5-1 peptide moieties as in Figure 12, it is seen that these angles are anticorrelated. This means that the allowed conformational states are related by peptide librations. It was this finding of the large amplitude peptide librations that suggested the dielectric relaxation studies (see Figure 3) and led to the finding and assignment of a peptide librational mode. The temperature dependence for the characteristic frequency of this relaxation is low of the order of one or a few kT (15) as would be expected for elasticity at physiological temperatures. Also experimental estimates of the dipole moment change at 40°C of about 2 Debye/pentamer (50) are of the same order of magnitude as the calculated dipole moment change of about 4 Debye/pentamer for the relaxed β-spiral structure using a 1.5 kcal/mole-residue cutoff energy. Thus it is not unreasonable to relate the two.

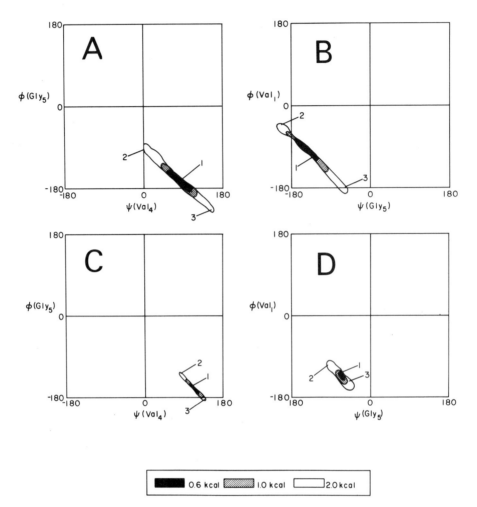

Figure 12. Lambda plots showing the anticorrelated relationship between changes in ψ_4 and ϕ_5 (A) and ψ_5 and ϕ_1 (B) for the β-spiral depicted in Figure 11 D and E. This indicates peptide librational motions within the β-spiral conformation. The dark area indicates the range of libration allowed within a 0.6 kcal/mole-residue cutoff energy; the shaded area goes out to 1.0 kcal/mole residue, and the outer contour shows the range of librational motion allowed within a 2.0 kcal/mole-residue cutoff energy. C and D demonstrate the decrease torsion angle space (decrease in configurational entropy) that results from a 130% extension of the β-spiral structure. Reproduced with permission from reference 43.

Much as a potential energy surface for a chemical reaction can give insight into the dynamic process of reaction rates, so too can consideration of the potential energy surface in a torsion angle representation of configuration space provide insight into the dynamic peptide librational process. The librational excursions possible within a 2 kcal/mole-residue cutoff energy are depicted in Figure 13 A for the central one of three pentamers, and the dampening of those librational motions on extension are apparent on comparing A and B of Figure 13. As also discussed above,

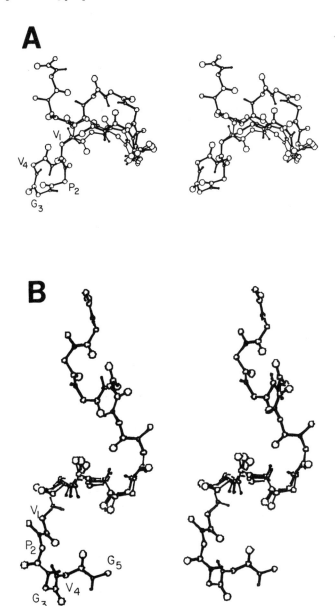

Figure 13. A. Stereo pair of three pentamer units in the β-spiral conformation of Figure 11 D and E in which the central pentamer is allowed excursion in conformation up to the 2 kcal/mole-residue cutoff energy. The 4-5 and 5-1 peptide moieties are seen to be capable of large rocking motions (peptide librations). It is this result that suggested the dielectric relaxation studies and that provides for a librational entropy. B. Stereo pair of the three pentamer units in which the end to end chain separation along the spiral axis has been extended 130%. Again the central pentamer is allowed conformational excursion up to 2 kcal/mole-residue. While librational processes are still observed the amplitudes are dramatically reduced. Thus this represents a marked decrease in configurational entropy and is proposed to be the source of the entropic restoring force of the elastomeric polypentapeptide. Reproduced with permission from reference 43.

comparisons of volumes in configuration space for the same basic structure in extended and relaxed states can provide insight into entropy changes.

Another formalism within which to consider the entropy of a structure arises from a normal mode analysis of the structure described by the valence force field approach. Dauber et al. (51) have calculated the Helmholtz free energies (A = E − TS) of a nonapeptide in four different helical conformations, designated as C_{7eq}, C_{7ax}, α_R and α_L. The TS term was obtained from a normal mode analysis where the eigenvalues were used to determine the vibrational frequencies, ν_i, which were then utilized in a vibrational partition function, i.e.

$$f_{vib} = (1 - e^{-h\nu_i/kT})^{-1} \tag{5}$$

to calculate the contribution of a single vibrational mode to the entropy, S_i for a mole of structures, i.e.

$$S_i = R \ln f_{vib} + \frac{Nh\nu_i}{T}(e^{h\nu_i/kT} - 1)^{-1} \tag{6}$$

Where N is Avogadro's number. The interesting qualititative statement is that the lower the frequency of the mode the larger its contribution to the entropy of the structure. In the case of the nonapeptide in the C_{7eq} structure, the calculated lowest frequency mode was 3.6 cm^{-1} or 10.8 × 10^{10}/sec, and the contribution of that single mode to the entropy of the structure was a calculated value for TS$_i$ at 298°K of 2.99 kcal/mole (51). Such low frequency modes contribute substantially to lower the free energy and thereby to increase the probability of the structure. In relation to the β-spiral of the polypentapeptide this loose helical structure would be expected to have stabilizing low frequency modes. In particular the 25 MHz peptide librational mode can be so considered. While it is necessary to consider the validity of the vibrational partition function for such low frequency modes, it is interesting to note that at 40°C by means of Eq. (6) the 25 MHz peptide librational mode would contribute some 8 kcal/mole to the stability of the structure. By these considerations, it is also apparent that the polypentapeptide is a structure of substantial entropy and that a decrease in entropy on extension is, within this context, also to be expected. In this regard it will be of interest to see what happens to the 25 MHz dielectric relaxation on stretching of the γ-irradiation cross-linked polypentapeptide coacervate.

Acknowledgment

This work was supported in part by NIH grant HL 29578. The author gratefully acknowleges the past and present members of the Laboratory of Molecular Biophysics who have contributed so significantly to work reviewed here.

References and Footnotes

1. P.J. Flory, A. Ciferi, C.A.J. Hoeve, *J. Polymer Sci. XLV*, 235, (1960).

2. P.J. Flory, *Rubber Chem. Technol. 41,* G41 (1968).

3. J.E. Mark, *Rubber Chem. Technol. 46,* 593, (1973).

4. J.E. Mark, *J. Polymer Sci., Macromolecular Reviews 11,* 135, (1976).

5. L.B. Sandberg, W.R. Gray, J.A. Foster, A.R. Torres, V.L. Alvarez, and J. Janata, *Adv. Exp. Med. Biol. 79,* 277, (1977).

6. L.B. Sandberg, N.T. Soskel, and J.B. Leslie, N. Engl. *J. Med. 304,* 566, (1981).

7. W.R. Gray, L.B. Sandberg and J.A. Foster, *Nature 246,* 461, (1973).

8. C.A.J. Hoeve and P.J. Flory, *Biopolymers 13,* 677, (1974).

9. K.L. Dorrington, and N.G. McCrum, *Biopolymers 16,* 1201, (1977).

10. A.L. Andrady and J.E. Mark, *Biopolymers 19,* 849, (1980).

11. D.W. Urry, S.A. Wood, R.D. Harris and K.U. Prasad in *Polymers as Biomaterials,* Eds. S.W. Shalaby, T. Horbett, A.S. Hoffman and B. Ratner, Plenum Publishing Corporation, New York, p.17 (1985).

12. D.W. Urry, T.L. Trapane, and K.U. Prasad, Biopolymers, in press.

13. D.W. Urry and K.U. Prasad in *Biocompatibility of Natural Tissues and Their Synthetic Analogues,* Ed. D.F. Williams, CRC Press, Inc., Boca Raton, Florida, 89-116, (1985).

14. D.W. Urry, R. Henze, R.D. Harris and K.U. Prasad, *Biochem. Biophys. Res. Commun. 125,* 1082 (1984).

15. R. Henze and D.W. Urry, *J. Am. Chem. Soc. 107,* 2991 (1985).

16. D.W. Urry, T.L. Trapane, M. Iqbal, C.M. Venkatachalam and K.U. Prasad, *Biochemistry, 24,* 5182-5189, (1985).

17. W. Kauzmann, Adv. *Protein Chem. 14,* 1 (1959).

18. C. Tanford in *The Hydrophobic Effect: Formation of Micelles and Biological Membranes,* Wiley, New York (1973).

19. H.S. Frank and M.W. Evans, *J. Chem. Phys. 13,* 507 (1945).

20. D.W. Urry, K.U. Prasad, T.L. Trapane, M. Iqbal, R.D. Harris and R. Henze, *Amer. Chem. Soc. Div. Polym. Mater.: Sc. Eng. 53,* 241-245, (1985).

21. D.W. Urry, M.M. Long, B.A. Cox, T. Ohnishi, L.W. Mitchell and M. Jacobs, *Biochim. Biophys. Acta 371,* 597 (1974).

22. D.W. Urry and M.M. Long, *CRC Crit. Rev. Biochem. 4,* 1 (1976).

23. R.W. Woody in *Peptides, Polypeptides and Proteins,* Eds. E.R. Blout, F.A. Bovey, M. Goodman and N. Lotan, Wiley, New York, p. 338 (1974).

24. P.Y. Chou and G.D. Fasman, *J. Mol. Biol. 115,* 135 (1977).

25. D.W. Urry, W.D. Cunningham and T. Ohnishi, *Biochemistry 13,* 609 (1974).

26. M.A. Khaled and D.W. Urry, *Biochem. Biophys. Res. Commun. 70,* 485, (1976).

27. W.J. Cook, H.M. Einspahr, T.L. Trapane, D.W. Urry and C.E. Bugg, *J. Am. Chem. Soc. 102,* 5502 (1980).

28. D.W. Urry, R.G. Shaw and K.U. Prasad, *Biochem. Biophys. Res. Commun. 130,* 50-57, (1985).

29. D.W. Urry in *Methods in Enzymology,* Eds. L.W. Cunningham and D.W. Frederiksen, Academic Press, Inc., New York 82, 673 (1982).

30. D.W. Urry, *Ultrastruct. Pathol. 4,* 227 (1983).

31. D.W. Urry, *J. Protein Chem. 3,* 403 (1984).

32. D.W. Urry, K. Okamoto, R.D. Harris, C.F. Hendrix and M.M. Long, *Biochemistry 15,* 4083 (1976).

33. D.W. Urry and M.M. Long in *Elastin and Elastic Tissue,* Eds. L.B. Sandberg, W.R. Gray and C. Franzblau, Plenum Press, New York, *Adv. Exp. Med. Biol. 79,* 685 (1977).

34. D. Volpin, D.W. Urry, I. Pasquali-Ronchetti and L. Gotte, *Micron 7,* 193 (1976).

35. D.W. Urry, T.L. Trapane, H. Sugano and K.U. Prasad, *J. Am. Chem. Soc. 103,* 1080 (1981).

36. C.M. Venkatachalam, M.A. Khaled, H. Sugano and D.W. Urry, *J. Am. Chem. Soc. 103,* 2372 (1981).

37. C.M. Venkatachalam and D.W. Urry, *Macromolecules 14,* 1225 (1981).

38. P.J. Flory, *Principles of Polymer Chemistry,* Cornell University Press, Ithaca (1953) and *Statistical Mechanics of Chain Molecules,* Interscience, New York (1969).

39. R.T. Deam and S.F. Edwards, *Phil Trans. Roy. Soc. Sec. A., (London) 280,* 317 (1976).

40. B. Lotz, F. Colonna-Cesari and G. Spach, *J. Mol. Biol. 106,* 15 (1976).

41. D.W. Urry, M.M. Long, M. Jacobs and R.D. Harris, *Ann. NY Acad. Sci. 264,* 203 (1975).

42. D.W. Urry, M.C. Goodall, J.D. Glickson and D.F. Mayers, *Proc. Natl. Acad. Sci. USA 68,* 1907 (1971).

43. D.W. Urry, C.M. Venkatachalam, S.A. Wood and K.U. Prasad in *Structure and Motion: Membranes, Nucleic Acids and Proteins,* Eds. E. Clementi, G. Corongiu, M.H. Sarma and R.H. Sarma, Adenine Press, Guilderland, NY p.185 (1985).

44. D.W. Urry, C.M. Venkatachalam, M.M. Long and K.U. Prasad in *Conformation in Biology,* Eds. R.

Srinivasan and R.H. Sarma, G.N. Ramachandran Festschrift Volume, Adenine Press, USA, p. 11 (1982).

45. L. DeBolt and J.E. Mark, *Polymer Preprints 25,* 193 (1984).
46. F.A. Momany, L.M. Carruthers, R.F. McGuire and H.A. Scheraga, *J. Phys. Chem. 78,* 1595 (1974).
47. F.A. Momany, R.F. McGuire, A.W. Burgess and H.A. Sheraga, *J. Phys. Chem. 79,* 2361 (1975).
48. N. Go and H.A. Sheraga, *Macromolecules 6,* 273 (1973).
49. N. Go and H.A. Sheraga, *Macromolecules 9,* 867 (1976).
50. R. Henze and D.W. Urry, unpublished results.
51. P. Dauber, M. Goodman, A.T. Hagler, D. Osguthorpe, R. Sharon and P. Stern, *ACS Symposium Series, No. 173 Supercomputers in Chemistry,* Eds. P. Lykos and I. Shavitt, p. 161 (1981).

Biomolecular Stereodynamics III, Proceedings of the Fourth Conversation in the Discipline Biomolecular Stereodynamics, State University of New York, Albany, NY, June 04-09, 1985, Eds., Ramaswamy H. Sarma & Mukti H. Sarma, ISBN 0-940030-14-4, Adenine Press, ©Adenine Press 1986.

Conformational Changes in Cyclic Peptides Upon Complexation with Alkali and Alkaline Earth Metal Ions

Isabella L. Karle
Laboratory for the Structure of Matter
Naval Research Laboratory
Washington, D.C. 20375

Abstract

The structures and conformations of a number of free cyclic peptides and the same peptides complexed with Li^+, Na^+, K^+, Mg^{++} or Ca^{++} ions have been established by x-ray diffraction analyses of single crystals. The different modes of complexation that have been found are: (1) an infinite sandwich in which the metal ion and the cyclic peptide alternate; (2) a discrete sandwich in which the metal ion lies between two cyclic peptide molecules; (3) incomplete encapsulation of the metal ion by one peptide molecule; (4) complete encapsulation by one peptide molecule; and (5) the metal ion partially *exo* to the polar cavity formed by the peptide.

The complexation of cyclic oligopeptides with light metal ions requires the formation of ligands between the metal ion and carbonyl oxygens. If the peptide is too small, even acting in pairs, then the incomplete coordination sphere about the metal ion is completed by ligands to the oxygens of water molecules or to O or N atoms of other solvent molecules such as CH_3CN, C_2H_5OH, $(CH_3)_2CO$, or to counterions such as SCN^-. The Mg^{++} and K^+ ions have shown an almost exclusive preference for octahedral coordination, whereas the Li^+ and Na^+ ions adjust their coordination to the local geometry, as for example, in the complexes with antamanide and antamanide analogs, the Li^+ and Na^+ ions have pentacoordination with ligands to four carbonyl O atoms in a square array and the fifth ligand to a polar atom of a solvent molecule at the apex of the coordination pyramid.

The free cyclic peptides are not in a conformation ready to accept and encapsulate a metal ion. They do not have preformed polar cavities. Severe conformational changes occur upon complexation, including folding and large rotations in the backbones. To demonstrate that complexation is responsible for the conformational changes rather than other factors in the crystal environment such as packing forces, intermolecular interactions, and/or solvent inclusion, a number of crystal structure analyses were performed on the uncomplexed peptides or closely related analogs of the peptide. To illustrate, the cyclic decapeptide antamanide and the biologically active analog [Phe4,Val6] antamanide were crystallized from solutions containing water and common organic solvents such as C_2H_5OH, CH_3CN, acetone, or DMFA and from completely nonpolar solvents such as n-hexane. The resulting crystals had different packing arrangements of the peptide, different solvent inclusions (both polar and nonpolar) and several different side chains on the peptides. Nevertheless, the folding of the backbone and the twisting of the side chains are almost identical in all the crystals. The conformations of the cyclic pentapeptides (Gly-Pro-Gly-D-Ala-Pro) and

(D-Phe-Pro-Gly-D-Ala-Pro) are superimposable despite a different side chain and different packing in the respective crystals. Similarly, uncomplexed valinomycin crystallized from different solvents and occurring in different crystalline packing arrangements has a unique elongated conformation.

The process of forming complexes with Li^+, Na^{++} and Mg^{++} involves major conformational changes in the peptides such as the rotation of one or more peptide units by as much as 180°. In addition, *cis/trans* isomerism of amide bonds may occur, intramolecular hydrogen bonds are broken, and elongated peptide ring backbones become folded.

Introduction

Transport of metal ions in a biological system is a function essential to life. One means of transporting a charged particle through a lipophilic system is to encapsulate the metal ion with a peptide molecule that has lipophilic side chains. The peptide molecule must have the property of folding in such a manner that the metal ion is surrounded by carbonyl oxygens and can form metal-oxygen ligands in an appropriate coordination sphere in the interior of the complex. For the metal ion-peptide complex to be active in a lipophilic membrane, the outer surface of the globule needs to be lipophilic, hence the peptide residues should have non-polar side chains that will extend outward and cover the complex. Accordingly, emphasis in crystal structure analyses has been placed on the naturally occurring and synthetic model peptides that are composed of residues having solely non-polar side chains, whether or not they perform as ion carriers.

One of the intentions of this paper is to illustrate the different modes of complexation with ions of light metals that have been observed as a function of the size of the cyclic peptide. They are: (1) an *infinite sandwich* in which the metal ion and the peptide alternate; (2) a *discrete sandwich* in which the metal ion lies between two cyclic peptide molecules; (3) *incomplete encapsulation* of the metal ion by one peptide molecule; (4) *complete encapsulation* by one peptide molecule; and (5) the metal ion partially *exo* to the polar cavity formed by the peptide. Significantly, the various peptide molecules in the uncomplexed state generally are not already in a conformation that is suitable to accept a metal ion, in contrast to polyethers such as grisorixin where essentially no conformational changes occur upon complexation (1). The severe conformational changes that occur in peptides upon complexation have been established by x-ray diffraction analyses of single crystals of free peptides and the same peptides complexed with Li^+, Na^+, Mg^{++}, Ca^{++} or K^+.

Invariability of Conformations of Uncomplexed Cyclic Paptides

The immediate question is whether the complexation process is primarily responsible for the change in conformation. Peptides are known to be extremely flexible molecules with rotations possible about each $N-C^\alpha$ and $C^\alpha-C'$ bond in the backbone, as well as about bonds in the side chains, Fig. 1 (2). Limiting values for rotations about the $N-C^\alpha$ and $C^\alpha-C'$ bonds were first derived by Ramachandran (3) in which he used

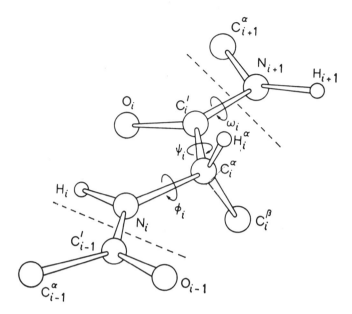

Figure 1. Labelling of atoms and torsional angles in peptide groups (IUPAC-IUB Commission on Biochemical Nomenclature (2)).

contacts between nonbonded atoms as a limiting criterion. The amide bond was assumed to be planar and in the *trans* conformation, a generally good assumption. Acceptable ranges for torsional angles ϕ and ψ about the N-C^a and C^a-C$'$ bonds for L-residues are shown in Fig. 2. Within the constraints shown in Fig. 2 there are still very many possible conformations, especially for linear peptides. Even for cyclic peptides where the number of degrees of freedom for the backbone is reduced by the additional constraint of having to close the backbone ring, many different conformational possibilities remain. Variability in backbone conformation has been demonstrated for oligopeptides containing 2-5 residues, for example. Differences in local environment in the crystal, in the availability of donors or acceptors for hydrogen bond formation, and in the nature and amount of cocrystallized solvent can have a significant effect on the conformation of small peptides. This result is not surprising in view of the fact that a small peptide generally does not have strong attractions within the individual molecule, such as internal hydrogen bonds or stacking interactions between rings in side-chains, that can overcome the attractions or repulsions for neighboring molecules.

Experimental results obtained from x-ray diffraction analyses of single crystals of *pseudo* polymorphs have established that the conformations of larger peptides are quite constant. Different crystal packing, substitution of polar for non-polar solvent molecules, inclusion or exclusion of solvent, and even, in some cases, substitution of side chains has had very little effect on conformation. Several examples are the following:

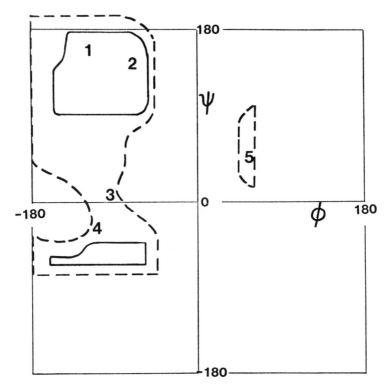

Figure 2. Allowable regions for torsional angles ϕ (rotation about N-C^a) and ψ (rotation about C^a-C′) for L-residues with *trans* amide bonds (Ramachandran and Sasisekharan (3)). The experimentally determined values for antamanide, a decapeptide with approximate two-fold rotation symmetry, are shown as points 1-5.

1) Valinomycin, a cyclic dodecadepsipeptide with the sequence (LVal-DHyv-DVal-LLac)$_3$, has been crystallized in a triclinic cell with two independent, but very similar, molecules in the cell from n-octane or acetone (4) and from ethanol/water (5). A monoclinic form has also been obtained from n-octane (5). In each case the conformation remained constant despite crystallizations from polar or from non-

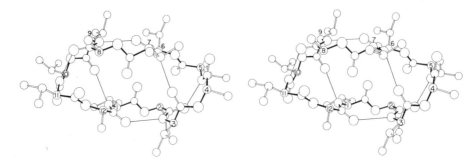

Figure 3. Stereodiagram for the conformation of uncomplexed valinomycin (4,5). Only the C^a atoms are numbered. The six transannular hydrogen bonds are depicted by light lines.

polar solvents, different local environments in the triclinic and monoclinic cells, and the inclusion of a n-octane molecule in the monoclinic form whereas the triclinic form was free of solvent. Figure 3 shows the conformation of the uncomplexed form (4).

2) Antamanide, a cyclic decapeptide with the sequence (Val-Pro-Pro-Ala-Phe-Phe-Pro-Pro-Phe-Phe), and the biologically active synthetic analogue, [Phe^4,Val^6] antamanide with the sequence (Val-Pro-Pro-Phe-Phe)$_2$, have nearly identical conformations for the backbone and for comparable side chains. The synthetic analogue crystallizes with large solvent channels between the peptide molecules. The continuous channels have a diameter ~5 A to 6 A after taking into consideration the van der Waals' radii of the atoms in the peptide. In one crystal form, the channel is filled with 12 water molecules per peptide molecule that are hydrogen-bonded to the peptide molecule and to each other to form an uninterrupted column of polar atoms (6). Crystallization from dioxane gives an identical result (7). In a *pseudo*-polymorphic crystal grown from n-hexane/methyl acetate, the large channel is occupied by unordered and nonpolar n-hexane and/or methyl acetate molecules (8) instead of water, however, the peptide conformation remains essentially undisturbed.

Natural antamanide crystallized from CH_3CN/H_2O solution (9) has quite a different arrangement of molecules in the cell than the [Phe^{4}Val^6] antamanide. Nevertheless, the conformations of the two different molecules are nearly identical. The only significant difference occurs for the torsion angle ϕ_6, a difference of 30°, where Val in the synthetic analogue replaces Phe in the natural molecule. The two molecules are compared in Fig. 4. Note that three H_2O molecules are intimately associated with the central region of the synthetic analogue by hydrogen-bonding, whereas the natural compound has four H_2O molecules in the same region. The 30° change in ϕ_6 has allowed space for the additional H_2O molecule.

3) The cyclic pentapeptides (Gly-Pro-Gly-DAla-Pro) and (DPhe-Pro-Gly-DAla-Pro) are superimposable despite a different side chain and different packing in the respective crystals (10,11).

4) The 18-membered ring formed in cyclic (Leu-Tyr-δAva-δAva) (where δAva = δ-aminovaleryl) mimics one of the conformations for cyclic (Gly)$_6$. In both the monoclinic crystal with one DMSO and one H_2O per peptide and the orthorhombic crystal with four H_2O molecules per peptide, the conformation of the peptide is almost identical. Furthermore, in each crystal four peptide molecules aggregate by interpeptide hydrogen bonding to form cavities of similar size and shape. In one case, the cavity is filled with water molecules that make strong hydrogen bonds with polar groups of the peptide, whereas in the other case, DMSO forms one hydrogen bond with the peptide using the sulfoxyl oxygen as a donor and the two methyl groups are directed toward the polar lining of the cavity (12,13).

5) Cycloamanide A, the natural cyclic hexapeptide (Pro-Val-Phe-Phe-Ala-Gly), crystallizes in two different crystal forms from H_2O/C_2H_5OH solutions (14,15). One

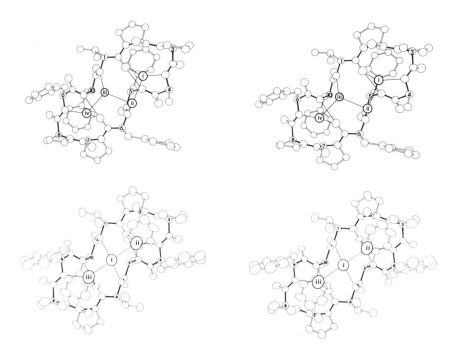

Figure 4. A comparison of the conformations of natural antamanide with four water molecules (i-iv) intimately bound to the peptide (9), upper diagram, with the conformation of the synthetic analog [Phe⁴Val⁶] antamanide (6-8) with three closely associated water molecules (i-iii), lower diagram.

crystal form contains four H_2O molecules as cocrystallized solvent, while the other form has one H_2O and three C_2H_5OH molecules. It is noteworthy that not only do the peptide molecules retain the same conformation in each *pseudo*-polymorphic crystal, but also retain a "forbidden" Type II′ beta-bend where a Type I had been expected for an L,L sequence.

The cyclic peptides discussed above contain five to twelve residues. In peptides of this size the internal forces that stabilize the conformation of the backbone and side chains appear to be strong enough to overcome any perturbations that may come from packing or solvent effects in the crystal. Constancy of conformation has, of course, been demonstrated in protein molecules that crystallize in several polymorphs.

Metal Ion Complexation

The formation of a complex between a metal ion and a cyclic peptide causes profound changes in the conformation of the cyclic peptide. Other factors, as shown above, can be effectively eliminated as being the causative agents for conformational change. In transforming from the uncomplexed conformation to that assumed in the metal ion complex, values for some pairs of torsional angles (ϕ_i, ψ_i) pass through a forbidden region of the ϕ, ψ map. Specific examples will be shown below.

As mentioned earlier, the type of metal ion peptide complex formed depends upon the size of the cyclic peptide and, of course, upon the size of the metal ion. In the present study, only the small metal ions are considered.

1) An example of the *infinite sandwich* is provided by the n:n complex of Mg^{++} and cyclic (DPhe-Pro-Gly-DAla-Pro) (16) in which the pentapeptide and the hydrated magnesium ion alternate in an infinite stack, Fig. 5. The Mg^{++} is coordinated to five O atoms and one N atom (from the NCS^- counterion) in an octahedral array. The Mg^{++} makes ligands directly with the peptide with only two of these O atoms, that is, the carbonyl O of the Phe residue from the peptide molecule below the Mg^{++} and the carbonyl O of the Gly residue from the peptide molecule above. The other three O atoms forming the Mg-O ligands belong to H_2O molecules. Three additional H_2O molecules, a total of six in the complex, make bridges between the H_2O molecules in the magnesium octahedron and the peptide by hydrogen-bonding to NH or CO moieties in the peptide. Obviously, the cyclic pentapeptide molecules are not large enough to encapsulate the Mg^{++} and its polar environment.

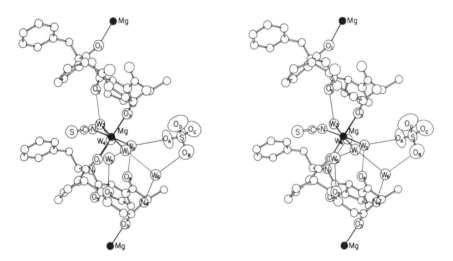

Figure 5. Stereodiagram of the n:n complex of Mg^{++} and cyclic (DPhe-Pro-Gly-DAla-Pro) (16). Six water molecules and the NCS^- ion are closely associated by ligand formation and hydrogen bonding.

A comparison of the conformation of the free peptide with the complexed form is snown in Fig. 6. All peptide bonds are in the *trans* conformation. The largest changes upon complexation can be described by the rotation of O_2 from inside to outside the backbone ring and the rotation of O_3 from outside to inside. The changes in the values of the torsional angles are shown in the ϕ,ψ map in Fig. 7. On this drawing it appears that the Gly[3] residue transverses a forbidden region near $\phi \sim 0$. However, taking the rotational path in the opposite direction, i.e. toward $\phi = 180$, all the ϕ,ψ values remain in an allowable region for Gly. The transformation between the free and the complexed conformation can be fairly readily achieved in space filling models.

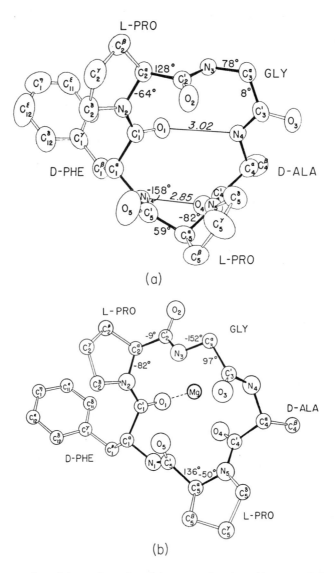

Figure 6. A comparison of the conformation of the uncomplexed peptide, top, and the same peptide as it appears in the complex with Mg^{++}, see Fig. 5.

A somewhat similar infinite stack of alternating peptide molecules and Mg^{++} ions has been found for the n:n complex of Mg^{++} and the cyclic hexapeptide $(GlyPro)_3$ (17). In this complex, water molecules are also needed to complete the oxygen octahedron about the Mg^{++} ion and to make bridges by means of hydrogen bonding between adjacent peptide molecules and Mg^{++} ions in a stack.

2) An illustration of a *discrete sandwich* is the 1:2 complex formed by Mg^{++} and the cyclic hexapeptide $(GlyProPro)_2$ (18) in which the MgO_6 octahedron is formed

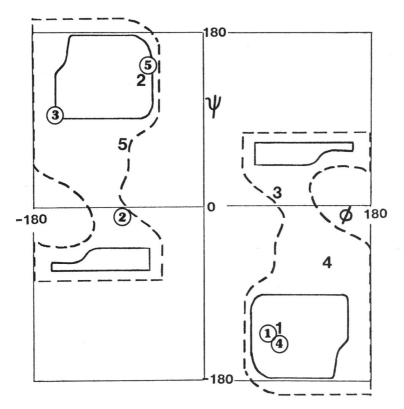

Figure 7. A comparison of the torsional angles ϕ and ψ determined for the uncomplexed cyclic pentapeptide, plain numbers, and the complexed form, circled numbers, shown in Fig. 6. Both L and D residues occur in this molecule. The allowable regions for L residues are on the left of the diagram and for D residues they are on the right. The Gly residue, 3 or ③ , is allowed on either the right or left side of the diagram (3).

entirely by using the carbonyl oxygens of the peptide molecules for the Mg-O ligands, Fig. 8, three carbonyl oxygens from the lower peptide molecule and three from the upper peptide molecule. The upper and lower peptide molecules are related approximately by a *pseudo* two-fold rotation axis that passes horizontally through the Mg^{++} atom in the orientation of the complex as shown in Fig. 8. In this complex, there are no water molecules. The polar environment of the Mg^{++} is shielded quite effectively by the non-polar side chains of the Pro residues. The upper and lower surfaces of the peptide sandwich are highly lipophilic, except for the exposed N_1H and O_3 moieties and comparable atoms in the lower peptide. However, recent experiments have shown that this complex does not carry Mg^{++} ions across polymeric membrances (19).

The individual peptide molecules in the complex do not possess any internal symmetry. There are *cis* peptide bonds between Pro^2-Pro^3 and Pro^5-Pro^6, Fig. 9. The uncomplexed cyclic $(GlyProPro)_2$ is also shown in Fig. 9 (20). Two major changes upon complexation are *cis-trans* isomerizations of two amide bonds: that

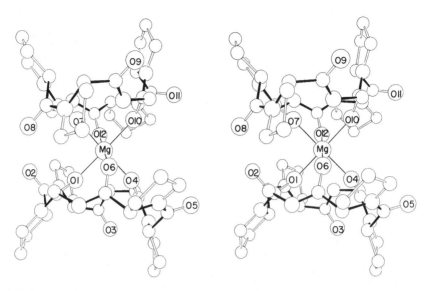

Figure 8. A stereodiagram of the 1:2 complex between Mg^{++} and cyclic(GlyProProGlyProPro) (18). Only the O atoms are numbered.

is, the *trans* C_2'-N_3 between Pro^2-Pro^3 becomes *cis* in the complex and the *cis* C_4'-N_5 between Gly^4-Pro^5 becomes *trans* in the complex. The isomerizations require a change of ~180° in ω_2 and ω_4 and are also accompanied by very large changes in ψ_2 and ψ_4, see Fig. 10. The ϕ,ψ values for residues Pro^3, Pro^5 and Pro^6 remain invariant in the free and complexed forms; in fact, ϕ values for all the residues remain essentially unchanged. The ψ value for Gly^1 readjusts somewhat in order to make the Mg-O_1 ligand. It should be noted that O_4 travels from the exterior of the backbone ring to the interior for complexation. Space-filling models of the peptide can be twisted from one conformation to the other fairly readily.

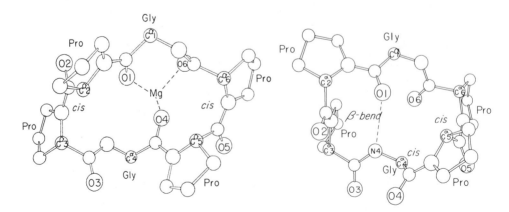

Figure 9. A comparison of one of the cyclic(GlyProProGlyProPro) moieties in the Mg^{++} complex in Fig. 8, left, (the other peptide moiety in the complex is almost identical) with the conformation of the uncomplexed peptide, right (20).

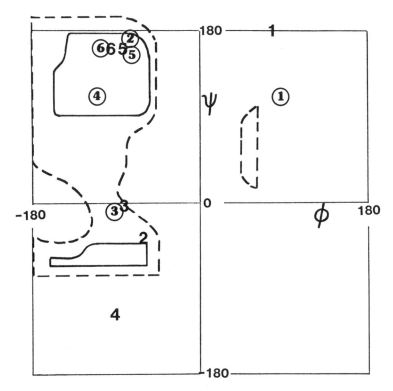

Figure 10. A comparison of the torsional angles ϕ and ψ determined for the uncomplexed cyclic-(GlyProProGlyProPro), plain numbers, and for the same peptide complexed with Mg^{++}, circled numbers, shown in Fig. 9. Residues 1 and 4 are glycyls, hence they have a much larger allowable region in ϕ,ψ space than residues with side chains. Note also that there is a *trans-cis* isomerization in the amide bond between Pro^2-Pro^3 and a *cis-trans* isomerization between Gly^4-Pro^5.

Other examples of discrete sandwiches for which crystal structure analyses have established the structure involve a number of variations of Ca^{++} and/or Na$^+$ complexes with the cyclic hexapeptide (GlyPro)$_3$ (17,21-23). Each of the complexes has true or approximate threefold symmetry and all peptide bonds are *trans,* although the free peptide has an asymmetric conformation with one *cis* peptide bond in a Gly residue (17). In the 1:2 complex with Ca^{++}, the ion has octahedral coordination with six carbonyl oxygens from the glycyl residues, three from each of the two peptides forming the sandwich. The carbonyl oxygens from the prolyl residues are directed outward. They attract either polar solvent molecules or additional ions, specifically Na$^+$ (21-23), to make stacked sandwich complexes of the type Solvent·Na$^+$·Peptide·Ca^{++}·Peptide·Na$^+$·Solvent, for example. The Na$^+$ ion makes ligands with the oxygens from three prolyl residues and completes tetrahedral coordination with one DMSO or octahedral coordination with three H$_2$O molecules. Such attractions for additional metal ions on the outside of the sandwich are not possible in the complex with cyclic (GlyProPro)$_2$, discussed above, because of the greater hydrophobicity of the exterior surface of the complex.

Structures of sandwich complexes containing heavy (large) ions have been reported for $2Rb^+ \cdot 2$ cyclo(GlyPro)$_4$ where both Rb^+ ions are sandwiched between the two peptide molecules (24) and for the intricate sandwich $(Bv \cdot Ba^{++} \cdot picrate_3 \cdot Ba^{++} \cdot Bv)^+$ where Bv is the hexadepsipeptide beauvericin, cyclo(LMePhe-DHyv)$_3$ (25).

3) *Incomplete encapsulation* of a metal ion by one peptide molecule is exhibited by the cyclic decapeptide antamanide. Complexes of $Li^+ \cdot antamanide \cdot CH_3CN$ (26) and $Na^+ \cdot [Phe^4Val^6]antamanide \cdot C_2H_5OH$ (27) have been made with entirely different packing in their respective crystals. Complexes with other solvents such as acetone, DMFA, and DMSO have made crystals that are isomorphous with one or the other of the above crystal forms. Complexation with Na^+ occurs selectively over K^+ in solution (28). No crystalline K^+ complexes have been found yet, presumably because the K^+ ion is too large to be accommodated in the cavity formed by the peptide.

Despite an entirely different mode of packing in the crystal, alkali ions of different radii, different side chains on residues 4 and 6 and different ligated solvent molecules, the Li^+ and Na^+ complexes are nearly isostructural. That is, none of the factors that are different, either in the complex or in the environment, affect the conformation of the cyclic peptide in the complexed form. However, the conformation in the complexed form is quite different than that in the uncomplexed form, compare Figs. 4, 11 and 12, and compare the changes in the ϕ_i, ψ_i values in Fig. 13. In the

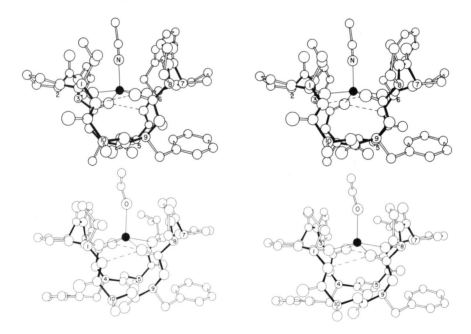

Figure 11. A comparison of the conformations of $Li^+ \cdot antamanide \cdot CH_3CN$ (26), upper diagram, and $Na^+ \cdot [Phe^4Val^6]antamanide \cdot C_2H_5OH$ (27), lower diagram. The black spheres represent the alkali metal ions. Note that the phenyl groups of residues 5 and 10 in both complexes have been omitted since they fold up against the body of the complex and obscure the folding of the backbone.

COMPLEXATION OF $\left[\text{Phe}^4\text{Val}^6\right]$ANTAMANIDE

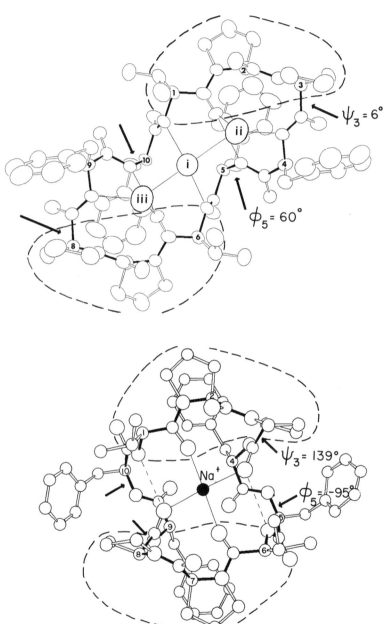

Figure 12. The change in conformation from uncomplexed [Phe⁴Val⁶]antamanide, upper diagram, to the Na⁺ complex, lower diagram. The regions enclosed by the dashed lines are not disturbed by the complexation. The largest rotations take place about C_3^α-C_3' and C_8^α-C_8', as well as N_5-C_5^α and N_{10}-C_{10}^α.

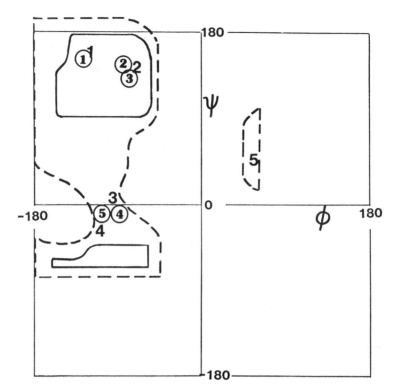

Figure 13. A comparison of the torsional angles ϕ and ψ determined for uncomplexed antamanide, plain numbers, and complexed antamanide, circled numbers, see Fig. 12. Since the antamanide backbone has an approximate two-fold rotation axis, only residues 1-5 are shown.

complexed form, the backbone is folded in a fashion somewhat similar to the seam in a tennis ball, the Pro²-Pro³ and Pro⁷-Pro⁸ peptide bonds have the *cis* configuration, the Li^+ or Na^+ ion is located in a shallow cup formed by the folding of the backbone, and four ligands between the metal ion and carbonyl oxygens O_1, O_3, O_6 and O_8 are formed. The fifth ligand involves a N or O atom from the solvent that may be CH_3CN, C_2H_5OH, or a number of other compounds that have a polar and a nonpolar end. The decapeptide is not large enough to completely encapsulate an ion, hence a solvent molecule not only provides an additional ligand for the metal ion in the interior, but also completes the lipophilic exterior of the complex created by the side chains of the Phe, Pro and Val residues of the peptide.

The uncomplexed form has an elongated, relatively planar ring. The segments between C_1^α to C_3^α and from C_6^α to C_8^α that contain the *cis* amide bonds between pairs of Pro residues have similar conformations in both the complexed and uncomplexed molecules, Fig. 12. However, the segments containing O_3 and O_4 (and O_8 and O_9) have been turned inside out upon complexation. In the complexed form, these O atoms are directed inward and O_3 and O_8 participate in the ligation with the metal ion, whereas in the uncomplexed form, these O atoms are directed outward with respect to the backbone ring. This major conformational change is accomplished mainly by rotations about the C_3^a-C_3' bond with a change in ψ_3 of ~140° and the N_5-C_5^a bond with a change in ϕ_5 of ~155°. Of course, similar rotations occur about C_8^a-C_8' and N_{10}-C_{10}^a since the backbones in both the complexed and uncomplexed form have approximate two-fold rotation symmetry.

The ϕ,ψ values for residues 5 and 10 for the uncomplexed molecule fall into the isolated allowable region for L-residues in the +,+ quadrant of the ϕ,ψ map. In order for the conformational change to take place during complexation, the ϕ,ψ values for residues 5 and 10 must traverse unallowable regions.

4) *Total encapsulation* is illustrated by the classic case of the complex of K$^+$ and valinomycin (29-31), the cyclic dodecadepsipeptide (LVal-DHyv-DVal-LLac)$_3$. The K$^+$ ion is coordinated to six carbonyl oxygen atoms of the six amide residues in a nearly regular octahedron with K$^+$-O distances near 2.75 A, Fig. 14. Each of the six carbonyl oxygens of the ester residues forms a hydrogen bond with the NH group of the n+3 residue, thus the backbone forms a bracelet of six β-bends, three up and three down, with non-crystallographic S_6 symmetry. In each of the three structure determinations the anions and the crystal packing have been quite different, but the conformation of the complex remained the same.

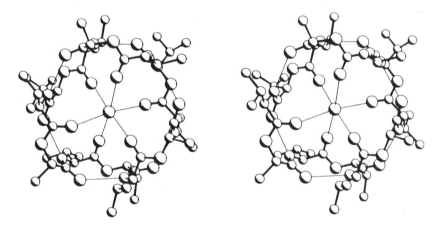

Figure 14. Stereodiagram of the K$^+$ complex of valinomycin (29-31).

A comparison with uncomplexed valinomycin, Fig. 3 shows that the circular bracelet backbone has been flattened in the uncomplexed form, thus eliminating the central cavity (4,5). Four of the *beta*-bends remain the same as in the K$^+$ complex, but the N_1H moiety of L-Val^1, instead of binding to O_{10} as in the complexed form, reaches across the ring to O_9 and forms a 5⇒1 hydrogen bond (13-membered loop). A similar 5⇒1 hydrogen bond is formed between N_7H and O_3. Owing to the threefold repetition of residues, the same conformation is obtained if N_3H and N_9H or N_5H and N_{11}H were to participate in 5⇒1 type hydrogen bonds instead of the 4⇒1 type. In the flattened bracelet, four of the carbonyl oxygen atoms that complex to the K$^+$ ion are still directed toward the interior of the ring, but the O_1-O_7 distance has decreased to 4.2 A while the O_3-O_9 distance has increased to 7.1 A from the 5.5 A value found in the complex. The other two carbonyl oxygens that ligate with the K^+ ion, O_5 and O_{11}, have been rotated downward and upward, respectively, with respect to the ring as viewed in Fig. 3. Atom O_{11} is more exposed to the solvent than O_5 because it is flanked by one methyl and one isopropyl side chain as compared to two isopropyl side chains for O_5. Atoms O_4 and O_{10} do not participate in ligand formation, however, they also are quite exposed to the solvent in the uncomplexed form. A comparison of ϕ_i,ψ_i angles in the complexed and free valinomycin are shown in Fig. 15.

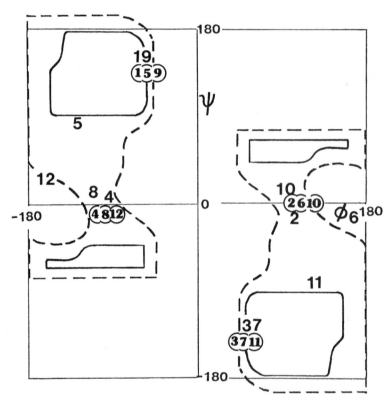

Figure 15. Torsional angles for uncomplexed valinomycin, plain numbers, and the complex with K^+, circled numbers.

The conformation of the backbone in the complex in Rb^+-prolinomycin, a synthetic peptide analog of valinomycin, is quite similar to the K^+-valinomycin complex (32). 5) For completeness, two other complexes of valinomycin that are at variance with the above categories should be mentioned. The bracelet form of the backbone in valinomycin is quite rigid and accommodates the K^+ or Rb^+ ion easily because these ions have an approximate radius to fit into the cavity and form six ligands of 2.75 or 2.90 A. The Na^+ ion is too small and since the bracelet cannot shrink, a different accommodation is made (33). The central position in the cavity is occupied by a H_2O molecule while the Na^+ is displaced by 2.3 A from the center of the cavity, slightly to the outside of the coordination cage, in the direction of the lactyl residues. The Na^+ forms only three ligands to carbonyl oxygens. The remaining ligands are provided by the H_2O molecule in the cavity and a picrate anion outside the complex.

When the complexing ion is too large, the backbone of valinomycin stretches into an extended chain forming a large oval, without any internal hydrogen bonds. Such a conformation has been established for the 2:1 complex with Ba^{++} ions, where the two Ba^{++} ions are located at the focii of the ellipse. Each Ba^{++} ion makes three ligands with carbonyl oxygen atoms and additional ligands with perchlorate anions (34).

Concluding Remarks

The several dozen crystal structure analyses of cyclic peptides and their complexes with alkali or alkaline earth metal ions show that large conformational changes occur in the peptide backbones upon complexation, regardless of the size of the peptide. The peptides fold and twist in such a manner as to provide carbonyl oxygens in proper orientations for the coordination sphere of the alkali or alkaline earth metal ions. These metal ions adopt the same environment of oxygen atoms, at similar interatomic distances, as found in minerals, for example, the octahedral arrangement of O atoms about the Mg^{++} or Ca^{++} ion in MgO (35) or CaO (36). In solution also, these metal ions are surrounded by O atoms from H_2O molecules comprising the hydration sphere. If a cyclic peptide molecule, singly or in pairs, cannot provide a sufficient number of carbonyl oxygens to surround a metal ion, then the coordination sphere is completed with H_2O molecules or with O or N atoms from other solvent molecules, as shown in Figs. 5 and 11.

An examination of the data presented in this paper for the conformational changes that occur upon complexation suggests a common feature for all the types of complexes that have been formed. In each case, in the uncomplexed peptide there is a carbonyl oxygen that is directed away from the body of the peptide and is unshielded by neighboring side groups. This oxygen appears to form an initial point of attraction for the metal ion that presumably exchanges one of the H_2O molecules in its hydration sphere for the extended carbonyl oxygen from the peptide. The intermediate steps are, of course, not observable by present means. However, after the complexation process is completed, the carbonyl bond of the "attracting" oxygen, along with the metal ion, has been rotated into the interior of the backbone ring. The examples of this process are: O_3 in the cyclic pentapeptide in Fig. 6 with an accompanying rotation of $+89°$ about C_3^a-C_3'; O_4 in cyclic $(GlyProPro)_2$ in Fig. 9 with an accompanying rotation of $-135°$ about C_4^a-C_4' and simultaneous *cis*$\Rightarrow$*trans* isomerization for residue 4; O_3 and O_8 in the decapeptide antamanide in Fig. 12 where the ψ_3 and ψ_8 values change by $+150°$; and O_5 and O_{11} in valinomycin in Figs. 3 and 14 where ψ_5 and ψ_{11} change by $+55°$ and $-55°$, respectively. In uncomplexed valinomycin (4,5), O_{11} is more exposed than O_5 owing to smaller side chains on that face of the molecule and, therefore, it may be O_{11} that is the point of attraction for K^+ or Na^+ ions. The somewhat *exo* positioning of Na^+ with respect to the valinomycin cavity (33), toward the less lipophilic face, supports the concept that O_{11} is the attracting atom. In each of these examples, internal hydrogen bonds in the uncomplexed peptide molecules have been broken during the complexation process and different hydrogen bonds have been formed within the complex.

Cyclic (GlyPro)$_3$ (17) needs a separate mention in that the total hydrophobicity of the side chains is considerably smaller than in the other peptides discussed in this paper. In the uncomplexed state, five of the six carbonyl oxygens are directed outward from the backbone and are quite exposed. The glycyl O atoms are selectively chosen for complexation. Two of the glycyl O atoms are approximately in proper positions to complex with the metal ions whereas the remaining glycyl residue has to isomerize from *cis* to *trans* to bring the carbonyl O to the proper orientation. Whether it is the *cis* glycyl oxygen that attracts the metal ion is not possible to determine from existing data. However, such a suggestion can be made from drawing a parallel with the other peptides discussed in this paper, where the carbonyl O that travelled the greatest distance in the complexation process was also the most probable point of attraction for the metal ion.

This research was supported in part by NIH grant GM30902.

References and Footnotes

1. M. Alléaume and D. Hickel, *Chem. Commun.* 1422 (1970); and in M. Dobler, *Ionophores and Their Structures,* J. Wiley and Sons, New York, pp. 85-89 (1981).
2. IUPAC-IUB Commission on Biochemical Nomenclature, *Biochemistry 9,* 3471 (1970).
3. G.N. Ramachandran and V. Sasisekharan, *Advances in Protein Chemistry,* Eds. C.B. Anfinsen, Jr., M.L. Anson, J.T. Edsall and F.M. Richards, Academic Press, New York, pp. 283-438 (1968).
4. I.L. Karle, *J. Amer. Chem. Soc. 97,* 4379 (1975).
5. G.D. Smith, W.L. Duax, D.A. Langs, G.T. deTitta, J.W. Edmonds, D.C. Rohrer and C.M. Weeks, *J. Amer. Chem. Soc. 97,* 7242 (1975).
6. I.L. Karle and E. Duesler, *Proc. Natl. Acad. Sci. USA 74,* 2602 (1977).
7. I.L. Karle, Unpublished. Submitted to Int. J. Peptide Protein Res.
8. I.L. Karle, *J. Amer. Chem. Soc. 99,* 5152 (1977).
9. I.L. Karle, T. Wieland, D. Schermer and H.C.J. Ottenheym, *Proc. Natl. Acad. Sci. USA, 76,* 1532 (1979).
10. I.L. Karle, *J. Amer. Chem. Soc., 100,* 1286 (1978).
11. I.L. Karle, *Perspectives in Peptide Chemistry,* Eds. A. Eberle, R. Geiger and T. Wieland, S. Karger, Basel, pp. 261-271 (1981).
12. I.L. Karle, *Macromolecules 9,* 61 (1976).
13. I.L. Karle and J.L. Flippen-Anderson, *Acta Cryst. B34,* 3237 (1978).
14. C.C. Chiang, I.L. Karle and Th. Wieland, *Int. J. Peptide Protein Res. 20,* 414 (1982).
15. I.L. Karle and C.C. Chiang, *Acta Cryst. C40,* 1381 (1984).
16. I.L. Karle, *Int. J. Peptide Protein Res. 23,* 32 (1984).
17. G. Kartha, K.I. Verughese and S. Aimoto, *Proc. Natl. Acad. Sci. USA 79,* 4510 (1982).
18. I.L. Karle and J. Karle, *Proc. Natl. Acad. Sci. USA 78,* 681 (1981).
19. W. Simon, ETH, Zurich, private communication.
20. M. Czugler, K. Sasvari and M. Hollosi, *J. Am. Chem. Soc. 104,* 4465 (1982).
21. K. Bhandary and G. Kartha, *Amer. Cryst. Assoc. Abstr. Series 2, 10(1),* 19 (1982).
22. G. Kartha and K. Bhandary, *Amer. Cryst. Assoc. Abstr. Series 2, 10(2),* 44 (1982).

23. G. Kartha and K. Bhandary, *Amer. Cryst. Assoc. Abstr. Series 2, 12(1),* 53 (1984).
24. Y.H. Chiu, L.D. Brown and W.N. Lipscomb, *J. Amer. Chem. Soc. 99,* 4799 (1977).
25. J.A. Hamilton, L.K. Steinrauf and B. Braden, *Biochem. Biophys. Res. Commun. 64,* 151 (1975).
26. I.L. Karle, *J. Amer. Chem. Soc. 96,* 4000 (1974).
27. I.L. Karle, *Biochemistry 13,* 2155 (1974).
28. T. Wieland and O. Wieland, *Microbial Toxins,* Eds. A. Ciegler and S.J. Ali, Academic Press, New York, Vol. 8, pp. 249-280 (1972).
29. M. Pinkerton, L.K. Steinrauf and P. Dawkins, *Biochem. Biophys. Res. Commun. 35,* 512 (1972).
30. K. Neupert-Laves and M. Dobler, *Helv. Chim. Acta 58,* 432 (1975).
31. J.A. Hamilton, M.N. Sabesan, and L.K. Steinrauf, *J. Amer. Chem. Soc. 103,* 5880 (1981).
32. J.A. Hamilton, M.N. Sabesan, L.K. Steinrauf, *Acta Cryst. B36,* 1052 (1980).
33. L.K. Steinrauf, J.A. Hamilton and M.N. Sabesan, *J. Amer. Chem. Soc. 104,* 4085 (1982).
34. S. Devarajan, M. Vijayan and K.R.K. Easwaran, *Int. J. Peptide Res. 23,* 324 (1984).
35. Natta and Passerini, *Gazz. Chim. Ital. 59,* 129 (1929).
36. Smith and Leider, *J. Appl. Cryst. 1,* 246 (1968).

Biomolecular Stereodynamics III, Proceedings of the Fourth Conversation in the Discipline Biomolecular Stereodynamics, State University of New York, Albany, NY, June 04-09, 1985, Eds., Ramaswamy H. Sarma & Mukti H. Sarma, ISBN 0-940030-14-4, Adenine Press, ©Adenine Press 1986.

Mobility and Heterogeneity in Protein Structure as Seen by Diffraction

Wayne A. Hendrickson, Janet L. Smith and Steven Sheriff*

Department of Biochemistry and Molecular Biophysics
Columbia University, New York, NY 10032 USA

Abstract

The picture of a protein that derives from a crystallographic analysis of its x-ray diffraction is an average over the lattice and over the time of measurement. This average can usually be represented by the mean atomic positions of the molecule and by parameters of the distribution of displacements about these atomic centroids. Displacements are due to thermally induced vibrations, dynamic conformational heterogeneity, and static variations among the molecules that compose the crystal. Thus, although it is not time resolved, the diffraction experiment does record the aggregate dynamic history of a structure and provides a measure of the individual atomic mobilities. In the event that conformational states are widely enough separated, such discrete disorder can be incorporated into the model. We have analyzed the atomic mobilities in several carefully refined protein structures. These include crambin, pancreatic trypsin inhibitor, myohemerythrin, lamprey hemoglobin and erabutoxin. Each shows substantial variation along the course of the molecule and the influence of lattice contacts is evident. In many cases the atomic distribution functions are found to be quite anisotropic. Discrete disorder is also prevalent. The adaptability to environmental context that these protein structures exhibit may have functional importance. A correlation is seen between atomic mobility and antigenicity. Flexibility is also involved in ligand binding and enzyme action.

Introduction

Analyses of the x-ray diffraction from crystals provide the most definitive of available pictures of the atomic structure in biological molecules. However, in contrast to many spectroscopic techniques where each observation can be assigned to a particular structural feature, in the diffraction experiment all atoms contribute to every measurement. Deductions about individual atoms require a comprehensive analysis of the complete diffraction pattern. This analysis averages over a long time (usually many hours) and the entire lattice (> 10^{14} molecules in a typical protein crystal); consequently, the images obtained are static and unsynchronized. The distributed character of structural information in diffraction data makes the technique ill-

*Present address: Laboratory of Molecular Biology, National Institute of Arthritis, Diabetes, Digestive and Kidney Diseases, Bethesda, MD 20205.

suited to time resolved studies. Although crystallographic results are usually static averages, this does not mean that diffraction data are insensitive to the dynamic properties of crystalline molecules. It is readily apparent from the basic theory described below that these data depend upon the individual atomic mobilities as well as the atomic positions. In particular, the diffracted intensity is exponentially related to the second moment of the component, u, of the distribution of atomic displacements in the direction normal to the Bragg plane of the x-ray reflection. These displacements from rest positions arise from several sources including thermally induced vibrations and dynamic or static disordering among discrete conformational states. In the case of biological macromolecules the displacements are often quite large and this has an important impact on the diffraction pattern.

Thus while the time course of motions in a protein molecule are not accessible to conventional crystallographic experiments, the integrated history of these motions— i.e. the distribution of atomic positions in the average structure—can be measured. The parameters of atomic mobility and conformational disorder can only be described with confidence when the structural analysis has reached an advanced stage of refinement. We describe here our analyses of thermal parameters and disorder in several carefully refined protein structures and discuss some implications of these results for the functional properties.

Theoretical Background

The sensitivity of crystallographic results to the dynamic properties of the crystalline molecules is immediately apparent from the well established theory of diffraction. The integrated intensity of a diffraction spot located by the indices *h, k* and *l* is related simply by

$$I(h,k,l) = K\,|F(h,k,l)|^2 \tag{1}$$

to a structure factor F_{hkl} that embodies all of the pertinent atomic parameters and a proportionality factor K that includes various physical and geometric factors of the experiment. The structure factor equation is

$$F(h,k,l) = \sum f_j\,(s_{hkl})\,\exp\,(-B_j s_{hkl}^2)\,\exp\,\{2\pi i(hx_j + ky_i + lz_j)\} \tag{2}$$

Here $s_{hkl} = \sin\theta/\lambda$ where θ is the Bragg scattering angle and λ is the wavelength of the radiation and f_j is the theoretically known atomic scattering factor of atom j. The pertinent structural parameters are the positional coordinates x_j, y_j and z_j of the jth atom and the "temperaturè" parameter B_j. In the case of disorder, an occupancy factor must also be incorporated into the equation. Clearly the intensity of diffraction spots will be exponentially diminished as the B values increase and as the scattering angles of these Bragg reflections increase. The B value cited above is related by

$$B = 8\,\pi^2\,\overline{u^2} \tag{3}$$

to the mean-square isotropic displacement $\overline{u^2}$ of the atom from its rest position. In the classic Debye-Waller treatment, these displacements are due to thermal vibrations and B is directly proportional to temperature above a certain characteristic temperature (below this point, B stays at a value that corresponds to the quantum zero-point energy of the system). This "temperature factor" theory applies rigorously only to certain simple systems. In the case of biological macromolecules, the atomic displacements include contributions from vibrations of groups as well as those of individual atoms, from dynamic disordering among thermally accessible conformational states, and from static lattice imperfections or conformational heterogeneity that is frozen into the crystal. These displacements are typically quite large since proteins and nucleic acids are intrinsically flexible and crystals of these molecules are highly hydrated and only loosely knit together by lattice interactions. This high degree of atomic mobility and disorder leads to large effective B values and this causes diffraction intensities to decrease with scattering angles.

Initial atomic models generated by crystallographic methods are generally rather inaccurate in atomic positions and have little if any quantitative information about thermal parameters. These models must be refined to optimize the agreement between observed diffraction values and those calculated from Eqs. 1 and 2. We have developed a refinement procedure that uses the method of least squares to drive the model to a fit with the diffraction data and at the same time to agreement with ideal stereochemical features (1). Since the equations are non-linear the process is iterative and manual intervention via molecular graphics (2) is an important ingredient. Stereochemistry has implications for thermal parameters as well as for the atomic position. Since bond distances and angles vary very little with time, atomic motions of bonded groups are highly correlated. We impose this stereochemical condition on the refinement by restraining the variances of time-dependent interatomic distances to suitably small values (3,4).

Correlations in the movement of atoms naturally imparts an anisotropic character to the motions, particularly when displacements are large. Moreover, large excursions of an atom will tend to bring it into collision with other atoms and this leads to anharmonicity. In cases where closely separated conformations are accessible, the distribution function might be bimodal. Such complications obviously require more elaborate models than obtain with simple isotropic parameters given by Eq. 3. This is possible with cumulant expansions of the distribution function (5), but only at the expense of adding variable parameters. Perversely, since large displacements mean fewer measurable data, it is just where higher order effects are most pronounced that they tend to become indeterminate. Thus, most refinements use an isotropic model, anisotropy is included when extensive data are available, and anharmonic effects are just beginning to be considered.

Atomic Mobility

The thermal parameters (B values) of a structure cannot be determined as accurately as can positional parameters. They can compensate for scaling errors, inappropriate absorption corrections and other systematic mistakes. They also take on erroneously high values if associated with wrongly placed atoms. However, conversely, the final structural details cannot be determined until the thermal parameters are refined. Despite the possible pitfalls, it is found that thermal parameters derived from carefully done studies carried to a high level of refinement appear to be meaningful (6-8), i.e. that they faithfully represent the distribution of atomic displacements. These displacement distributions can be identified with the freedom of a given protein atom to move within the crystal. This atomic mobility has various components, as discussed above, and to some extent these can be distinguished (9,10).

We have recently carried out thorough refinements of several protein structure including careful analyses of thermal parameters. These protein studies include crambin (11) now refined at 0.945 Å resolution, bovine pancreatic trypsin inhibitor (12) refined anew (13) against the 1.5 Å Deisenhofer and Steigemann data, myohemerythrin refined with data at 1.3/1.7 Å (8,14), erabutoxin (15) refined to 1.4 Å (16), and lamprey hemoglobin at 2.0 Å resolution (17). The results of this work coupled with experience in refinement of other structures help to characterize atomic mobility behavior and we cite some examples here.

Atomic mobilities in proteins are rather variable but in general quite large. The average isotropic thermal parameter, B, for the atoms in a typical protein crystal is about 20 Å². From Eq. 3, this corresponds to an r.m.s. displacement of 0.5 Å which is large on the atomic scale where bond lengths are about 1.5 Å. The average B values vary considerably for different protein crystals and so do the individual B values for the different atoms within any one protein. For example, the average isotropic equivalents of the B values in three of our highly refined protein structures are 8.8 Å² for crambin, 16.3 Å² for trypsin inhibitor and 22.9 Å² for myohemerythrin (solvent included in averages). Within trypsin inhibitor the refined values range from 6.7 to 36.1 Å². Some of the individual variation can be appreciated from the stereodrawings of thermal ellipsoids in the peptide segments shown in Figures 1 and 2. These are 50% probability ellipsoids; that is, an atom would be expected to be found within its envelope half of the time. (Note in the trypsin inhibitor example, a constant "lattice disorder" contribution has been subtracted for clarity.) In these examples, as is true generally, the atomic mobility in main chain atoms is appreciably lower than in side chain atoms.

The models shown in Figures 1 and 2 are from anisotropic refinements made with rather tight restraints to correlate the motions on adjacent atoms. The rapid increases in mobility out along some side chains are a reflection of this anisotropy—such increases with isotropic atoms would violate stereochemical restrictions on bond fluctuations. Appreciable anisotropy is evident in these structures and is particularly noticeable in the carbonyl oxygens. However, the anisotropic model used in this work, which fixes the principal axes of displacement, appears to be unsatisfactory. Tests of this model on known small molecule structures fail to match the observed data whereas a full 6-parameter model does (J.H. Konnert, personal communication). A comparison between the trypsin inhibitor refinement results and a molecular

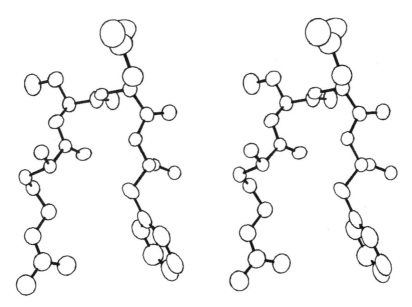

Figure 1. Stereodrawing of peptide Arg 10-Ser 11-Asn 12-Phe 13 from the anisotropically refined structure of crambin at 0.945 Å resolution. Thermal ellipsoid surfaces are drawn at the 50% probability contour.

dynamics simulation likewise indicates that the fixed axis model is too restrictive (13). Correlations of the kind employed in the thermal restraints are also evident in the dynamic simulations.

The variations in atomic mobilities tend to be highly correlated with solvent accessibility (8), but just as surface exposure changes quickly from residue to residue so too do the mobilities. Main chain atoms on the other hand feel a dampened influence of environment and usually are more smoothly varying. The pattern of main chain B values in the myohemerythrin crystal is shown in Figure 3. The low frequency rise and fall in this pattern clearly corresponds to strong regional distinctions in mobility. Perhaps more interesting are the striking higher frequency features. Many of these, particularly in the B and C helix regions correspond precisely in period and phase to the surface exposure of these helices. Remarkably in light of such evidence of reasonable behavior, the B value pattern in myohemerythrin proved to differ markedly from that in octameric hemerythrin. However, the differences proved to be associated mainly with intersubunit contacts and it was possible to use correlation to surface accessibility to adjust the crystalline atomic mobilities to ones that might pertain to the protomeric units free in solution (8). The influence of lattice interactions on thermal parameters is also strongly present in crystals of cytochrome c' (18).

Movement of the kind involved in conformational change during enzymatic reactions or allosteric control is clearly a crucial aspect of biochemical activity. Whether the more rapid and continual mobility that is reflected in thermal factors is biologically significant is a more open question. There are increasing indications that it is. For

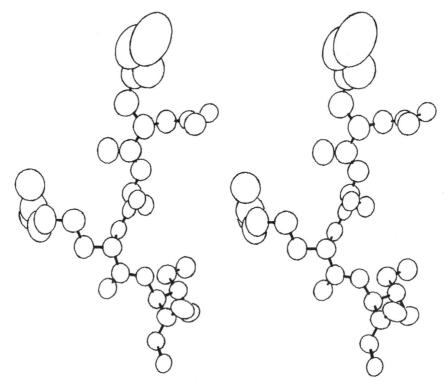

Figure 2. Stereodrawing of peptide Lys 15-Ala 16-Arg 17-Ile 18 from the anisotropically refined high-pH structure of bovine pancreatic trypsin inhibitor at 1.5 Å resolution. A constant isotropic "lattice disorder" component of 5 Å^2 has been subtracted from each thermal parameter prior to plotting the reduced thermal ellipsoids at the 50% probability level. Lysine 15 binds at the active site of trypsin.

example, in *Streptomyces griseus* protease A the three stretches of residues that comprise the polypeptide substrate binding site of this enzyme have some of the highest atomic mobilities in the molecule (7). When substrate tetrapeptides are soaked into these crystals, the mobilities of the binding site regions are reduced substantially in the complex (19). At a minimum the entropic changes involved have implications for binding energies, but it is plausible that flexibility of binding regions is also involved in the kinetics of the capture process. Perhaps in a related way, the antigenic properties of peptides and proteins are highly correlated with atomic mobilities. Sequential antigenic determinants in protein correspond to high B-value stretches (20) and antipeptide antibodies against mobile stretches of myohemerythrin bind strongly to the native protein whereas those against the least mobile regions do not (21). Other factors such as surface accessibility must also be involved, but flexibility is likely to be a significant ingredient in immunological recognition.

Conformational Heterogeneity

Atomic mobility as represented by thermal parameters necessarily pertains to the distribution of positions centered at a single atomic center. It can also happen that

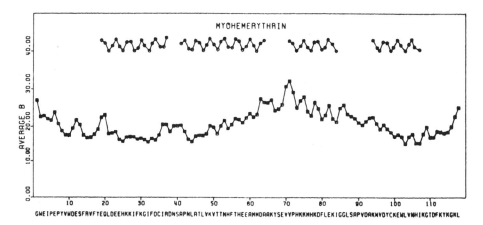

GWEIPEPTYWDESFRVFTEQLDEEHKKIFKGIFOCIRDNSRPNLRTLYKVTTNHFTHEERMHDRRKYSEVYPHKKMHKDFLEKIGGLSRPVDRKNVDYCKEWLYNHIKGTDFKYKGNL

Figure 3. Plot of the average B values for the main chain atoms (N, C$_a$, C and O) of myohemerythrin for the residues of the polypeptide chain. Average B values (□—□) are shown below. Projections of the α-helical segments (○—○) are shown above with the external surfaces facing upward. The helical periodicity is also evident in the pattern of B values, especially for helices B and C.

the potential energy surface in which an atom or group of atoms resides contains multiple minima of similar energy. In such a case, atomic populations for the group might be centered about two or more discrete sites. If the energy barrier can be surmounted by thermal kinetic energy then these multiple states will be in dynamics equilibrium. Alternatively, a heterogeneous population of states might be frozen into the system during folding or crystallization events.

During the course of our recent refinements of protein structures we have found that the final stages must involve the modeling of discrete disorder into multiple conformation states. This is especially true for side chains, but ramifications in the solvent structure are also pervasive. The results of our disorder refinements (22) are summarized in Table I. Crambin was the first of these structures to be refined and it is one of the most highly ordered and stable proteins known to exist. Yet there is appreciable disorder. The full extent of this heterogeneity could not be

Table I
Summary of Disordered Protein Refinements

Feature	Crambin	Erabutoxin	Myohemerythrin	Lamprey Hemoglobin
Resolution	0.945	1.4	1.3/1.7	2.0
Amino-acid residues	46	62	118	149
Disordered side chains	7	11	7	10
Bound solvent (sites per residue)	1.9	2.0	1.3	1.6
Average B value (protein atoms)	7	15	23	16
R value	0.111	0.152	0.159	0.142

appreciated until the structure was highly refined at high resolution. Then conformers separated by only a few degrees in torsion angles could be resolved. The clear existence of disorder in the well defined crambin case served as a model for the interpretation of the other structures, but the internal evidence is also clearcut in those cases. Tightly held thermal restraints were helpful in the visualization of disorder and geometrical restraints were essential for its inclusion in the refinement.

One particular example of conformational heterogeneity is shown in Figure 4. Tyr 29 was initially refined with anisotropic temperature factors, at 1.5 Å resolution, and while it exhibited the highest apparent atomic mobility in this molecule, it was still well within the bounds of typical proteins. The final electron density map through the plane of the ring is shown in the lower left panel. When the refinement was extended to beyond 1 Å resolution, the first difference density map, shown at upper left, gave clear evidence of two discrete conformers. In retrospect, there is evidence for this disorder in the 1.5 Å map as well. However, while the disorder is certainly clear in the final density map at 0.945 Å as shown in the lower right panel, the contrast with the density for a well ordered tyrosine as shown at upper right is very striking. In the case of Tyr 29, the heterogeneity is connected via a lattice contact to compositional heterogeneity at another position. In other cases the conformational disorder is isolated. Most often the residues involved are at the protein surface, but internal heterogeneity (e.g. at Phe 13 in crambin) is also seen.

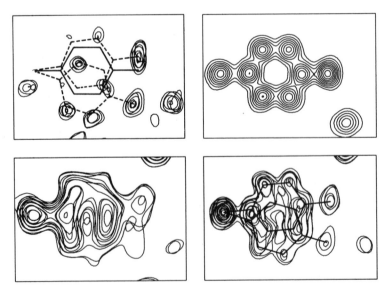

Figure 4. Electron density distributions associated with structures of tyrosine residues in crambin. Clockwise from lower left: 2 Fo-Fc density of Tyr 29 at 1.5 Å resolution, ΔF density at 0.945 Å resolution with the 1.5 Å refine average model (solid lines) and 0.945 Å refined disorder mates (dashed lines) superimposed, 2 Fo-Fc density for Tyr 44 at 0.945 Å resolution, 2 Fo-Fc density for Tyr 29 at 0.945 Å resolution.

The frequency of occurrence and widespread prevalence of conformational heterogeneity in carefully refined protein structures suggests that such multiplicity of states is a common feature of protein structure. In the four proteins cited in Table I, from 6 to 18% of the residues are found to be disordered. Clearly the line between atomic mobility as seen in unimodal distributions and heterogeneity requiring multimodel descriptions depends on the fineness of separation of the conformers and on the resolution of the technique. In the case of crambin we discern discrete states of Phe 13 that are separated by only 9° and 4° respectively in χ_1 and χ_2 torsion angles. At 1.5 Å resolution this residue appeared to be satisfactorily fit by a single conformation with B values of 9 Å^2 or less. Many protein structures are only known at lower resolution and have higher B values. It is quite probable that many of the residues in such structures might actually have more than one conformational state. One indication that this might be the case comes from the distribution of torsion angles in aliphatic side-chains. In crambin these angles have an r.m.s. deriative of only 9° from the staggered positions. By contrast in lamprey hemoglobin and erabutoxin, which are known at lower resolution, we obtain 19.2° and 16.8° for the breadth of these distributions. Some of this greater breadth may reflect a compromise fitting to unresolved conformers.

In some cases, the heterogeneity at one site is linked to that at another through hydrogen bonds or other interactions. This is particularly true in the solvent where alternative networks of water structure can often be resolved. However many of the sites of disorder in the proteins we have studied appear to be independent. One can then ask how many stable conformational states of the protein might exist. If one ignores energetically indistinguishable states such as can be produced by aromatic ring flips or methyl rotations, then a lower bound on the number of states is given by the product of the number of conformers observed at a site and the number of independent sites. Here our estimates range from $3 \times 2^6 = 192$ for crambin to $2^{10} = 1024$ for erabutoxin. Each of these states would be populated similarly. As these are lower bounds it seems clear that the proliferation of conformational states in a protein is quite large. However, it should be emphasized that the states described here are very similar to one another and are only distinguished by fine detail.

Concluding Remarks

Our experience leads us to conclude that diffraction studies can reveal considerable solid information about mobility and heterogeneity in protein structure. Here we have emphasized flexibility as observed directly. There also have been a number of diffraction studies where flexibility can be inferred either because the essentially same structure is crystallized in more than one state or because segments of a protein known to exist chemically are not evident from the diffraction results (23). Such domain flexibility also clearly has considerable significance in biochemical action.

We should also remark that although results presented here are refined as thoroughly as possible with present methods, the calculated diffraction data deviate from the observed values by amounts beyond the errors in the data and by amounts beyond what is routinely achieved in small molecule crystallography. For example, the crambin model at 0.945 Å resolution gives a crystallographic agreement factor of R=0.11 whereas a value of R=0.04 should be obtained if the observation were matched at the level of measurement error. It is mainly limitations in our description of dynamic properties that are responsible for this deficiency. As described above, improvements on current methods can be made but advances will probably not come easily.

Acknowledgements

This work was supported in part by a grant, No. PCM-84-09658, from the National Science Foundation.

References and Footnotes

1. W.A. Hendrickson and J.H. Konnert in *Biomolecular Structure, Function, Conformation and Evolution,* Ed., R. Srinivasan, Pergamon, Oxford, pp. 43-57 (1980).
2. T.A. Jones, in *Computational Crystallography,* Ed. D. Sayre, Clarendon Press, Oxford, pp. 303-317 (1982).
3. J.H. Konnert and W.A. Hendrickson, *Acta Crystallogr. sect. A 34,* 344-350 (1980).
4. W.A. Hendrickson and J.H. Konert, *Biophys. J. 32,* 645-647 (1980).
5. C.K. Johnson, *Acta Crystallogr. 25,* 187-194 (1969).
6. P.J. Artymiuk, C.C.F. Blake, D.E.P. Grace, S.J. Oatley, D.C. Phillips and M.J.E. Sternberg, *Nature (London) 280,* 563-568 (1979).
7. A. Sielecki, W.A. Hendrickson, C. Broughton, L.T.J. Delbaere, G. Brayer and M.N.G. James, *J. Mol. Biol. 134,* 781-804 (1979).
8. S. Sheriff, W.A. Hendrickson, R.E. Stenkamp, L.C. Sieker and L.H. Jensen, *Proc. Natl. Acad. Sci. U.S.A. 82,* 1104-1107 (1985).
9. H. Frauenfelder, G.A. Petsko and D. Tsernoglou, *Nature (London) 280,* 558-563 (1979).
10. F. Parak, E.W. Knapp and D. Kucheida, *J. Mol. Biol. 161,* 177-194 (1982).
11. W.A. Hendrickson and M.M. Teeter, *Nature (London) 290,* 107-113 (1981).
12. J. Deisenhofer and W. Steigemann, *Acta Crystallogr. sect. B 31,* 238-250 (1975).
13. H.-a. Yu, M. Karplus and W.A. Hendrickson, *Acta Crystallogr. sect. B 41,* 191-201 (1985).
14. S. Sheriff, W.A. Hendrickson and J.L. Smith, in preparation.
15. M.R. Kimball, A. Sato, J.S. Richardson, L.S. Rosen and B.W. Low, *Biochem. Biophys. Res. Commun. 88,* 950-959 (1979).
16. J.L. Smith, W.A. Hendrickson, P.W.R. Corfield and B. Low, in preparation.
17. R.B. Honzatko, W.A. Hendrickson and W.E. Love, *J. Mol. Biol. 184,* 147-164 (1985).
18. B.C. Finzel and F.R. Salemme, *Nature (London) 315,* 686-688 (1985).
19. M.N.G. James, A. Sielecki, G. Brayer and L. Delbaere, *J. Mol. Biol. 144,* 43-88 (1980).
20. E. Westhof, D. Altschuh, D. Moras, A.C. Bloomer, A. Mondragon, A. Klug and M.H.V. VanRegenmortel, *Nature (London) 311,* 123-126 (1984).
21. J.A. Tainer, E.D. Getzoff, H. Alexander, R.A. Houghten, A.J. Olson, R.A. Lerner and W.A. Hendrickson, *Nature (London) 312,* 127-134 (1984).
22. J.L. Smith, W.A. Hendrickson, R.B. Honzatko and S. Sheriff, in preparation.
23. W.S. Bennett and R. Huber, *CRC Crit. Rev. Biochem. 15,* 291-384 (1984).

Biomolecular Stereodynamics III, Proceedings of the Fourth Conversation in the
Discipline Biomolecular Stereodynamics, State University of New York,
Albany, NY, June 04-09, 1985, Eds., Ramaswamy H. Sarma & Mukti H. Sarma,
ISBN 0-940030-14-4, Adenine Press, ©Adenine Press 1986.

Ligand Binding:
New Theoretical Approaches to Molecular Recognition

J. Andrew McCammon and Terry P. Lybrand
Department of Chemistry
University of Houston—University Park
Houston, TX 77004

Stuart A. Allison
Department of Chemistry
Georgia State University
Atlanta, GA 30303

Scott H. Northrup
Department of Chemistry
Tennessee Technological University
Cookeville, TN 38505

Abstract

Two new methods for the theoretical study of ligand-receptor interactions are described and illustrated. The first method is concerned with the dynamics of diffusional encounters between ligands and receptors. In this method, one simulates the diffusional motion of the interacting molecules on a computer, and then determines the mechanism and rate constant for association by analysis of the resulting trajectories. Application to increasingly detailed models of superoxide dismutase shows that this enzyme achieves its high catalytic rate in part by electrostatic steering of substrate molecules into its active sites. The second method is concerned with the thermodynamics of ligand-receptor binding. Here, one uses computer simulations to sample thermally-accessible configurations of ligand, receptor, and solvent while making specified changes in the chemical structures of the ligand or receptor. The results are analyzed to provide the relative free energy of binding of related ligand-receptor pairs. The method is being used to determine the relative stabilities of complexes of the macropolycyclic amino-ether SC-24 with two different anions.

Introduction

Complex formation is an essential part of the action of enzymes, hormones, antibodies, drugs and other biological molecules. Within the past couple of years, two new theoretical methods have been introduced that will permit detailed and realistic studies of key aspects of such molecular interactions. These methods are expected to be useful in the molecular interpretation of experimental data and in the design of new molecules with specified activities.

The first step in complex formation is the diffusional encounter of ligand and receptor molecules. Here, we use the terms ligand and receptor generically to include any molecules or other chemical species that recognize and bind to one another (e.g., substrate and enzyme, hormone and receptor). The rate of initial diffusional encounter determines or controls the overall rates of a number of biological processes. For example, certain enzymes are known to catalyze reactions of their substrates at diffusion-controlled rates (1). Existing theoretical methods for studying diffusion-controlled reactions have been limited to rather idealized models, e.g., spherical reactants interacting by centrosymmetric forces (2,3). Although such methods have helped enormously to shape our general understanding of bio-molecular kinetics, it is important to develop more quantitative methods that recognize the details of biomolecular structure, flexibility and interactions. Such methods could be required, for example, to predict the increases in efficiency resulting from proposed modifications in enzyme molecules that operate with diffusion-controlled kinetics.

From a practical point of view, the thermodynamic aspects of ligand-receptor associations are often even more important than the kinetic aspects. The specificity of action of antibodies, hormone receptors, and other types of receptors is determined by the relative affinity of different ligands for these molecules. Where the structures of the ligands and receptors are known, it has been possible to form qualitative explanations of binding affinities in terms of interactions (steric, hydrogen-bonding, hydrophobic, etc.) suggested by examination of these structures (4). Somewhat more detailed analyses are possible by application of approximate potential energy functions of the molecular mechanics type (5). As with the diffusional phenomena, however, it is important to develop truly quantitative methods that take account of solvation, entropy, and other features of real molecules and their surroundings.

In what follows, we discuss methods that are specifically designed to deal with the aspects of ligand binding described above. These methods are new elements in the remarkable flowering of theoretical work in structural molecular biology that has been underway for several years (3,5,6-9).

Kinetics of Encounter

Simulation Method. Computer simulations of reactant trajectories have provided some of our most detailed insights to gas-phase bimolecular reactions (10). Thus, it would seem desirable to have available corresponding techniques for the study of diffusion-controlled reactions. In principle, such techniques would allow theoretical studies of biomolecular associations that include such realistic features as detailed distributions of electrostatic charge, rough molecular surfaces, geometrically restricted binding sites, internal structure fluctuations of the reactants, and so forth. Although algorithms for simulating Brownian motion have been available for some time (11-17), the formalism needed to extract rate constants from Brownian trajectories has only recently been derived (18-20). This formalism was designed to take into account an important difference between diffusion-influenced and gas-phase reactions.

In diffusion-influenced reactions (unlike their gas-phase counterparts), there is always a chance that collisions with the solvent will drive the reactant molecules together, even if these molecules have moved out of their range of mutual interaction. The contribution of such trajectories to the biomolecular rate constant is accounted for by an exact analytic correction in the new technique.

Here, we will describe the simulation method for the special case of a spherical, uniformly reactive ligand subject to central forces interacting with a receptor which has none of these restrictions. This special case is sufficient to treat the superoxide dismutase-superoxide system (see below). It is straightforward to extend the method to more complicated cases involving structured ligands (19).

In dilute solution, it is sufficient to consider the relative motion of one reactant molecule in the field of a fixed reaction partner (21). Here, we choose the receptor to be the fixed target. As shown in Figure 1, one defines a spherical surface of radius b centered on the receptor. This radius should be chosen large enough that the direct and hydrodynamic forces between ligand and receptor are centrosymmetric to a good approximation for ligand-receptor distances R > b. Then, under steady state conditions, the bimolecular rate constant k is given by

$$k = k_D(b)p \qquad (1)$$

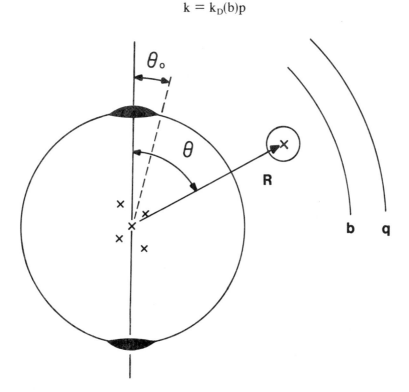

Figure 1. Schematic of the superoxide dismutase (receptor)—superoxide (ligand) system. Dark caps represent the two active sites of the enzyme. Crosses represent electrostatic charges in the enzyme.

where $k_D(b)$ is the rate constant for ligands with R initially greater than b to first achieve a separation $R = b$, and p is the probability that a ligand starting at $R = b$ will ultimately react rather than diffuse away. Because of the restriction placed on b, $k_D(b)$ can be determined analytically and is given by

$$k_D(b) = (\ \int_b^\infty dR[\frac{exp[u(R)/k_B T]}{4\pi R^2 D}])^{-1} \tag{2}$$

where $u(R)$ is the solvent-averaged potential felt by the ligand, $k_B T$ is the Boltzmann constant multiplied by temperature, and D is the relative diffusion coefficient of ligand and receptor. The quantity p depends on the behavior of trajectories in the "complicated" region $R < b$ and is evaluated by trajectory simulations. Trajectories are initiated at random points on the b surface and are continued until they either react (hit a reactive point on the receptor surface) or hit a spherical surface of radius $q > b$. Let β denote the computed probability of reaction in this finite reaction region. As discussed above, there is always a chance that a trajectory terminated at q would, if continued, have returned to hit an active part of the receptor. Because the ligand-receptor interaction has a simple form for $R > b$, this probability of return can be determined analytically and the probability of reaction in an infinite reaction region shown to be (18)

$$p = \frac{\beta}{1-(1-\beta)\Omega} \tag{3}$$

where

$$\Omega = k_D(b)/k_D(q) \tag{4}$$

The desired rate constant is thus given by Equations (1)-(4).

Any of a variety of stochastic or Brownian dynamics algorithms could be used to simulate the trajectories from which the key quantity β is obtained (11-17). We have used the Ermak-McCammon algorithm, which is suitable for overdamped dynamics (11). For the special case considered here, and neglecting hydrodynamic interactions, the algorithm takes a very simple form. If the ligand is currently at the position R°, its position a time Δt later is taken to be

$$\underset{\sim}{R} = \underset{\sim}{R}^\circ + (k_B T)^{-1} D\underset{\sim}{F}^\circ \Delta t + \underset{\sim}{S} \tag{5}$$

where F° is the current solvent-averaged force on the ligand and the components S_i of the vector S are random numbers with the properties

$$\langle \underset{\sim}{S} \rangle = 0$$

$$\langle S_i S_j \rangle = 2D\delta_{ij}\Delta t \tag{6}$$

The time step must be small enough that the solvent-averaged force changes only slightly during Δt. Also, small steps should be taken near the receptor surface and the surface at $R = q$.

The methods described here have been presented in their simplest forms for purposes of clarity and to emphasize the key ideas involved. These methods are readily generalized to handle structured ligands with noncentrosymmetric interactions, hydrodynamic interactions, finite reactivity of active sites, and other features as desired (18-20).

Application to the Enzyme Superoxide Dismutase. Superoxide dismutase (SOD) is known to catalyze reaction of superoxide (O_2^-) with diffusion-controlled kinetics (22). Chemical and structural studies suggest that the high efficiency of SOD is due in part to the distribution of electrostatic charges in this enzyme (23,24). The enzyme is a dimer with identical subunits. Although the net charge of the enzyme is -4 at normal pH, there are concentrations of positive charge near the two active sites on opposite sides of the dimer, and it has been suggested that the resulting electric field could steer substrates into the active sites (23,24).

To investigate the possibility of steering effects in the diffusion-controlled kinetics of SOD, several simulation studies have been carried out. In the first of these (25,26), the SOD dimer was modeled as a sphere of 30A radius (Figure 1). Two small reactive patches corresponding to the active sites were defined on opposite sides of the enzyme. A cluster of five charges was placed within the sphere to reproduce approximately the monopole, dipole and quadrupole electric moments deduced from the X-ray structure of SOD. Representative trajectories of an O_2^- ligand were computed as described above. In these initial studies, typical values of b and q were 300 and 500A, respectively; it is, however, possible to use much smaller values of b (20). Analysis of the trajectories showed that the charge distribution does indeed bias substrate trajectories toward the active sites. The computed rate constants for the five charge SOD model are about 40% larger than those found for a model containing only a single central charge of -4.

In subsequent studies, the model of the SOD-O_2^- system has been generalized to include approximate descriptions of effects due to solvent electrolytes (25), the recessed character of the active sites (20), and the detailed distribution of charge on SOD (i.e., explicit charges on all non-hydrogen atoms) (27). All of these studies support the substrate steering picture, while providing interesting additional details concerning the enzyme kinetics.

Thermodynamics of Binding

Simulation Method. Generally, thermodynamic considerations govern the overall ligand-receptor binding process. Since relative effectiveness (e.g. biological response, catalytic efficiency, etc.) for a series of ligands often correlates with the relative binding affinities of those ligands at a receptor site, computer simulation techniques that can reliably predict relative free energies of binding for structurally similar ligands at a receptor are of great interest and possible utility.

The relative free energy of binding for two ligands, L and M, to a receptor R can be computed by calculating the free energy of binding for each reaction (7)-(8) and taking the difference of these results.

$$L + R \Rightarrow L:R \quad \Delta A_1 \tag{7}$$

$$M + R \Rightarrow M:R \quad \Delta A_2 \tag{8}$$

$$\Delta\Delta A = \Delta A_2 - \Delta A_1$$

Computational methods such as the umbrella sampling technique may in principle be used to calculate these free energy values (28-31). However, this approach consumes large amounts of computer time and may be very difficult to implement in the present context.

A new computer simulation method to predict relative free energies of binding for ligand-receptor complexes employs a perturbation theory technique applied to a set of reactions that form a thermodynamic cycle (32). In this procedure, two hypothetical reactions are considered:

$$L + R \Rightarrow M + R \quad \Delta A_3 \tag{9}$$

$$L:R \Rightarrow M:R \quad \quad \Delta A_4 \tag{10}$$

Reactions (7)-(10) form a complete thermodynamic cycle, so that $\Delta\Delta A = \Delta A_4 - \Delta A_3$, since A is a thermodynamic state function. Standard perturbation techniques can be used to compute the free energy changes for reactions (9) and (10) by slowly changing (i.e. "perturbing") ligand L into ligand M (32,33).

The procedure is straightforward. First, define a potential energy function for the L/R/solvent system, V_L, and a similar potential energy function for the M/R/solvent, V_M. Next, define a "mixed" potential energy function V_λ, such that

$$V_\lambda = \lambda V_M + (1-\lambda)V_L \tag{11}$$

The free energy change for a reaction can be computed by performing small, stepwise perturbations of the system. For reaction (9), the potential energy function parameters for L are changed to those of M in small steps. For each step, a molecular dynamics simulation yields the free energy for values of λ about λ_i

$$A_3(\lambda) - A_3(\lambda_i) = -\beta^{-1}\ln\langle \exp[-\beta(V_\lambda - V_{\lambda_i})]\rangle_{\lambda_i} \tag{11}$$

where $\beta^{-1} = kT$ (k is Boltzmann's constant, T is temperature in Kelvin) and $\langle\,\rangle_{\lambda_i}$

is a canonical ensemble simulation average for V_{λ_i}. The stepsize, $\lambda_{i+1}-\lambda_i$, is sufficiently small so as to assure good statistics in the region of λ where the results from simulation i overlap those from i+1. The free energy change for reaction (9), $\Delta A_3 = A_3(\lambda=1)-A_3(\lambda=0)$, is obtained by piecing together the individual steps. The free energy change for reaction (10), ΔA_4, is computed in a similar manner and the relative free energy difference for reactions (7) and (8) is then $\Delta\Delta A = \Delta A_4 - \Delta A_3$. In contrast to alternate techniques for computing relative free energies of binding (e.g. umbrella sampling methods), the thermodynamic cycle-perturbation method typically requires much less computer time and is much easier to employ.

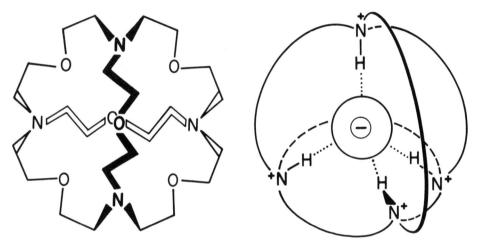

Figure 2. Schematic of the SC-24 (receptor)—halide (ligand) system (38). a. SC-24 structure; b. SC-24—halide complex

Application to the Ionophore SC-24. This novel simulation approach has been used to compute the relative free energy of binding for Cl^- and Br^- to the macropolycyclic compound, SC-24 (see Figure 2), in aqueous solution. The thermodynamic cycle involves the hypothetical free energy changes ΔA_3 and ΔA_4.

$$
\begin{array}{ccc}
 & \Delta A_1 & \\
Cl^- + SC\text{-}24 & \Rightarrow & Cl^-{:}SC\text{-}24 \\
\Delta A_3 \Downarrow & & \Delta A_4 \Downarrow \\
Br^- + SC\text{-}24 & \Rightarrow & Br^-{:}SC\text{-}24 \\
 & \Delta A_2 &
\end{array}
$$

The relative free energy of binding is $\Delta\Delta A = \Delta A_4 - \Delta A_3 (= \Delta A_2 - \Delta A_1)$.

The calculations were performed using the molecular modeling package AMBER (34,35) and SPC potential functions (36). Parameters for Br^- were so derived as to reproduce reasonably the relative ion-single water interaction energies for Br^- versus Cl^-. Note that ΔA_3 in the reaction scheme represents the relative free energy of solvation for Cl^- and Br^-. The perturbation method predicted a value for ΔA_3 of 3.4 kcal/mol, in good agreement with the tabulated result of 3.3 kcal/mol (37). The computed relative free energy for binding of Cl^- and Br^- to SC-24 ($\Delta\Delta A = 4.1$ kcal/mol) is also in good agreement with the experimental value of 4.3 kcal/mol (38).

Concluding Remarks

Both of the methods described here are very new and will surely be refined and modified with time. Nevertheless, the preliminary results from both methods are encouraging. The diffusional method will be applied to SOD with amino acid sequence changes and to other biological molecules. As for the thermodynamic method, perturbations of charge and changes in the total number of atoms are possible in addition to the steric perturbation explored here. Many applications can be envisioned for this method. For example, it may be helpful in computer-assisted design of ligands (e.g. drugs) for specific biomacromolecular receptors or prediction of amino acid modifications to yield proteins with specific characteristics and properties.

Acknowledgments

We thank Professor Peter A. Kollman and Dr. Georges Wipff for helpful discussions concerning SC-24. This work has been supported in part by grants from the NIH, NSF and Robert A. Welch Foundation to the University of Houston. J.A.M. is the recipient of an NIH Research Career Development Award and a Camille and Henry Dreyfus Teacher-Scholar Award. S.A.A. is the recipient of an NSF Presidential Young Investigator Award and a Camille and Henry Dreyfus Grant for Newly Appointed Faculty in Chemistry. S.H.N. is the recipient of an NIH Research Career Development Award.

References and Footnotes

1. J.R. Knowles and W.J. Albery, *Acc. Chem. Res. 10,* 105 (1977).
2. D.F. Calef and J.M. Deutch, *Annu. Rev. Phys. Chem. 34,* 493 (1983).
3. O.G. Berg and P.H. von Hippel, *Annu. Rev. Biophys. Biophys. Chem. 14,* 131 (1985).
4. C.R. Cantor and P.R. Schimmel, *Biophysical Chemistry I,* Freeman, San Francisco (1980).
5. P.A. Kollman, *Acc. Chem. Res. 18,* 105 (1985).
6. J.A. McCammon, *Rep. Progr. Physics 47,* 1 (1984).
7. S.C. Harvey, *Comm. Mol. Cell. Biophys.,* to be published.
8. R.M. Levy and J.W. Keepers, *Comm. Mol. Cell. Biophys.,* to be published.
9. M. Karplus and J.A. McCammon, *Sci. Amer.,* to be published.
10. D.G. Truhlar, Ed., *Potential Energy Surfaces and Dynamics Calculations,* Plenum, New York (1981).
11. D.L. Ermak and J.A. McCammon, *J. Chem. Phys. 69,* 1352 (1978).
12. M. Fixman, *J. Chem. Phys. 69,* 1527, 1538 (1978).
13. E. Helfand, Z.R. Wasserman and T.A. Weber, *J. Chem. Phys. 70,* 2016 (1979).
14. M.R. Pear and J.H. Weiner, *J. Chem. Phys. 71,* 212 (1979).
15. G.T. Evans and D.C. Knauss, *J. Chem. Phys. 71,* 2255 (1979).
16. W.F. van Gunsteren and H.J.C. Berendsen, *Mol. Phys. 45,* 637 (1982).
17. G. Lamm and K. Schulten, *J. Chem. Phys. 78,* 2713 (1983).

18. S.H. Northrup, S.A. Allison and J.A. McCammon, *J. Chem. Phys. 80,* 1517 (1984).
19. S.A. Allison, N. Srinivasan, J.A. McCammon and S.H. Northrup, *J. Phys. Chem. 88,* 6152 (1984).
20. S.A. Allison, S.H. Northrup and J.A. McCammon, *J. Chem. Phys. 83,* 2894 (1985).
21. G. Wilemski and M. Fixman, *J. Chem. Phys. 58,* 4009 (1973).
22. I. Fridovich in *Advances in Inorganic Biochemistry,* Ed., G.L. Eichhorn and L.G. Marzilli, Elsevier North Holland (1979).
23. A. Cudd and I. Fridovich, *J. Biol. Chem. 257,* 11443 (1982).
24. E.D. Getzoff, J.A. Tainer, P.K. Weiner, P.A. Kollman, J.S. Richardson and D.C. Richardson, *Nature 306,* 287 (1983).
25. S.A. Allison and J.A. McCammon, *J. Phys. Chem. 89,* 1072 (1985).
26. S.A. Allison, G. Ganti and J.A. McCammon, *Biopolymers 24,* 1323 (1985).
27. G. Ganti, J.A. McCammon and S.A. Allison, *J. Phys. Chem. 89,* 3899 (1985).
28. C. Pangali, M. Rao and B.J. Berne, *J. Chem. Phys. 71,* 2975 (1979).
29. G. Ravishankar, M. Mezei and D.L. Beveridge, *Faraday Symp. Chem. Soc. 17,* 79 (1982).
30. M. Berkowitz, O.A. Karim, J.A. McCammon and P.J. Rossky, *Chem. Phys. Lett. 105,* 577 (1984).
31. J. Chandrasekhar, S.F. Smith and W.L. Jorgensen, *J. Amer. Chem. Soc. 107,* 154 (1985).
32. B.L. Tembe and J.A. McCammon, *Comput. Chem. 8,* 281 (1984).
33. D.A. McQuarrie, *Statistical Mechanics,* Harper and Row, New York (1976).
34. P.K. Weiner and P.A. Kollman, *J. Comput. Chem. 2,* 287 (1981).
35. U.C. Singh and P.A. Kollman, *J. Comput. Chem. 5,* 129 (1984).
36. H.J.C. Berendsen, J.P.M. Postma, W.F. van Gunsteren and J. Hermans, in *Intermolecular Forces,* Ed., B. Pullman, Reidel, Holland, p. 331 (1981).
37. H.L. Friedman and C.V. Krishnan in *Water, A Comprehensive Treatise,* Ed., F. Franks, Plenum, New York (1972) Vol. 3.
38. E. Graf and J.M. Lehn, *J. Amer. Chem. Soc. 98,* 6403 (1976); E. Kauffman, Thesis, Universite Louis Pasteur, Strasbourg (1979).

*Biomolecular Stereodynamics III, Proceedings of the Fourth Conversation in the
Discipline Biomolecular Stereodynamics, State University of New York,
Albany, NY, June 04-09, 1985, Eds., Ramaswamy H. Sarma & Mukti H. Sarma,
ISBN 0-940030-14-4, Adenine Press, ©Adenine Press 1986.*

Solvent Effects on Conformational Stability
in the ALA Dipeptide: Full Free Energy Simulations

D.L. Beveridge, G. Ravishanker, M. Mezei and B. Gedulin
Chemistry Department
Hunter College of the City University of New York
695 Park Avenue
Hunter College, CUNY,
New York, NY 10021.

Abstract

We report herein theoretical studies of the conformational stability of the Ala dipeptide as
an isolated molecule in water. The C_7, C_5, α_R and P_{II} conformational forms are considered.
The thermodynamics of hydration were computed by liquid state computer simulation on a
solute and 202 waters molecules under periodic boundary conditions at T=25°C. The free
energy of hydration was computed by determining the relative probabilities of conformations that lie on a line between conformations in (ψ,ϕ) space. The intramolecular thermodynamics was worked out by Monte Carlo methods in the quasiharmonic approximation, with
intramolecular entropies determined from the covariance matrix of atomic displacements.
The total free energy, intramolecular plus hydration, is found to be similar for C_7, α_R and P_{II},
and thus it is reasonable to expect that all forms are thermally populated at room temperature.

Introduction

The secondary structure of the Ala dipeptide, N-acetylalanyl-N-methylamide
(AcAlaNHMe), in aqueous solution is a matter of fundamental interest in structural
biochemistry. The sterically allowed regions of conformation in (ψ,ϕ) space for
AcAlaNHMe are prototypical for the polypeptide backbone, and thus a knowledge
of the conformational stability of this molecule is quite relevant to numerous
aspects of protein and enzyme structural analysis. At the same time this molecule is
small enough to be treated with considerable rigor using modern methods of theoretical chemistry, and for some years now AcAlaNHMe and the "dipeptide model"
(21) have played an important historical role in the elucidation of the conformational
possibilities in polypeptides.

The intramolecular energetics of AcAlaNHMe and related molecules have been
studied extensively by means of empirical energy functions and molecular quantum
mechanics. However a complete account of structural chemistry of a potentially
flexible molecule in solution requires a consideration of the manifold structures of

lowest free energy, including both intramolecular and intermolecular contributions. We describe herein some new theoretical studies of AcAlaNHMe as an isolated molecule and in aqueous solution at 25°C, in which the statistical thermodymanics of various molecular conformations are obtained numerically using Monte Carlo computer simulation techniques. The results obtained are compared with those of previous theoretical calculations and with experiment. The original articles from this Laboratory summarized herein are those of Mehrotra et al. (22) and Mezei et al. (23) on the liquid state Monte Carlo studies of hydration of AcAlaNHMe, and of Ravishanker et al. (24) and of Ravishanker and Beveridge (25) on the quasiharmonic Monte Carlo determinations of the intramolecular (vibrational) thermodynamics. The general area of free energy simulations in which these studies are carried out has been reviewed recently by Mezei and Beveridge (26).

Background

The conformation of AcAlaNHMe can be specified in terms of the angles of torsion made by the intersecting CNH and NCH planes (ψ) and by the NCO and CCH planes (ϕ). Structural studies to date indicate four conformations from the allowed regions of (ψ,ϕ) space for specific consideration (27): $C_7(90°,-90°)$, $C_5(150°,-150°)$, $a_R(-50°,-70°)$ and $P_{II}(150°,-80°)$; c.f. Figure 1. The C_7 and C_5 structures are characterized by 7- and 5-atom ring structures completed by an intramolecular hydrogen bond between a CO and NH group. The C_7 form can exist in both axial and equatorial forms; we consider herein the equatorial form. The a_R and P_{II} conformers have ψ and ϕ angles similar to those found in the right-handed polypeptide α-helix and the poly(L-proline)-II helix. However, the a_R and P_{II} forms of AcAlaNHMe are not fully representative of helical polypeptides, which are further stabilized by intramolecular hydrogen bonds between subunits.

The structure of AcAlaNHMe in CCl_4 was established by Avignon and Lascomb by infrared spectroscopy (28) to be predominantly C_7. The crystal structure, reported by Harada and Iitaka (29), involved two conformers: P_{II} and the structurally similar poly-(L-proline)-I form. Avignon et. al. subsequently extended their study to AcAlaNHMe in water (30) and proposed on the basis of an analysis of depolarized Rayleigh scattering that the C_7 conformer is maintained in water, stabilized by a water molecule bridging exterior CO and NH groups. Recently Madison and Kopple (31) studied the alanine dipeptide in CCl_4 and in H_2O using CD spectra and NMR spin coupling analysis. They confirmed earlier proposals of C_7 for non-polar solvents, but found evidence for contributions from both a_R and P_{II} conformations of AcAlaNHMe in water.

The results of diverse theoretical calculations of the relative stability of the C_7, C_5, a_R and P_{II} conformers in isolated AcAlaNHMe (32-38) are collected in Table I. Calculations using empirical energy functions (EPF) by Brant and Flory (32) and by Scheraga and coworkers (38) show that C_7, C_5 and a_R structures correspond with energy minima of comparable depth, with P_{II} in a sterically allowed region likely to be thermally accessible. The Consistent Force Field developed by Hagler et al. (39)

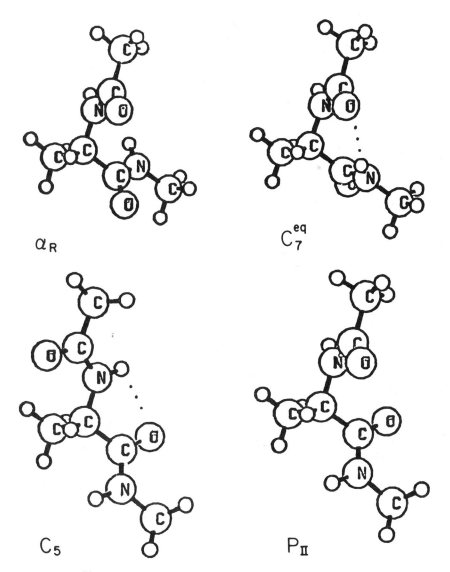

α_R

C_7^{eq}

C_5

P_{II}

Figure 1. AcAlaNHMe in the C_7, C_5, α_R and P_{II} conformations.

applied to a related molecule N-formylalanylamide by Robson et al. (36) shows considerable destabilization of the α_R form; Karplus and Rossky (37) used a similar prescription for the intramolecular energy. Quantum mechanical calculations (33-36) show sharper differences in conformational energies, but generally support C_7 as the energy minimum in the free space approximation. Inclusion of electron correlation, as in the PCILO calculation (21), tends to flatten the surface. The intramolecular thermodynamics of peptides is more fully dealt with in a paper by Hagler et al. (40).

Table I

Calculated intramolecular energies of the C_5, C_7, α_R and P_{II} conformations of AcAlaNHMe in kcal/mol, relative to C_7

Authors	Method	C_7	C_5	α_R	P_{II}
Brant and Flory (32)	EPF	0	0	1	0
Momany et al. (33)	CNDO	0	5	5	5
Hoffman and Imamura (34)	EHT	0	−2	0	0
Pullman et al. (21)	PCILO	0	2	3-4	4
Robson et al. (35,36)	3s2p	0	−3.2	7.8	
	4-31G	0	−2.1	7.5	
	CFF	0	1.6	7.7	
Karplus and Rossky (37)	EPF	0	6	8	8
Scheraga et al. (38)	ECEPP	0	.38	1.13	1.0

Studies of solvent effects on the conformational stability of AcAlaNHMe were first reported by Venkatachalam and Krimm (41), who computed a (ψ,ϕ) map with specific water molecules fixed in hydrogen bonding positions to the interior CO and NH groups. Rein and coworkers (42) reported calculations on AcAlaNHMe in CCl_4 and in H_2O using an Onsager continuum model for the solvent. The minimum in CCl_4 was near C_7 in accord with experiment and the minimum in water was found at $\psi=120°$, $\phi=60°$. Recently Scheraga and coworkers, using a new extended version of the hydration shell approach, obtained C_7 as a minimum for AcAlaNHMe in CCl_4 and C_5 in H_2O (38,43). Their calculated conformational map is however remarkably flat in the sterically allowed region, and thus other conformations are indicated to be thermally accessible. In summary, one finds general agreement between experiment and theory on C_7 as the conformational preference of the alanine dipeptide in nonpolar solvents, but diverse results on the nature of the structure in water.

Liquid state computer simulation has recently been used to study the molecular dynamics of $[AcAlaNHMe]_{aq}$ in the C_7 form by Rossky and Karplus (44). Hagler (45) has reported Monte Carlo calculations of the internal energy and structure of C_7 and α_R forms of AcAlaNHMe and found a dipeptide-water energy difference of 5.6 kcal/mol favoring α_R. Quite recently, Pettitt and Karplus (46) reported dynamical simulations with normal mode calculations and some intramolecular entropy determinations on the Ala dipeptide. They demonstrated that the vibrational entropy of the dipeptide is not sensitive to the electrostatic charge parameters in the energy functions, and that the entropies the C_7^{eq}, C_7^{ax} and C_5 forms were very similar. Brady and Karplus (47) have recently carried out a molecular dymanics study of the internal entropy of the dipeptide in water. In vacuum the hydrogen bonded C_7 form exhibited a lower entropy than the C_5 due presumably to the better hydrogen bond. On dissolving C_7, the entropy dropped slightly. The C_7 and α_R form were found to have similar entropies in solution with P_{II} somewhat larger. We compare further on in this article our results on intramolecular entropies of AcAlaNHMe with those of

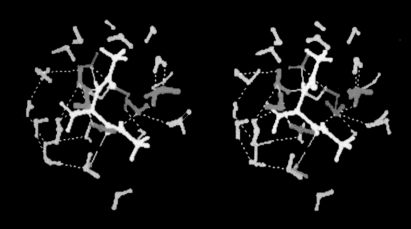

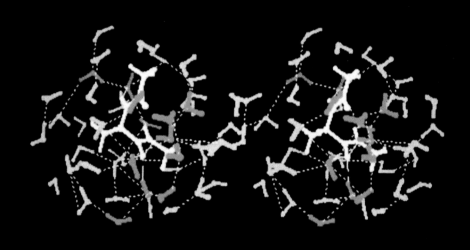

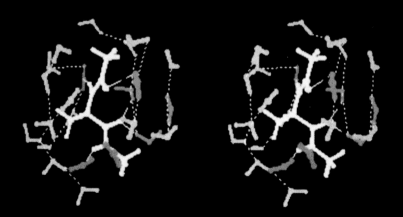

Stereographic views of water-solute hydration complexes for the Ala dipeptide in the C_7 (top), α_R (middle) and P_{II} (bottom) conformations. The middle panel shows an entire first hydration shell; in the others the number of waters is reduced to those nearest to the solute for clarity. Water molecules are color coded with respect to mode of hydration based on the Proximity Criterion: hydrophobic — yellow, hydrophilic (C=0) — blue, hydrophilic (N-H) — green. Solute water hydrogen bonds are indicated by dashed lines, and water-water hydrogen bonds by dotted lines.

Brady and Karplus, and add in the contributions from the internal energy to give the full intramolecular free energy of the various conformers and the solvation thermodymanics. Concluding this section, we note the particularly interesting study on the potential of mean torsion for the dipeptide in water in progress by Pettitt and Karplus (48), using the extended RISM method of Hirata, Pettitt and Rossky (49).

Theory and Methodology

A conformationally flexible molecule in solution can in principle assume various structural forms, and the equilibrium state of the system must be described in terms of these structures and their corresponding statistical weights. The relative free energy of the various structures is simply related to their respective probability of occurrence in the ensemble average of structures representative of the statistical state of the system, considering all intramolecular and intermolecular degrees of freedom. A complete theoretical study of this type of problem is not feasible at present for AcAlaNHMe in water, and thus various assumptions and approximations must be introduced in order to reduce the problem to tractable form.

First, we identify specific conformations of the dissolved molecule which are expected on the basis of other information, experimental or theoretical, to be of particular structural interest. We then direct our calculations to the difference in free energy and related quantities between the selected structures in solution. We assume for two specific conformations i and j with conformational free energies A_i and A_j, that

$$\Delta A_{ij} = \Delta A_{ij}^{int} + \Delta A_{ij}^{hyd} \tag{1}$$

where ΔA_{ij}^{int} is the difference in free energy due to intermolecular factors and ΔA_{ij}^{hyd} is the free energy difference due to hydration. The coupling of hydration to internal modes and vice versa is neglected.

Intramolecular Contributions: Intrinsic Differences and Vibration

The quantities A_i and A_j include in principle averages over the respective vibrational manifolds of conformational states i and j. Preliminary estimates of ΔA_{ij}^{int} are typically constructed with the vibrational contributions to A_i^{int} and A_j^{int} assumed to cancel,

$$\Delta A_{ij}^{int} = \Delta E_{ij}^{int} \tag{2}$$

where ΔE_{ij}^{int} is the intrinsic difference in intramolecular energy of a molecule in conformational states i and j. The AcAlaNHMe system, where structures with intramolecular hydrogen bonds are compared with open forms, is vulnerable in this approximation. Procedures for treating ΔA_{ij}^{int} more rigorously have become available, and are incorporated in our most recent studies on the Ala dipeptide (24,25). In brief, the configurational internal energy for the intramolecular modes in the vicinity of a local minimum is computed by numerical integration of the expectation value

$$U_i^{int} = \langle E(\mathbf{q}) \rangle \qquad \mathbf{q} \in i \tag{3}$$

where $E(\mathbf{q})$ is the intramolecular potential energy as a function the $\mathbf{q}$ internal degrees of freedom and the brackets indicate a Boltzmann configurational average. The energy function $E(\mathbf{q})$ is evaluated from an analytical force fields supplied as indicated in the following section. For the determination of entropy, our procedure follows the quasi-harmonic formalism as it evolved in the series of papers by Karplus and Kushick (50) and Levy et al. (51,52) based on an effective potential energy of the form

$$E^{QH} = (\tfrac{1}{2}) \, \mathbf{q} \cdot \mathbf{F}^{QH} \cdot \mathbf{q} \tag{4}$$

with the elements of the force constant matrix $\mathbf{F}^{QH}$ given by the expression

$$F_{kl} = kT \, [\sigma^{-1}]_{kl} \tag{5}$$

where σ is the covariance matrix of the internal coordinate fluctuations

$$\sigma_{kj} = \langle (q_k - \langle q_k \rangle)(q_j - \langle q_j \rangle) \rangle. \tag{6}$$

Assuming that the difference in the Jacobians evaluated at the two conformations i and j is negligible, the quasiharmonic conformational entropy difference is given by

$$\varDelta S_{ij}^{int} = (\tfrac{1}{2}) \, k\ln \, (\sigma_i/\sigma_j) \tag{7}$$

where

$$\sigma_i = \det \boldsymbol{\sigma}_i \tag{8}$$

for displacements in the vicinity of the i th conformation. The relative internal energies and free energies follow straightforwardly from

$$\varDelta U_{ij}^{int} = U_j^{int} - U_i^{int} \tag{9}$$

$$\varDelta A_{ij}^{int} = \varDelta U_{ij}^{int} - T \varDelta S_{ij}^{int}. \tag{10}$$

In a quasiharmonic molecular simulation, the force-constant matrix can be used in a Wilson FG analysis and the calculated thermodymanic quantities can be decomposed into contributions from the various internal coordinates i.e. bond stretches, angle bends, and torsional or improper displacements for analysis and interpretation.

Intermolecular Contributions: Hydration

In a computer simulation on aqueous hydration (53) involving a dissolved molecule A and W water molecules, the total internal energy of solvation is determined by numerical integration of the expectation value

$$U^{hyd} = \langle E(\mathbf{X}^A, \mathbf{X}^W) \rangle \tag{11}$$

where $E(\mathbf{X}^A, \mathbf{X}^W)$ is the many-particle configurational energy for the system in configuration $(\mathbf{X}^A, \mathbf{X}^W)$ and the brackets indicate Boltzmann averaging. The difference in the internal energy between two conformations i and j of A is determined from separate simulations, one involving A in conformation i the other A in j, each subject to statistical uncertainty (noise) in the simulation. In taking this difference, we neglect the effect of the partial molar volume of A on conformation. The calculated U_i^{hyd}, U_j^{hyd}, ΔU^{hyd} and the various molecular distribution functions of the system can be partitioned into various contributions for analysis and interpretation of results (53,54).

The calculation of the free energy difference between i and j is carried out in a separate series of simulations. Here we define a "conformational transition coordinate" λ connecting i and j in conformation space by any desired path. The free energy as a function of this coordinate, a potential-of-mean force (55), is

$$
\begin{aligned}
A^{hyd} &= -kT \ln g(\lambda) \\
&= -kT \ln P(\lambda) + C
\end{aligned}
\tag{12}
$$

where $g(\lambda)$ is a spatial correlation function for conformations along λ in solution, $P(\lambda)$ is the probability of occurrence of the conformation λ in solution, and C is a constant. If we now define λ such that $\lambda=0$ corresponds to structure i and $\lambda=1$ corresponds to j, then

$$
\begin{aligned}
\Delta A_{ij}^{hyd} &= A^{hyd}(\lambda)_{\lambda=1} - A^{hyd}(\lambda)_{\lambda=0} \\
&= kT \ln [P(\lambda)_{\lambda=0}/P(\lambda)_{\lambda=1}]
\end{aligned}
\tag{13}
$$

the defining equation for the "probability ratio method" for free-energy simulations (26). The determination of ΔA_{ij}^{hyd} from computer simulation is thus reduced to the determination of $P(\lambda)$ for a conformational transition coordinate λ. Generally, we cannot expect the entire range of λ from 0 to 1 to be adequately sampled in a standard computer simulation involving associated liquids, and one finds only a fraction of the λ coordinate covered in a single simulation of reasonable, O(1000K), length. An approach to this problem was developed by Patey and Valleau (56) and Torrie and Valleau (57) and implemented for the study of molecular associations by Pangali et al. (58), and Ravishanker et al. (59), and for conformations of organic molecules in the neat liquid as reviewed by Jorgensen (60). The idea, called "umbrella sampling", is to use separate simulations to develop $P(\lambda)$ about points $\lambda_1, \lambda_2, \ldots$ such that the $P(\lambda)$ for successive points overlap. Then $P(\lambda)$ for the entire range of λ desired can be determined by a straightforward matching procedure.[58]

In a given simulation for point λ_k, the coordinate λ can be restricted to motion about λ_k by adding to the configurational energy $E(\mathbf{X}^A, \mathbf{X}^W)$ a constraining function $U_H(\lambda, \lambda_k)$, chosen here to be harmonic (58,59). This is essentially a non-Boltzmann sampling procedure, but Valleau and coworkers (56,57) showed that unbiased estimates of $P(\lambda)$ can be extracted by forming

$$P(\lambda) = \mathcal{N}\langle N(\lambda) \exp(U_H(\lambda)/kT)\rangle_H/\langle \exp(U_H(\lambda)/kT)\rangle_H \qquad (14)$$

where $N(\lambda)$ is the frequency of occurence of structure λ, N is a normalization factor, and the subscript H denotes an ensemble average determined with the harmonic constraining function added to the conventional expression for the configurational energy. Note that the harmonic form is only chosen for convenience. In fact, the non-Boltzmann bias can be introduced in tabular form (57) and its actual values can be determined in a self-consistent manner from the simulation itself (61,62). With $P(\lambda)$ thus determined, ΔA^{hyd} can be computed from Eq. (13), and in our problem relative free energies of hydration for the various conformations of AcAlaNHMe can be determined in computer simulation by umbrella sampling on conformational transition coordinates linking the various structures under consideration.

Various conformers of AcAlaNHMe differ in ψ and ϕ values, which suggest a two dimensional version of umbrella sampling. We initially explored this possibility, but found that developing overlapping distributions for the entire (ψ,ϕ) space would require a minimum of 36 separate simulations and would be prohibitively expensive. The additional difficulties of matching multidimensional umbrella distributions have been discussed by Mezei et al. (23). We subsequently chose a one dimensional procedure with the defining equation for λ taken to be

$$(\psi,\phi) = \lambda \, (\psi_j,\phi_j) + (1-\lambda) \, (\psi_i,\phi_i) \qquad (15)$$

Thus λ becomes a reduced, correlated conformational coordinate (63), analogous to the virtual bonds defined in complex conformational studies of nucleic acids (64).

In many instances a correlated change as described here may at some points place the solute in unphysical, sterically forbidden conformations. However in the calculation of the thermodynamics of hydration, the intramolecular energy is not directly included, i.e. temporarily assumed to be zero for this aspect of the calculation (65). Therefore intramolecular repulsions will not hinder the sampling of the system as a function of λ. The required free energy, a state function, can be obtained from any path on which overlapping $P(\lambda)$ can be constructed. Thus for the determination of differences in thermodynamic indices of hydration, the most convenient and economical correlated conformational coordinate can be chosen. Clearly the correlated conformational coordinate approach can be generalized to any number and any types of internal coordinates (63), and may be useful for diverse problems in structural chemistry and biochemistry.

In this approach, the internal coordinates of the solute not explicitly considered become rigid constraints, and a metric correction as described by Fixman (66) is required. The correction is expected to be within the statistical uncertainty of energy expectation values. Also, in taking conformational energy differences constraint errors tend to cancel (60). Thus we neglect herein errors due to constraints.

Calculations

The intramolecular thermodynamics of the Ala dipeptide were determined by a Monte Carlo calculation on the intramolecular degrees of freedom. Our procedure follows the essential ideas set forth by Levy, Rojas and Friesner (52) for sampling the intramolecular potential surface in the region of a minimum energy conformation. Details differ in that our Metropolis method is carried out based on cartesian displacements of atoms and our covariance matrix is collected in internal coordinates. The energy functions used in the calculations reported herein are those included in the program CHARMM (67) with local terms associated with bond lengths, bond angles, torsion angles and improper torsion angles together with non-local terms associated with van der Waals, electrostatic and hydrogen bonding interactions. The initial geometries for the AcAlaNHMe calculations for C_7, C_5, α_R and P_{II} are given above.

The Monte Carlo sampling involved uniform cartesian displacement along x, y and z coordinates for each atom of .02 Å, found by trial and error to produce ~50% acceptance rate in the Metropolis procedure. Calculation each of the conformers were limited to an area of ± 30 degrees on the (ψ, ϕ) conformational map with a constraint similar to that used for the gauche and trans conformers of butane by Levy et al. Simulation on each conformer involved at least 2000K configurations, where each configuration involved a displacement of all atoms of the molecule. Fuller details of the MC calculations on the internal modes of the dipeptide will be described in a forthcoming article by Ravishanker and Beveridge (25).

In our studies of the hydration of the Ala dipeptide, separate (T,V,N) ensemble liquid state Monte Carlo simulations were carried out on $[AcAlaNHMe]_{aq}$ in the conformations C_7, C_5, α_R and P_{II}, respectively. A modified Metropolis procedure incorporating the force bias method and preferential sampling for convergence acceleration was used. The system for study in each case comprised of 202 rigid molecules, one AcAlaNHMe and 201 water molecules. The temperature in the simulation was 25°C and the volume of the simulation cell was determined based on the density measurment of Bose and Hudt (68). The condensed phase environment of the system was provided by means of periodic boundary conditions. Convergence was monitored by the method of batch means (69). Full details of the Monte Carlo methodology used in our programs are given in a recent article by Mehrotra et al. (70).

The N-particle configurational energy of the system were in each case calculated using potential functions determined from quantum-mechanical calculations. For the water-water interactions, the MCY-CI(2) potential developed by Matsuoka et al. (71) was used. For the solute-water interactions, a potential function constructed from the 12-6-1 functional form and transferable parameters of Clementi et al. (72)

was obtained. Net atomic charges for atoms of AcAlaNHMe for the various conformations were determined by LCAO-SCF-MO calculations using Gaussian-80 (73) with the basis set described by Matsuoka et al. (74) and consistent with the potential function determination. Our strategy in choosing potential functions was to assume the best functions available to us at the time the study began for both intramolecular and intermolecular sets, and thus obtain the best available well defined estimates of the thermodynamic quantities under the circumstance. We note without further apologies certain incompatibilities between the intramolecular and intermolecular sets of functions, such as the electrostatic charges. At the time of this writing, new and improved energy functions have become available for both the intramolecular and intermolecular parts of the problem and our future studies on this problem, described briefly in the discussion section (vide infra), will deal with improvements in the quality of the simulations and the sensitivity of the simulations and results to choice of potential.

In the computer simulation, all potential functions for water-water interactions were truncated at a spherical cutoff of 7.75 Å, and solute-water interaction were treated under the minimum image convention. Solute-solute interactions are neglected and thus the system represents a state of infinite dilution. Each complete simulation involved an equilibration period of at least 500K configurations with ensemble averages formed over a succeeding 1500K segment of the realization.

The free energy determinations were carried out on correlated conformational coordinates—between the C_7 and α_R conformations and the C_7 and P_{II} conformations of the alanine dipeptide. Details of the simulations were maintained fully consistent with the internal energy calculations described above. The atomic charges on solute atoms used in the solute-water interaction potential were varied accordingly with λ:

$$q_n(\lambda) = \lambda \, q_{nj} + (1-\lambda) \, q_{ni} \qquad (16)$$

where q_n denotes the net atomic charge of atom n and q_{ni} and q_{nj} are the corresponding quantities for states i and j. In the umbrella sampling, the configurational energy of the system was supplemented with a harmonic constraining function $U_H(\lambda,\lambda_k)$ whose parameters were determined from trial runs to sample λ in overlapping segments.

Results and Discussion

The calculated thermodynamic quantities for the C_7, C_5, α_R and P_{II} conformations of the Ala dipeptide are collected in Table II, with all results presented relative to the internally hydrogen bonded C_7 form. The error estimates given represent two standard deviations computed with the method of batch means (69) (except for the intermolecular free-energy contributions which is discussed below). The intramolecular contributions are given in Table IIa. The transitions from C_7 to the C_5, α_R or P_{II} forms all involve disrupting the intramolecular hydrogen bond in C_7, and are indicated to be endothermic. The entropies for the corresponding transitions are

Table II

Calculated Thermodynamic Quantities for the C_7, C_5, α_R and P_{II} Conformation of AcAlaNHMe, relative to C_7

	C_7	C_5		α_R	P_{II}
a) Intramolecular					
ΔE_{ij}^{int}	0	+4.5	(3.5)	+4.7	+4.8
ΔU_{ij}^{int}	0	+4.1±.3	(4.1)	+4.2±.3	+3.9±.3
ΔS_{ij}^{int}	0	+1.0±1.3		+1.1±1.6	+1.6±1.2
$-T\Delta S_{ij}^{int}$	0	−0.3	(−0.5)	−0.3	−0.5
ΔA_{ij}^{int}	0	+3.8±.5	(3.6)	+3.9±.6	+3.4±.5
b) Intermolecular					
ΔU^{hyd}	0	+8.5±5.3		−6.6±5.6	−5.5±8.0
ΔU_{ij}^{SW}	0	−1.0±.8		−5.1±1.2	−6.2±1.2
ΔU_{ij}^{WW}	0	+9.5±5.3		−1.5±5.6	−6.2±0.7
ΔS_{ij}^{hyd}	0			−10.0±18.8	−7.7±26.7
$-T\Delta S_{ij}^{hyd}$	0			+3.0	+2.3
ΔA_{ij}^{hyd}	0			−3.6±.6	−3.2±.3
c) Total, intramolecular + intermolecular					
ΔU_{ij}	0			−2.4±5.6	−1.6±8.
ΔS_{ij}	0			−8.9±18.8	−6.1±26.7
$-T\Delta S_{ij}$	0			+2.7	+1.8
ΔA_{ij}	0			+0.3±.9	+0.2±.6

Legend: Energies are in Kcal.mol, entropies are in cal/mol deg.

all positive, indicating more disorder in weakly hydrogen bonded (if at all) C_5 form and in the open forms, α_R and P_{II}. The P_{II} form is indicated to be slightly more disordered than α_R. Examination of the (ψ,ϕ) map for AcAlaNHMe clearly reveals that the region of sterically allowed configuration space in the vicinity of α_R is distinctly smaller than that for P_{II}, consistent with the calculated entropies. The temperature-weighted entropy $-T\Delta S$ is seen in each case to be small, and the free energy changes are dominated by the internal energy change. Here as well, the statistical contribution is small relative to the intrinsic energy change. Thus the free energies of transition from C_7 to α_R, C_5 and P_{II} are all endergonic, with magnitudes corresponding primarily to the energy of disrupting the intramolecular hydrogen bond in the C_7 form.

We compare our results where possible with those obtained from the molecular dynamics study on AcAlaNHMe reported by Brady and Karplus (47) and Karplus et al. (75). Their calculated values for terms in Table I are given in parenthesis beside our corresponding entries. For C_5 relative to C_7, the results of the two studies are in reasonable agreement, allowing for possible slight discrepancies in characteristics such as molecular geometry, etc. between the respective studies. The agreement is in fact remarkably good considering the fact that independent methodologies, Monte Carlo and molecular dynamics, and independently written lenghty computer programs are involved. There are minor discrepancies in the relative statisical contribution to the internal energy for C_5 with respect to C_7, with ours going down by 0.4 kcal/mol while theirs goes up by 0.6 kcal/mol. This may be due to the slight discrepancies in the way in which conformations and conformational domains are defined in the two studies. Molecular dynamics values for the entropy of AcAlaNHMe in solution are also reported by Karplus et al., whereas our values refer only to the isolated molecule. However, their comparison of the intramolecular entropy differences shows that values for the isolated molecule and the molecule in ST2 water are of the same magnitude. We concur with Karplus et al. that the origin of the conformational entropy differences is not only due to fluctuations in the (ψ, ϕ) torsion angles, but from contributions from other internal coordinates and in their correlations with (ψ, ϕ) displacements.

The calculated internal energies of $[AcAlaNHMe]_{aq}$, total and partitioned into solute-water and water-water contributions, are given in Table IIb. In our results the calculated energetics can be expected to be somewhat sensitive to choice of intermolecular potential, an aspect to be dealt with in more detail in forthcoming work. Also, the statistical uncertainties in the relative energies are rather high since they are obtained as small differences between large relatively noisy numbers.

As described in Section II, the C_7 conformation seems to be well established as the predominate form of AcAlaNHMe in nonpolar solvents. For ease of interpretation, we thus present our results for aqueous solutions of AcAlaNHMe relative to those for the C_7 conformation, and consider those aspects of aqueous hydration which lead to stabilization and destabilization. Overall, we find the internal energy of the C_5 conformation of AcAlaNHMe to be destabilized and the α_R and P_{II} conformers to be stabilized in water. The α_R and P_{II} conformations in solution are relatively close energetically, 1.1 kcal/mol, with a preference for α_R indicated. The partitionings of total internal energy into solute-water and water-water contributions indicate that the origin of the destabilization of C_5 lies in the water-water term whereas the stabilization effects for α_R and P_{II} are found mainly in the solute-water term. The P_{II} conformation is in fact preferentially favored by solute-water interaction. The ultimate preference for α_R over P_{II} is due to the favorable contributions from the water-water solvent reorganization term. The calculated pressure values show a variation of 450 atm. Since this can give a pV correction of 0.2 kcal only or less, the effect of keeping the partial molar volumes constant is well within the statistical noise in the simulation.

To explore in more detail the structural origins of these results, the calculated first shell coordination numbers for C_7, C_5, α_R and P_{II} are collected in Table III. The average numbers of water molecules in the first shell hydration complex for $[AcAlaNHMe]_{aq}$ is found to range between 32 and 33.8, with the variation relative to C_7 only 1.5 molecules. The coordination can be partitioned into contributions from the various functional groups by the proximity criterion (54). Here we find ~85% of the first shell waters to be engaged in hydrophobic hydration of the methyl groups of AcAlaNHMe, with the remainder assigned to the hydrophilic hydration of the CO and NH groups. Overall there seems to be little change in the first shell

Table III

Calculated first shell coordination numbers for the C_7, C_5, α_R and P_{II} conformations of AcAlaNHMe, relative to C_7

	C_7(abs)	C_7	C_5	α_R	P_{II}
Total	33.5	0.	−1.5	−1.4	.3
CH_3	29.	0.	−2.3	−2.	−1.
CO	2.6	0.	.7	.6	1.2
NH	1.9	0.	.1	0.	.1

hydration numbers with conformation and the nature of the hydration effects on structure must depend on the way in which waters interact with the solute and each other rather than the numbers involved.

Since the major stabilizing effect for the α_R and P_{II} conformations appears in the solute-water term, we partitioned to solute water binding energies according to functional groups also by means of the proximity criterion. The results are shown in Table IV. The major factor contributing to the stabilization of the internal energy of hydration for α_R and P_{II} in aqueous solution is clearly due to energetic factors originating in the hydration of the carbonyl groups of the alanine dipeptide. Stereo views of the structures of a selected hydration complex for the C_7, α_R, and P_{II} forms of the AcAlaNHMe are shown in Figure 3.

Table IV

Calculated solute-water binding energies for the C_7, C_5, α_R and P_{II} conformations of AcAlaNHMe, relative to C_7

	C_7	C_5	α_R	P_{II}
CH_3	0.	4.7±0.3	0.5±0.5	1.3±0.5
CO	0.	−1.1±0.5	−8.2±0.7	−5.9±0.9
NH	0.	−2.0±0.8	2.8±0.8	−2.2±1.0

The original proposal by Avignon et al. (30) of the C_7 structure stabilized by a water bridge has not held up in subsequent experimental and theoretical studies. However, we have examined specific structures in the C_7 simulation to see the extent to which the characteristic bridging structure they proposed can be found. We observed a structure markedly similar to that of Figure 14.6 of their 1973 paper (30). Thus, the structure seems to be involved, but does not confer a special stability relative to that of α_R and P_{II}.

From the free energy simulations, the ΔA_{ij}^{hyd} between the C_7 and α_R conformations was found to be −3.6 kcal/mol (76) and the ΔA_{ij}^{hyd} between C_7 and P_{II} conformations was −3.2 kcal/mol. Our assessment of the noise level in these results is as follows:

observation of the λ ranges covered by successive 25K segments of the runs showed a grand cycle of period $\sim$500K, similar to the long-range correlations in the energy found for liquid water (77,78). Observation of the $P(\lambda)/P(\lambda')$ ratios on the longest run for λ and λ' used in matching this run show it to be accurate within 10-20% after 500K. Depending on the number of runs required, this results in an uncertainty factor of 1.5-2.0 in the $P(0)/P(1)$ values, which translates to a statistical error of 0.3-0.6 kcal/mol in ΔA_{ij}^{hyd}. Note in particular that the noise level in the free-energy difference is much reduced over the internal energy difference.

Combining the results on the free energy and internal energy of hydration leads to estimates of the entropy difference between C_7 and α_R to be $\Delta S_{ij}^{hyd} = -10.1$ cal/mol°C and for C_7 and P_{II}, $\Delta S_{ij}^{hyd} = -7.72$ cal/mol deg. The uncertainty in the calculated entropy of hydration, computed here as $(\Delta A_{ij} - \Delta U_{ij})/T$ is necessarily quite high, but the value indicates that the entropy change for hydration is negative for both C_7-α_R and C_7-P_{II} transitions, presumably due to the ordering of water in the regions of hydrophilic hydration in the open forms of AcAlaNHMe.

A comparison of the calculated results with experimental data for $[AcAlaNHMe]_{aq}$ requires knowledge of the total free energy change for the various transitions, the intramolecular contribution plus that due to hydration. Our estimates of the total free energy change for the various conformational transitions are given in Table IIc. The endergonic free energy change attributed to disrupting the intramolecular hydrogen bond in going from C_7 to α_R and P_{II} is seen to be compensated by the exergonic free energy of hydration for α_R and P_{II}, attributed via proximity analysis of the simulation results to the hydration of the exposed carbonyl oxygen in the open forms of AcAlaNHMe and mainly due to the internal energy rather than the entropy of hydration. Madison and Koppel (31) in spectroscopic studies cited above found evidence for both α_R and P_{II} conformations of AcAlaNHMe in water, with C_7 not necessarily excluded. Our calculations indicate that the conformational free energy of the C_7, α_R and P_{II} conformations are very similar, and that it is reasonable to expect them all to be thermally populated at ambient temperatures.

Thus the Ala dipeptide is indicated, now from both experimental studies and theoretical calculations, to be indeed a conformationally flexible molecule. The hydration of the carbonyl oxygens in aqueous solution stabilizes the open conformational forms α_R and P_{II} relative to C_7, and renders them thermally accesible at 25°C. It is likely that the well-known conformational flexibility of numerous small peptides in water arises as a consequence of hydration competing successfully with intramolecular hydrogen bonds to stabilize open conformational forms.

Acknowledgements

This research was supported by NIH Grant GM 24914 and by NSF Grant CHE-8203501 and a CUNY Faculty Research Award. Computer graphics was prepared on the NIH prophet System. The assistance of Ms. Yvonne Lesesne and Mr. Arald Jean-Charles is gratefully acknowledged.

References and Footnotes

21. B. Pullman and B. Maigret, in *Conformation of Biological molecules and Polymers*, E. Bergman and B. Pulman, eds., Academic Press Inc., New York (1973).
22. P.K. Mehrotra, M. Mezei and D.L. Beveridge, *Int. J. Quantum Chem.: Quantum Biology Symposium, 11,* 301 (1984).
23. M. Mezei, P.K. Mehrotra and D.L. Beveridge, *J. Am. Chem. Soc., 107,* 2239 (1985).
24. G. Ravishanker, M. Mezei and D.L. Beveridge, *J. Comp. Chem.,* submitted.
25. G. Ravishanker and D.L. Beveridge, in preparation for the *Israel Journal of Chemistry.*
26. M. Mezei and D.L. Beveridge, *Annals of the N.Y. Acad. Sci.,* in press.
27. C.R. Cantor and P.R. Schimmel, *Biophysical Chemistry*, Vol. 1, W.H. Freeman and Co., San Francisco (1980).
28. M. Avignon and J. Lascombe, in *Conformation of Biological Molecules and Polymers*, E. Bergman and B. Pulman, eds., Academic Press Inc., New York (1973).
29. Y. Harada and Y. Iitaka, *Acta Cryst., B30,* 1452 (1974).
30. M. Avignon, C. Garrigou-Lagrange and P. Bothorel, *Biopolymers, 12,* 1651 (1973).
31. V. Madison and K.D. Kopple, *J. Am. Chem. Soc., 102,* 4855 (1980).
32. D.A. Brandt and P.J. Flory, *J. Mol. Biol., 23,* 47 (1967).
33. F.A. Momany, R.F. McGuire, J.F. Yan and H.A. Scheraga, *J. Phys. Chem., 75,* 2286 (1971).
34. R. Hoffman and I. Imamura, *Biopolymers, 7,* 207 (1969).
35. I.H. Hillier and B. Robson, *J. theor. Biol., 76,* 83 (1979).
36. B. Robson, I.H. Hillier and M.F. Guest, *J. Chem. Soc. (Far. Trans. II) 74,* 1311 (1978).
37. P.J. Rossky, M. Karplus and A. Rahman, *Biopolymers, 18,* 825 (1979).
38. Z.I. Hodes, G. Nemethy and H.A. Scheraga, *Biopolymers, 18,* 1565 (1979).
39. A.T. Hagler, G. Nemethy and S. Lifson, *J. Am. Chem. Soc., 96,* 5319 (1974).
40. A.T. Hagler, P.S. Stern, R. Sharon, J.M. Becker, and F.J. Naidler, *J. Am. Chem. Soc., 101,* 6842 (1979).
41. C.M. Venkatachalam and S. Krimm, in *Conformation of Biological molecules and Polymers*, E. Bergman and B. Pulman, eds., Academic Press Inc., New York (1973).
42. V. Renugopalakrishnan, S. Nir, and R. Rein, *Biochimica and Biophysica Acta, 434,* 164 (1976); Renugopalakrishnan and R. Rein, in *Environmental Effects on Molecular Structure and Propertes*, B. Pullman, ed., D. Reidel, Dodrecht-Holland (1976).
43. G. Nemethy, Z.I. Hodes and H.A. Scheraga, *Proc. Natl. Acad. Sci., 75,* 5760 (1978);
44. P.J. Rossky and M. Karplus, *J. Am Chem. Soc., 101,* 1913 (1979).
45. A.T. Hagler, D.J. Osguthorpe and B. Robson, *Science, 208,* 599 (1980).
46. B.M. Pettitt and M. Karplus, *J. Am. Chem. Soc., 107,* 1166 (1985).
47. J. Brady and M. Karplus, in *Molecular Dynamics and Protein Structure*, J. Hermans, ed., Polycrystal Book Service, Western Springs, Ill. (1985); *J. Am. Chem. Soc.,* submitted.
48. M. Pettitt and M. Karplus, private communication; *Abstracts of the American Chemical Society Conference on Theoretical Chemistry,* Jackson Hole, Wy. June (1984).
49. F. Hirata, B.M. Pettitt and P. Rossky, *J. Chem. Phys., 78,* 4133 (1983) and references therein.
50. M. Karplus and J.N. Kushick, *Macromolecules, 17,* 1370 (1984).
51. R.M. Levy, M. Karplus, J. Kushick and D. Perahia, *Macromolecules, 17,* 1370 (1984).
52. R.M. Levi, O.L. Rojas and R. Freisner, *J. Phys. Chem., 88,* 4233 (1984).
53. S. Swaminathan, S. Harrison and D.L. Beveridge, *J. Am. Chem. Soc., 100,* 5705 (1978).
54. P.K. Mehrotra and D.L. Beveridge, *J. Am. Chem. Soc., 102,* 4287 (1980).
55. D.A. McQuarrie, *Statistical Mechanics*, Harper and Row, Inc., New York, (1976).
56. G.N. Patey and J.P. Valleau, *Chem. Phys. Letters, 21,* 297 (1973).
57. G.M. Torrie and J.P. Valleau, *J. Comp. Phys., 28,* 187 (1977).
58. C. Pangali, M. Rao and B.J. Berne, *J. Chem. Phys., 21,* 2982 (1979).
59. G. Ravishanker, M. Mezei and D.L. Beveridge, *Faraday Symp. Chem. Soc., 17,* 79 (1982).
60. W.L. Jorgensen, *J. Phys. Chem., 87,* 5304 (1983).
61. G.H. Paine and H.A. Scheraga, *Biopolymers 27,* 1391 (1985).
62. M. Mezei, *J. Comp. Phys.,* submitted.
63. D.L. Beveridge and M. Mezei, in *Molecular Dynamics and Protein Structure*, J. Hermans, ed., Polycrystal Book Service, Western Springs, Ill. (1985).

64. W. Olson, in *Topics in Nucleic Acid Structure*, Part II, S. Neidle, ed., MacMillan, London (1982).
65. Computing ΔA_{ij} with $E^{int}(\mathbf{X}^A)=0$ is formally equivalent to computing ΔA_{ij} with $E(\mathbf{X}^A,\mathbf{X}^w) = E(\mathbf{X}^w) + E^{int}(\mathbf{X}^A)$ and performing an umbrella sampling with $U'=U_H-E^{int}(\mathbf{X}^A)$.
66. M. Fixman, *Proc. Nat. Acad. Sci. USA, 71,* 3050 (1974); D. Chandler and B.J. Berne, *J. Chem. Phys.,* 5386 (1979).
67. B.R. Brooks, R.E. Bruccoleri, B.D. Olafson, D.J.States, S. Swaminathan and M. Karplus, *J. Comp. Chem., 4,* 187 (1983).
68. L. Bose and A. Hudt, *J. Chem. Thermody., 3,* 663 (1971).
69. W.W. Wood, in *Physics of Simple Liquids*, H.N.V. Temperley, J.S. Rowlinson, and G.S. Rushbrooke, Ed., North-Holland Publishing Co., Amsterdam, (1968); R.B. Blackman and J.W. Tuckey, *The measurement of power spectra*, Dover, New York (1958); R.B. Blackman, *Data smoothing and prediction*, Addison-Wesley, Reading, Mass. (1965).
70. P.K. Mehrotra, M. Mezei and D.L. Beveridge, *J. Chem. Phys., 78,* 3156 (1983).
71. O. Matsuoka, E. Clementi and M. Yoshemine, *J. Chem. Phys., 64,* 1351 (1976).
72. E. Clementi, F. Cavallone and R. Scordamaglia, *J. Amer. Chem. Soc., 99,* 5531 (1977).
73. J.S. Binkley, R.A. Whitehead, P.C. Hariharan, R. Seeger, J.A. Pople, W.J. Hehre and M.D. Newton, GAUSSIAN-80, Quantum-Chemical Program Exchange, adapted for IBM by S. Topiol and R. Osmond.
74. O. Matsuoka, C. Tosi and E. Clementi, *Biopolymers, 17,* 33 (1978).
75. M. Karplus, J. Brady, B. Brooks, J. Kuschick and M. Pettitt, in *Molecular Dynamics and Protein Structure*, J. Hermans, ed., Polycrystal Book Service, Western Springs, Ill. (1985).
76. Recent calculations using the adaptive procedure described in Ref.62 indicated that this value may be reduced, due to unsatisfactory convergence of the run around the minimum of $P(\lambda)$, where the convergence difficulties were already recognized (23). Work is in progress to check this point.
77. M. Rao, C.S. Pangali and B.J. Berne, *Mol. Phys., 37,* 1779 (1979).
78. M. Mezei, S. Swaminathan and D.L. Beveridge, *J. Chem. Phys., 71,* 3366 (1979).

Biomolecular Stereodynamics III, Proceedings of the Fourth Conversation in the Discipline Biomolecular Stereodynamics, State University of New York, Albany, NY, June 04-09, 1985, Eds., Ramaswamy H. Sarma & Mukti H. Sarma, ISBN 0-940030-14-4, Adenine Press, ©Adenine Press 1986.

Water-Hydronium Interaction:
A Computer Simulation

S.L. Fornili*†, M. Migliore† and M.A. Palazzo*
†Istituto Applicazioni Interdisciplinari Fisica, C.N.R.
*Istituto di Fisica, Universitá di Palermo
Via Archirafi 36, I-90123 Palermo, Italy

Abstract

Atom-atom pair potentials for water-hydronium interaction have been evaluated by SCF-LCAO-MO calculations using an extended basis set. Results of Monte Carlo simulations, which have been carried out for an infinitely dilute aqueous solution of hydronium at 300 K, agree with experimental data and shed light on structural characteristics of the water molecules surrounding the ion.

Introduction

Hydronium ion (H_3O^+) plays an important role in many proton transfer processes (1-6), which, in aqueous solutions, most likely are affected by the structural properties of the water molecules surrounding hydronium (1).

Experimental data concerning the condensed-phase behaviour of hydronium (7,8) suggest that hydronium, Na^+ and K^+ have similar hydration characteristics. Neutron and X-ray diffraction results (8), however, suggest that hydronium is hydrogen-bonded to four water molecules, whereas computer simulations indicate that a larger number of water molecules are in the first coordination shell of Na^+ and K^+ (9-12). Therefore, the anisotropic hydrogen bonding charcteristics of hydronium may play a remarkable role in its hydration features.

Extensive ab initio molecular orbital calculations (13) carried ˉout on gas-phase hydration of hydronium led to an O . . . O distance in $H_3O^+ \cdot 3H_2O$ of 2.53 A, in excellent agreement with the experimental value of 2.52 A (8), but the addition of a fourth water molecule into the first solvation shell did not yield any significant stabilization of $H_{11}O_5^+$ as compared with $H_9O_4^+ + H_2O$ at infinite separation.

In the present work the water-hydronium interaction at 300 K is studied by Metropolis-Monte Carlo (MC) simulation, based on ab initio atom-atom pair potentials, with the aim of gaining a reasonably accurate and statistically significant overall picture of the condensed-phase hydration features of hydronium, rather than a very accurate information for small clusters of water molecules surrounding the ion. Since the average lifetime of hydronium is 10^{-12}-10^{-11} sec (14), and the MC method is concerned with static properties of systems, the assumption is made that the water molecules in the ion field can reach equilibrium configurations before the solvated proton migrates from hydronium to neighbouring water molecules. Comparison of MC findings with experimental results validates this assumption.

Atom-atom pair potentials

The ab initio and simulation computations reported in the present work were carried out on a DEC VAX11/750 computer. The Hartree-Fock (SCF-LCAO-MO) calculations were performed using IBMOL program (15) with an extended gaussian basis set including polarization—(11s,7p,1d) for oxygen atoms and (6s,1p) for hydrogen atoms (16). Total energies of −76.336412 a.u. and −76.051314 a.u. have been obtained for hydronium ion and water molecule, respectively.

In Table I we report atomic coordinates, net charges (NC) (17), and molecular orbital valency state (MOVS) energies (18) for hydronium ion. Atomic coordinates were computed on the basis of optimized geometrical parameters reported in Ref. 19. Net charges and MOVS energies together with atomic numbers and number of bonds determine the class grouping of atoms (20,21). In the case of the hydronium ion, there is one class of oxygen and one class for the three equivalent hydrogen atoms.

Table I

Cartesian coordinates (in a.u.), MOVS energies (in a.u.) and net charges, NC, (in electrons) for hydronium ion

	x^a	y	z	MOVS	NC
O	0.0	0.0	0.0	1.0422	−0.4167
H1	1.528	−0.882	0.491	0.44615	0.4722
H2	0.0	1.764	0.491	0.44615	0.4722
H3	−1.528	−0.882	0.491	0.44615	0.4722

(a) Based on optimized geometrical parameters from Ref. 19.

The interaction energy surface between water and hydronium ion has been sampled at 342 points corresponding to different positions and/or orientations of a water molecule relatively to the hydronium ion. The counterpoise technique (20) was used to avoid basis set superposition error. At O . . . O distance of 4.67 a.u. an interaction energy minimum of −31.06 kcal/mol has been found, which agrees with the experimental enthalpy change, $\Delta H = -31.6$ kcal/mol, for the hydration reaction of hydronium ion (22).

The ab initio interaction energies for the H_3O^+ . . . H_2O system were expressed in terms of atom-atom pair potentials $E_k = \Sigma V(i,j;a,b)$, where k represents a specific calculation, i and j refer to oxygen or hydrogen atoms of the hydronium ion and water molecule, respectively, and a and b are the class number of i and j atom, respectively. We assumed as fitting functions several simple analytical forms consisting of combinations of powers of the distance $r(i,j)$ between atoms i and j, since more common choices (e.g., 12-6-1 combination) gave no satisfactory results. The best fitting was obtained with the following expression

$$V(i,j;a,b) = -A(a,b)/r(i,j)^3 + B(a,b)/r(i,j)^8 - C(a,b)/r(i,j)^4 + q(i)q(j)/r(i,j) \quad (1)$$

where $q(i)$ and $q(j)$ denote the net charges of atoms i and j, respectively. The constants $A(a,b)$, $B(a,b)$ and $C(a,b)$ are the fitting parameters reported in Table II; E_k, $r(i,j)$, and $q(i)$ and $q(j)$ are expressed in kcal/mol, A and electrons, respectively. Ion symmetry was taken into account in the fitting procedure. In Fig. 1 we report ab initio *vs.* fitted values of the interaction energy. The corresponding standard deviation is 1.5 kcal/mol, i.e., about 5% of the lowest interaction energy value (-31.06 kcal/mol).

Isoenergy contour maps

Isoenergy contour maps for the $H_3O^+ \ldots H_2O$ system were computed for a further check of the analytical potentials obtained by the above mentioned fitting procedure. As it is well known (20), the isoenergy contour maps show the interaction energy between solute and one water molecule, whose oxygen atom is placed at the intersection points of a fine grid (in the present case, $\Delta x = \Delta y = 0.25$ a.u.). The orientation of the hydrogen atoms of water is optimized by minimizing the water-solute interaction energy. In Fig. 2, typical isoenergy contour maps are reported corresponding to the planes $z = 0.5$ a.u. (approximately, the z-coordinate of the hydrogen atoms of hydronium) and $x = 0.0$ a.u., respectively. Three energy minima are clearly shown in Fig. 2A, which correspond to the positions of the hydronium hydrogen atoms. Further, Fig. 2B shows that in the region close to the oxygen atom of hydronium, the ion-water attraction is much weaker than in the region close to the hydrogen atoms. These findings, which agree with previous calculations (13), are confirmed by analyses of MC results presented in the next section.

Table II
Fitting parameters A, B and C for atom-atom pair potentials

$i-j^a$	A^b	B	C
O-O	-479.34	15546.	893.61
O-H	379.47	5103.60	-760.50
H-O	157.44	8.8981	-144.06
H-H	-76.349	217.85	99.885

(a) Atoms under i and j belong to hydronium ion and water molecule, respectively;

(b) A, B and C constants are expressed in units such that $V(i,j)$ of eq. 1 is in kcal/mol if $r(i,j)$ is expressed in A, and $q(i)$ and $q(j)$ in electrons.

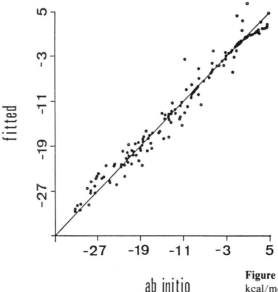

Figure 1. Fitted *vs.* ab initio energy values in kcal/mol for water-hydronium ion interaction.

Monte Carlo simulation

Metropolis-Monte Carlo simulations (23) were carried out at a temperature of 300 K. One hydronium ion and 215 water molecules were involved, which were confined within a cubic volume of 18.6 A side length. The condensed phase environment was simulated by periodic boundary conditions. The configurational energy of the system was computed by assuming MCY-CI potentials (24) for water-water interaction and the above described atom-atom pair potentials for the ion-water interaction. Both water-water and ion-water interaction energies were computed under the minimum image convention.

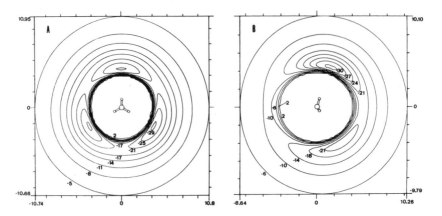

Figure 2. Isoenergy contour maps for water-hydronium ion pair-potentials in the planes A) z=0.5 and B) x=0.0. Coordinates and interaction energies are expressed in a.u. and kcal/mol, respectively.

After equilibration (about 2000 k configurations), energy and coordinate values for 1000 k configurations were stored and subsequently analyzed. The average values of water-water and water-ion interaction energies are -35.76 ± 0.07 kJ/mol and -5.09 ± 0.02 kJ/mol, respectively.

In Fig. 3 we report $g_{OO}(r)$ and $g_{OH}(r)$ radial distribution functions concerning the distribution of the oxygen and hydrogen atoms of water around the oxygen atom of hydronium. In the top of Fig. 3, the total radial distribution functions are shown. In the middle and in the bottom insets, the partial radial distribution functions are reported, which correspond to $z \geq 0$ (where the hydrogen atoms of hydronium are located) and $z < 0$, respectively. By comparing the total radial distribution functions with analogous findings from previous simulations on aqueous solutions of Li^+, Na^+ and K^+ ions (9-12), we can make the following remarks: i) as for these alkali cations, oxygen atoms of water are closer to hydronium ion than hydrogen atoms, due to the net positive charge of hydronium. ii) Three hydration shells are clearly shown in Fig. 3. This feature is also present in K^+ case, but it has not always been found for Li^+ (11). iii) The maximum of $g_{OO}(r)$ occurs for hydromium ion at $r = 2.48 \pm 0.02$ A, which is an intermediate value between those previously reported for Na^+ and K^+ ions (9-12). This result agrees within 1.6% with the experimental value obtained for the O . . . O distance in aqueous solutions of HCl at 20 C (8). The first minimum of $g_{OO}(r)$ occurs at $r = 3.0$ A. The corresponding running integration number is 4.1, which is also in good agreement with experimental data (8). These water molecules are preferentially located in the $z \geq 0$ region, as it is indicated by the partial radial distribution functions reported in the middle and bottom insets of Fig. 3. This feature is further evidenced by Fig. 4, in which the water molecules closest to hydronium are shown, as selected from a representative MC configuration. As one can see, the competition of the fourth water molecule for bonding to the hydronium hydrogens, causes relevant hydrogen-bond distortion.

Conclusions

The following conclusions can be drawn from the present work. i) Although the hydronium ion is unstable, still the MC simulation approach, which is concerned with static properties of systems, gives valuable results to be compared with experimental findings, as it has been shown in the discussion of the previous section. This situation is probably brought about by a reorientation rate of water molecules in the strong ion field, which is higher than the average rate of molecular reorientation in liquid water (25). A similar observation has been made for a previous molecular dynamics simulation (26). ii) The analyses of the MC results show that the size of the hydronium ion, as seen by water molecules, is intermediate between the sizes of Na^+ and K^+ ions, in agreement with experimental results (7,8). iii) About four water molecules are shown to compete for bonding with hydrogen atoms of hydronium, in agreement with X-ray and neutron diffraction data (8). The arrangement of the water molecules around the hydronium ion is quite different from what could be expected by space-filling considerations. Further analysis of MC data is in progress to study the structure and energetics of the water molecules surrounding

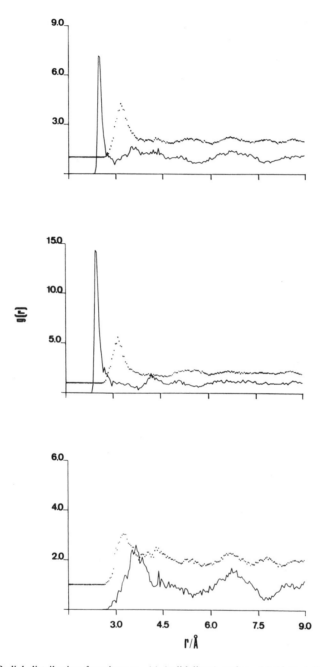

Figure 3. Radial distribution functions $g_{OO}(r)$ (solid lines) and $g_{OH}(r)$ (dotted lines) of oxygen and hydrogen atoms of water around the oxygen atom of the hydronium, for an infinitely dilute aqueous solution of hydronium ion at 300 K. The $g_{OH}(r)$ curves are shifted upwards one unit for legibility. Top inset, total distribution functions; middle and bottom insets, partial distribution functions for $z \geq 0$ and $z < 0$, respectively.

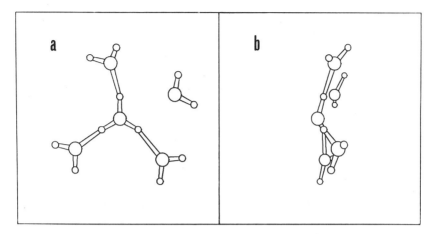

Figure 4. The closest water molecules surrounding hydronium, as selected from a MC representative configuration: a) top view, b) side view.

hydronium, their influence on proton migration from hydronium to neighbouring water molecules and on proton transport via the connectivity pathways of hydrogen-bonded water molecules (27).

We wish to thank Dr. G. Corongiu for the VAX version of IBMOL, and Prof. M.U. Palma, Prof. M.B. Palma-Vittorelli and Prof. L. Cordone for stimulating discussions. Technical assistance by Mr. M. Lapis and Mr. A. La Gattuta is acknowledged. The present work was performed at Istituto Applicazioni Interdisciplinari Fisica (CNR). Fundings from MPI 60% and CRRNSM are also acknowledged.

References and Footnotes

1. M. Eigen, *Angew. Chem., Int. Ed. Engl. 3,* 1 (1964).
2. A. Dalgarno and J.H. Black, *Rep. Progr. Phys. 39,* 573 (1976). Amsterdam, p. 195 (1967).
4. L.J. Denes and J.J. Lowke, *App. Phys. Lett. 23,* 130 (1973).
5. R.K. Sen, E. Yeager and W.E. O'Grady, *Ann. Rev. Phys. Chem. 26,* 287 (1975).
6. a) F.R. Salemme in Biochemistry, Ed., G. Zubay, Addison-Wesley, Reading, Massachusetts, p. 115 (1983); b) R. Breslow, *ibid.,* p. 158; c) J. Marmur, *ibid.,* p. 681.
7. a) R.W. Gurney, *Ionic Processes in Solutions,* McGraw-Hill, New York, p. 168 (1953); b) T. Erdey-Gruz, *Transport Phenomena in Aqueous Solutions,* Adam Hilger, London, p. 121 (1974).
8. R. Triolo and A.H. Narten, *J. Chem. Phys. 63,* 3624 (1975).
9. E. Clementi and R. Barsotti, *Chem. Phys. Lett. 59,* 21 (1978).
10. P. Bopp, W. Dietz and K. Heizinger, *Z. Naturforsch. 34a,* 1424 (1979).
11. M. Mezei and D.L. Beveridge, *J. Chem. Phys. 74,* 6902 (1981).
12. R.W. Impey, P.A. Madden and I.R. McDonald, *J. Phys. Chem. 87,* 5071 (1983).
13. M.D. Newton, *J. Chem. Phys. 67,* 5535 (1977).
14. T. Erdey-Gruz, *Transport Phenomena in Aqueous Solutions,* Adam Hilger, London, p. 327 (1974).
15. a) E. Clementi and D.R. Davis, *J. Comp. Phys. 1,* 223 (1966); b) E. Ortoleva, G. Castiglione and E. Clementi, *Comput. Phys. Commun. 19,* 337 (1980).

16. G. Bolis, E. Clementi, D.H. Wertz, H.A. Scheraga and C. Tosi, *J. Am. Chem. Soc. 105,* 355 (1983).

17. R.S. Mulliken, *J. Chem. Phys. 23,* 1833 (1955).

18. E. Clementi and A. Routh, *Intl. J. Quantum Chem. 6,* 525 (1972).

19. H.Z. Cao, M. Allavena, O. Tapia and E.M. Evleth, *Chem. Phys. Lett. 96,* 458 (1983).

20. E. Clementi, in *Lecture Notes in Chemistry, Vol. 19,* Springer, Berlin (1980).

21. a) G. Corongiu, S.L. Fornili and E. Clementi, *Intl. J. Quantum Chem.: Quantum Biol. Symp. 10,* 277 (1983); b) S.L. Fornili, D.P. Vercauteren and E. Clementi, *Biochim. Biophys. Acta 771,* 151 (1984); c) K.S. Kim, D.P. Vercauteren, M. Welti, S.L. Fornili and E. Clementi, *Biochim. Biophys. Acta,* in press.

22. A.J. Cunningam, J.D. Payzant and P. Kebarle, *J. Am. Chem. Soc. 94,* 7627 (1972).

23. a) N. Metropolis, A.W. Rosenbluth, M.N. Rosenbluth, A.H. Teller and E. Teller, *J. Chem. Phys. 21,* 1087 (1953); b) S. Romano and E. Clementi, *Gazz. Chim. Ital. 108,* 319 (1978).

24. O. Matsuoka, E. Clementi and M. Yoshimine, *J. Chem. Phys. 64,* 1351 (1976).

25. a) D. Eisenberg and W. Kauzmann, *The Structure and Properties of Water,* Oxford University Press, Oxford, p. 205 (1969); b) J.A. Glasel in *Water A comprehensive treatise,* Ed., F. Franks, Plenum, New York, Vol. 1 (1972), Chap. 6; c) J.B. Hasted, ibid., Chap. 7.

26. M. Rao and B.J. Berne, *J. Phys. Chem. 85,* 1496 (1981).

27. a) A. Geiger, F.H. Stillinger and A. Rahman, *J. Chem. Phys. 70,* 4185 (1979); b) F.H. Stillinger, *Science 209,* 451 (1980).

Biomolecular Stereodynamics III, Proceedings of the Fourth Conversation in the Discipline Biomolecular Stereodynamics, State University of New York, Albany, NY, June 04-09, 1985, Eds., Ramaswamy H. Sarma & Mukti H. Sarma, ISBN 0-940030-14-4, Adenine Press, ©Adenine Press 1986.

The Effect of Different Force Fields on the Probable Hydration Sites of Urea*

Françoise Vovelle[a] and Julia M. Goodfellow[b]

[a]Centre de Biophysique Moleculaire, et Université d'Orléans,
1A Avenue de la Recherche Scientifique,
45045 Orléans Cedex, France
[b]Department of Crystallography, Birkbeck College,
University of London, Malet Street, London, WC1E 7HX, UK

Abstract

Energy minimization techniques have been employed to find the primary hydration sites around solutes. We have used the urea molecule as a model system to investigate the effects of different force fields on the geometry and energy of interaction of water molecules with the carbonyl and amino groups. Analysis of our results (obtained with three different sets of both non-bonded interaction coefficients and partial atomic charges for urea) has shown that (i) the water model plays the major role in determining which of the possible primary hydration sites are occupied; (ii) the size of the partial atomic charges of the solute molecule affects the magnitude of the interaction energies and relative energies of different sites; (iii) the non-bonded interaction coefficients appear to correlate with the distances of the hydrogen bond between the primary hydration site and the solute molecule; (iv) although many combinations of parameters (water model, partial atomic charges and non-bonded interaction coefficients) lead to primary hydration sites with stereochemically acceptable hydrogen bonding geometries, a few do not. We have also calculated the energy of interaction of a hydrocarbon (methane) with urea, with the four water models and with another hydrocarbon in an initial attempt to understand more about the properties of the urea molecule in denaturation and increasing the solubility of hydrocarbons.

Introduction

Several computational methods are commonly used to study the hydration of both small and large solutes. These include molecular dynamics and Monte Carlo simulations of aqueous solution (1,2) and crystal hydrates of amino acids (3), peptides (4), proteins (5) and nucleic acids (6). Such calculations take account of both the individual water solute interactions and the surrounding water . . . water interactions. A simpler technique is that of energy minimization (7) in which the optimum energy and geometry of interaction between one water molecule and the solute is calculated albeit for many possible water molecule positions on the surface of the solute. This

*This paper was presented at the Fourth Conversation in Biomolecular Stereodynamics, SUNY at Albany, June 04-07 1985.

method is useful in that it enables a large number of such calculations to be undertaken using relatively small amounts of processor time compared with the computationally intensive molecular dynamics or Monte Carlo simulations. However, the drawback is that only the primary hydration sites can be predicted i.e. those which depend mainly on the interaction with the solute and not on interactions with other solvent molecules. We have chosen to use the energy minimization method in order to undertake many calculations with different parameters.

However, all these methods have a common element namely the calculation of the potential energy between one molecule and the atoms of the solute molecule. Both the energy and geometry of the optimum hydration sites will depend on the combination of models used to describe the interaction. There are many models for water itself (8) based usually on properties of the water dimer and we have chosen representative models from the four-point charge, three-point charge and polarisable electropole categories. The properties of the solute atoms are often described in terms of a partial atomic charge and two non-bonded interaction coefficients (one repulsive and one attractive). The values of these parameters come from quantum mechanical calculations and least squares fitting to crystal data. However, there are several different sets of values available in the literature (9).

In order to look at the effects of these different models, we have chosen the urea molecule (10) as the solute molecule. This is because of its small size and because it contains two types of potential hydrogen bonding groups found in macromolecules i.e. carbonyl and amine groups. However, the interactions of urea are of great interest in themselves because of its well-known denaturing effect on proteins (11) and the fact that the solubility of hydrocarbons in aqueous solutions can be increased in the presence of urea (12). There is still no agreement on the mechanism of action but many hypotheses involve the possible structure breaking or structure promoting effect that urea has on the water . . . water interactions (13). Recently, Tanaka et al (14) and Kuharski and Rossky (15) have independently undertaken molecular dynamics simulations of aqueous solutions of urea. The former authors find that the urea molecule enters aqueous solution without any appreciable distortion to the water structure. Kuharski and Rossky (15) come to similar conclusion in that the properties of the water molecules in the solvation region show only small differences compared with water molecules in the bulk and no substantial perturbations were found in the water . . . water interactions. Thus, the interaction of water molecules at the primary hydration sites of urea may be directly relevant to its mechanism of action.

Thus the main aim of this study is to look at the effect of different combinations of models (that might be used to represent the water . . . water interaction) on the energy and geometry of the optimum (i.e. minimum energy) hydration sites around the urea molecule. Further calculations have indicated the interaction energy of urea with a simple hydrocarbon in order to look at the relative energies of interaction of the various components when urea is added to aqueous solutions of hydrocarbons.

Materials and Methods

Water Models

We have used five different models to represent the interaction of the water molecule. These fall into two categories namely point charge and polarisable electropole models. Of the former, we have employed the three point charge model (TIPS2) of Jorgensen (16), the four point charge model (ST2) of Stillinger and Rahman (17) and a modified version of the four point charge model known as EMPWI (18). The latter model includes hydrogen . . . hydrogen non-bonded interactions and an oxygen . . . hydrogen repulsive term. The polarisable electropole model is that developed by Gellatly et al (19) and includes electropole terms up to the dipole-octopole interaction and an inverse 6/9 dependence for the non-bonded interaction (PE[DO]). As this model attempts to account for non-pair additive effects via a polarisable dipole formalism, our calculations have included a range of dipole moments from 1.85 to 3.05 Debye which is consistent with experimental and calculated data (20). Data presented in this paper is based on a dipole of 2.45 Debye. In the present calculations it is not possible to compute the increment to the dipole moment because only the field due to the urea molecule is known and no contribution to the field due to surrounding solvent molecules is included. Some characteristics of the water dimer energies are given for these models in Table I.

Table I

Water Model	Energy[a] Kcal·mol^{-1}	Dimer Characteristics Distance[b] (Å)	Angle[c] (degrees)
ST2	−6.84	2.85	54
TIPS2	−6.20	2.79	47
PE(DO)	−4.5	2.9	60
EMPWI	−7.8	2.75	51

[a]Minimum energy.
[b]Separation of water oxygen at minimum energy configuration.
[c]Angle between the direction of the dipole moment of the donor water molecule and the vector between the donor and acceptor water molecules.

Solute Models

The form of the force field for the urea atoms has been kept fixed and includes partial atomic charges and non-bonded interaction coefficients for each of the eight urea atoms. However, various parametrizations have been used for both the charges and the non-bonded interaction coefficients. The latter parameters have been taken from three main sources namely Momany et al (21), Lifson et al (22) and Weiner et al (23) and are given in Table II. The partial atomic charges have been calculated by using (a) the relatively simple semi-empirical CNDO/2 method,

Table II

Atom	(i)[b]		(iia)[c]		(iib)[c]		(iii)[d]	
	VR	VA	VR	VA	VR[e]	VA	VR	VA
N	732,540.	363.	2,271,000.	1230.	86,900.	2020.	540,675.	588.
H	8,430.	45.5	0.0	0.0	0.0	0.0	82.0	2.6
C	1,048,980.	767.	3,022,000.	1340.	12,500.	355.	789,954.	612.
O	170,190.	369.	275,000.	502.	45,800.	1410.	250,584.	429.

Non-Bonded Interactions Coefficients[a]

[a]Values given in Kcal·mol^{-1} Å^{-12} for repulsive coefficient (VR) and Kcal·mol^{-1} Å^{-6} for attractive coefficient (VA).
[b]Taken from Momany *et al.* (21).
[c]Taken from Lifson *et al.* (22).
[d]Taken from Weiner *et al.* (23).
[e]Kcal·mol^{-1} Å^{-9} units.

(b) taken from an intrinsically more accurate calcualtion by Orita and Pullman (24) using the SCF LCAO procedure with an (7s,3p/3s) atomic Gaussian basis contracted to a minimum basis set and (c) calculated by J.O. Baum (Birkbeck College) from the SCF electrostatic potential using ATMOL with STO-3G basis set and corrected to give the experimental dipole moment for urea. The latter charges are presumably the more realistic. These charges are listed in Table III.

The water . . . solute interaction is presented as (i) a coulombic interaction between the point charges or electropole terms on the water molecule and the partial atomic charges on the urea and (ii) a non-bonded Lennard Jones interaction (normally with an inverse 6-12 distance dependence).

Energy Minimization

The energy minimization subroutine is based on the method of Powell (25) and comes from the Harwell subroutine library. Initially, iso-energy contour maps are

Table III

Atom	I[b]	II[c]	III[d]
O	−0.42	−0.42	−0.59
C	0.46	0.54	0.81
N	−0.28	−0.98	−0.73
H$_c$	0.12	0.42	0.29
H$_t$	0.14	0.50	0.33
Dipole moment (Debye)	3.14	4.94	4.56

Partial Atom Charges[a]

[a]Fractional electronic charges.
[b]Calculated using CNDO/2 method.
[c]Taken from Orita and Pullman (24).
[d]Calculated using ATMOL with STO-3G basis set and corrected to the experimental dipole moment (J.O. Baum, Birkbeck College).

calculated in the plane of the urea molecule by calculating the energy of interaction for a water molecule and the urea molecule placed at fixed grid points and contouring between these points. The approximate positions of the solvent oxygen atoms of the local minima can be estimated from such maps and input to a full minimization program which optimizes the interaction energy with respect to three rotational and three translational degrees of freedom. The geometry of the water . . . urea interaction is then analysed at each of the final local minima.

Results and Discussion

Iso-Energy Contour Maps

Comparison of the iso-energy contour maps with each other enable us to recognise differences in the energy of interaction between the solvent and solute molecules. Each map is calculated using a combination of models which involve a choice of three parameter sets: (1) partial atomic charges from sources I, II or III (see Table III), (2) non-bonded interaction coefficients from three sources (i), (ii) or (iii) (see Table II) and, of course, (3) a water model (see Table I). In order to look at the effects of any one of these parameter sets, we have performed a series of calculations with the other parameter sets constant. The differences in the iso-energy contour maps due to the different choices of parameter sets are illustrated in Figures 1 to 3. Figure 1 shows the effects of changing the non-bonded interaction coefficients while (a) the partial atomic charges are those from CNDO/2 calculations and the ST2 water model is used (Figure 1A) and (b) the partial atomic charges are those from reference (24) and the TIPS2 water model is used (Figure 1B). Figure 2 illustrates the differences to the iso-energy contour maps as a function of changing the water model for calculations performed with coefficients from Momany et al (21) and (a) CNDO/2 charges (Figure 2A) and (b) charges from set III (Figure 3B). Finally we looked at changes in the iso-energy contour maps due to a change in partial atomic charges for the solute atoms using (a) the EMPWI water model and non-bonded interaction parameters from Momany et al (21) (Figure 3A) and (b) the PE(DO) water model with non-bonded coefficients from set (iib) (Figure 3B).

Inspection of these maps shows quite distinct differences in the position, shape and depth of the local minima in the plane of the molecule. The combination of parameters used in the calculations make it difficult to find simple correlations between the use of one set and the effects on the iso-energy contour maps. However, it seems that the main effect is due to the choice of water model. For example, in Figure 1 map (i), (ii) and (iii) calculated using the ST2 water model from Figure 1A are quite similar to each other. Similarly maps (i), (ii) and (iii) from Figure 1B calculated using the TIPS2 model are more similar to each other than to those from Figure 1A. The most striking differences can be seen in Figure 3 where the maps calculated with EMPWI water model (Figure 3A[i], [ii] and [iii] are quite distinguishable from those calculated with the PE(DO) water model (Figure 3B[i], [ii] and [iii]. In Figure 2, it is maps calculated with similar water models that look alike e.g. Figure 2A(i) and Figure 2B(i). Because of the difficulties in making quantitative

1 A   **1 B**

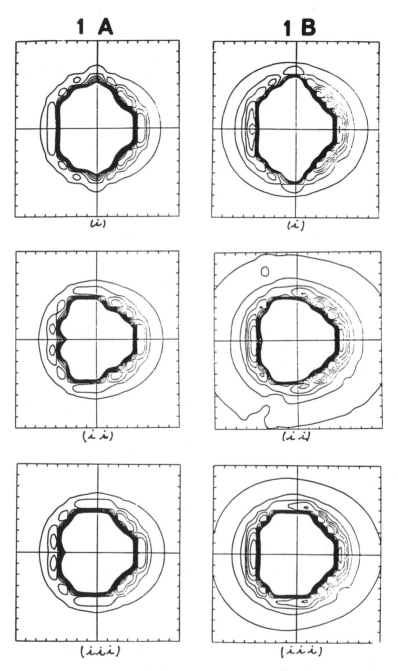

Figure 1. Iso-energy contour maps showing the interaction of one water molecule with an urea molecule using (A) partial atomic charges calculated using the CNDO/2 method and the ST2 water model and (B) partial atomic charges taken from the studies of Orita and Pullman (1977) and the TIPS2 water model. The figure illustrates the changes in the energy contours due to the use of different non-bonded interaction parameters taken from (i) Momany et al (1974), (ii) Lifson et al (1979) and (iii) Weiner et al (1984).

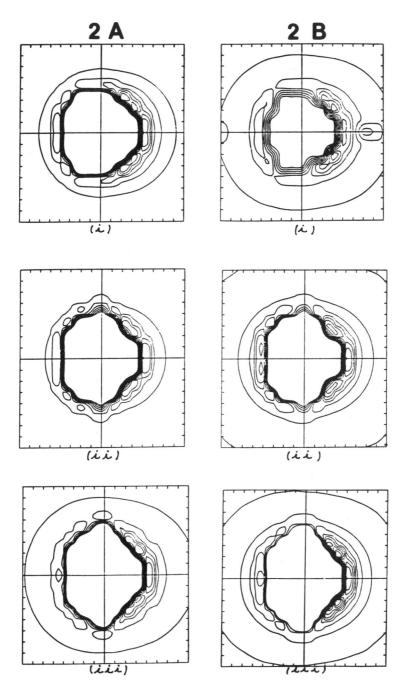

Figure 2. Iso-energy contour maps showing the interaction of one water molecule with an urea molecule using non-bonded interaction coefficients from Momany et al (1974) and (A) partial atomic charges calculated using the CNDO/2 method and (B) partial atomic charges from set III (table 3). This figure illustrates the changes in the energy contours due to the use of different water models including (i) the EMPWI model, (ii) the ST2 model and (iii) the TIPS2 model.

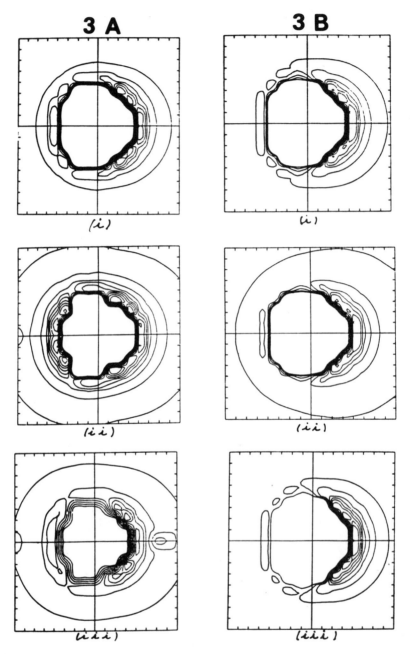

Figure 3. Iso-energy contour maps showing the interaction of one water molecule with an urea molecule using (A) the EMPWI water model with non-bonded interaction coefficients from Momany et al (1974) and (B) the PE(DO) water model and the non-bonded interaction coefficients from Lifson et al (1979). This figure illustrates the changes in the energy contours due to the use of different partial atomic charges on the atoms of the urea molecule from (i) CNDO/2 calculations, (ii) the studies of Orita and Pullman (1977) and (iii) the ab initio calculations corrected to give the correct experimental dipole moment for urea.

comparisons of iso-energy contour maps, it is necessary to proceed to full three-dimensional minimization as described in the next section.

Primary Hydration Sites

The positions of local minima were estimated from these iso-energy contour maps and were input to a full energy minimization procedure which allowed for translational and rotational freedom relative to the urea molecule. The resultant minima, from this full minimization procedure, were called primary hydration sites. Possible sites are illustrated in Figure 4 and are labelled S_0, S_1, S_2, S_3 and S_4 to indicate the position of the oxygen atom of the water molecule. S_0 and S_1 would correspond to interactions with the carbonyl group while S_2 and S_4 refer to interaction with the amino group. S_3 is a bridging site with the water molecule within hydrogen bonding distance of both the carbonyl and amine groups. Symmetry related sites have been omitted for clarity. Stereochemical constraints would prohibit some sites being occupied simultaneously (e.g. S_3 and S_4 on the same side of the carbonyl group).

Results of these full minimizations are given in terms of both the minimum energy of interaction between the water and urea molecules (Table IVA, B, and C) and the

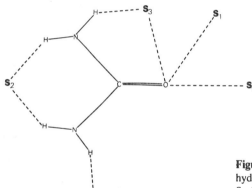

Figure 4. A schematic illustration of the possible hydration sites of urea. Each site is labelled S_0, S_1, S_2, S_3 and S_4 and the possible hydrogen bonds are indicated by dashed lines.

hydrogen bond distances in Table VA, B and C. Several points can clearly be seen from these tables.

1. The range of values of the energy is large from as high as -2.92 to as low as -18.40 Kcalmol^{-1}. On average, the highest values occur with the CNDO/2 charges set (I) (Table IVA). The lowest values occur with set III charges for site S_0, S_1 and S_3 (Table IVC) and with set II charges for site S_2 and S_4 (Table IVB). The latter sites are hydrogen bonded only to the amine group while the former sites (S_0, S_1 and S_3) involve interaction with the carbonyl oxygen atom of urea.

Table IVa

	Water model	Energy (Kcal mole^{-1}) S_0	S_1	S_2	S_3	S_4
(A)	ST2					
	(i)	—	−6.03	(−2.31)	—	—
	(ii)	—	—	−3.82	−5.56	—
	(iii)	—	—	−4.04	−5.95	—
(B)	TIPS2					
	(i)	−6.04	−6.43	(−4.21)	—	—
	(ii)	−5.65	−5.97	−2.92	—	—
	(iii)	−5.54	—	−3.24	−6.12	—
(C)	EMPWI	—	—	−4.31	−8.55	—
(D)	PE(DO)	—	−7.25	—	—	—

Table IVb

	Water model	Energy (Kcal mole^{-1}) S_0	S_1	S_2	S_3	S_4
(A)	ST2					
	(i)	—	—	−5.89	−8.59	—
	(ii)	—	—	−12.01	—	−18.4
	(iii)	—	—	−10.20	−12.22	—
(B)	TIPS2					
	(i)	−7.11	—	(−4.37)	−8.86	—
	(ii)	−6.46	—	−5.86	−8.83	—
	(iii)	−6.56	—	−6.30	−9.68	—
(C)	EMPWI	—	—	−10.39	−11.61	—
(D)	PE(DO)	—	−8.49	—	—	—

Table IVc

	Water model	Energy (Kcal mole^{-1}) S_0	S_1	S_2	S_3	S_4
(A)	ST2					
	(i)	—	—	−4.93	−9.14	—
	(ii)	—	—	−8.14	−9.06	—
	(iii)	—	—	−7.99	−9.85	—
(B)	TIPS2					
	(i)	−8.82	−9.63	(−3.88)	—	—
	(ii)	−8.04	—	−5.06	−9.00	—
	(iii)	−8.20	—	−5.47	−9.57	—
(C)	EMPWI	—	—	−8.33	−13.78	—
(D)	PE(DO)	—	−10.77	—	—	—

Table Va

	Water model	S_0	S_1	S_2	S_3	S_4
				Hydrogen bond distance (Å)		
(A)	ST2					
	(i)	—	2.55	(3.52)	—	—
	(ii)	—	—	3.08	2.63;3.34	—
	(iii)	—	—	3.06	2.63;3.02	—
(B)	TIPS2					
	(i)	2.67	2.67	(3.71)	—	—
	(ii)	2.74	2.75	3.35	—	—
	(iii)	2.73	—	3.28	2.74;3.26	—
(C)	EMPWI	—	—	3.08	2.75;3.41	—
(D)	PE(DO)	—	2.85	—	—	—

Table Vb

	Water model	S_0	S_1	S_2	S_3	S_4
				Hydrogen bond distance (Å)		
(A)	ST2					
	(i)	—	—	3.35	2.53;3.15	—
	(ii)	—	—	2.50	—	2.37
	(iii)	—	—	2.84	2.83;2.69	—
(B)	TIPS2					
	(i)	2.65	—	(3.59)	2.65;3.31	—
	(ii)	2.72	—	3.18	2.78;3.05	—
	(iii)	2.71	—	3.15	2.73;2.96	—
(C)	EMPWI	—	—	2.85	2.73;2.95	—
(D)	PE(DO)	—	2.82	—	—	—

Table Vc

	Water model	S_0	S_1	S_2	S_3	S_4
				Hydrogen bond distance (Å)		
(A)	ST2					
	(i)	—	—	(3.40)	2.48;3.23	—
	(ii)	—	—	3.25	2.68;2.82	—
	(iii)	—	—	2.91	2.58;2.80	—
(B)	TIPS2					
	(i)	2.59	2.59	(3.62)	—	—
	(ii)	2.66	—	3.22	2.67;3.29	—
	(iii)	2.65	—	3.18	2.66;3.09	—
(C)	EMPWI	—	—	2.91	2.66;3.22	—
(D)	PE(DO)	—	2.74	—	—	—

2. The energy of interaction with the carbonyl group (S_0 and S_1 sites) is lower than that for interaction with amine group (site S_2) with set I and set III charges. When a bridging (S_3) site occurs, its energy is lower than the other sites.

3. Not all combinations of models lead to the same hydration sites. For example, all calculations involving the TIPS2 water model give rise to the S_0 site with the water molecule out of the plane of the urea molecule.

4. Some combinations of models lead to sites which are too far away (greater than 3.4Å) from a urea atom for conventional hydrogen bonding. In five out of six examples of the use of the non-bonded coefficients from set (i) with TIPS2 and ST2 water models, we find distances of greater than 3.4Å for the S_2 hydration site and the sixth example is 3.35Å. One case of unacceptable short hydrogen bond (2.37Å) is found (Table VB).

5. Most combinations of models lead to acceptable hydrogen bond distances between the oxygen atom of the water molecule and the non-hydrogen atoms of the urea molecule. In these cases, the distances to the hydrogen atom of the water molecule were distributed around 2.05Å and the angles subtented at the water molecule hydrogen atom were within 30° of linearity.

6. The hydrogen bonds between the water molecule and the carbonyl group were always shorter than those to the amino group. This can be seen most clearly for the bridging hydration site (S_3) for which two distances are given of which the first refers to the interaction to the carbonyl group.

7. The length of the hydrogen bonds seems to correlate with the non-bonded interaction parameter set used in the calculation. For example, using non-bonded parameter set (i), with any combination of water models and charges, leads to, on average, the shortest distances from the primary hydration site to the carbonyl group on urea and the longest distances to the amine groups. In contrast, when non-bonded interaction coefficients from set (iii) are used, we find, on average, the longest hydrogen bond distances to the carbonyl group and the shortest of those to the amine group.

Interaction with Hydrocarbons

Similar calculations to those described above for the primary hydration sites of urea, were performed in order to look for possible sites of interaction between urea and a hydrocarbon (methane). Firstly iso-energy contour maps were produced and then full minimizations were undertaken. The results are given in Table VIA. In both cases, the non-bonded interaction parameters of Lifson et al (22) have been used with charge set I and III. The methane was modelled with CNDO/2 charges and non-bonded interaction coefficients from reference (22). These energies were found to be higher for interaction with the carbonyl group on urea (-0.88 or -0.84 Kcalmol^{-1}) and lower for interaction with the amino group (-0.98 and -0.97 Kcalmol^{-1}).

Table VI

	UREA[a]		UREA[b]	
	carbonyl	amine	carbonyl	amine
(A) ENERGY[c] (kcalmol^{-1})	−0.88	−0.98	−0.84	−0.97
	ST2	TIPS2	EMPWI	PE(DO)	Methane
(B) ENERGY[c] (kcalmol^{-1})	−0.20	−0.29	−1.05	−0.43	−0.41

[a]partial atomic charges from set III
[b]partial atomic charges from set I
[c]energy of interaction with methane

For comparison, we have also calculated the primary hydration sites of methane for the four different water models (Table VI B). The energy of interaction was very small for all water models being between −0.20 and −1.05 Kcalmol^{-1}. The energy of methane interacting with methane was only −0.41 Kcalmol^{-1}.

Conclusion

Any technique involving the calculation of potential energies depends on the use of realistic force fields to model each type of interaction between atoms. The development of suitable force fields for use in the study of water-solute interactions is not without problems (26) especially as it involves the combination of models representing both the water molecule and the atoms of the solute molecule. Although a variety of such models are available, it is difficult to choose between them. For this reason, we have attempted to look at the effects of the different models on the energy and geometry of the hydration sites around the urea molecule.

One of the major influences on these hydration sites is the choice of model for the water molecule itself. At present, a large number of possible models exist (8,27). These models, usually based on the interaction between water dimers, can be classified into categories such as four-point charge, three-point charge, central force, polarisable electropole. It is impossible to decide which is the optimum model because they predict various properties of water with varying degrees of success (8,27). Our calculations, using four of these water models, show that the greatest similarity between any two iso-energy contour maps occurs when they are both calculated using the same water model. Moreover, analysis of the primary hydration sites indicates that the main influence on which of the stereochemically reasonable sites are occupied seems to be the choice of water model.

One of the parameters used to model the atoms of the solute is the atomic charge. Estimates of partial atomic charges can be obtained from quantum mechanical methods of which a variety are available from the most sophisticated *ab initio* calculations to the semi-empirical methods such as CNDO/2. Even if the number of atoms in the solute molecule is small enough to allow the use of high-level *ab initio* calculations, the values obtained can depend on the choice of basis sets and the method of attributing continuous charge density to atom sites (28,29). We found, using three different sets of atomic charges, that the main consequences of different values was to change the magnitude of the interaction energy of the primary hydration sites and on their relative energies. Thus, the values of the partial atomic charges can have an effect on whether bridging sites (with one water molecule within hydrogen bonding distance of two solute atoms) are found.

The values for the non-bonded interaction coefficients are also difficult to obtain accurately. They are usually estimated from least squares fitting procedures of data from small organic molecules (12). Again a variety of values are found in the literature. The main effect appears to be on the distance of the hydrogen bond between the primary hydration sites of the water molecule and the solute atom. This seems to be intuitively reasonable as these coefficients define the position of the closest contact distance between neutral atoms.

Thus we find, on average, three relatively simple correlations between the choice of water model, partial atomic charges and non-bonded interaction coefficients with the occupancy of the primary hydration sites, the potential energies of interaction and the hydrogen bond distances respectively. However, these effects are obviously not independent as, for example, the relative size of the charges on the solute atoms determines, together with the water model, whether the bridging sites are at a local minimum. We find that some combinations of parameters lead to sites which are not within reasonable hydrogen bonding distance of solute atoms, i.e. between 2.5 and 3.2 Å (from crystal hydrate studies). When we look at most of the combinations which lead to reasonable hydration site geometries, we find a spread in energies and geometries.

At present, it is not possible to compare these results directly with experimental data for urea as the crystal structure does not contain any water of hydration and the structure of water around urea in solution is only just being determined using neutron scattering techniques (30). However, comparison with quantum mechanical calculations (24) and molecular dynamics (14,15) simulations is possible. The supermolecular quantum mechanical approach of Orita and Pullman (24) shows the presence of three main hydration sites with energies of -10.2 Kcalmol^{-1} to the carbonyl (similar to site S_1), and -7.4 Kcalmol^{-1} and -8.4 Kcalmol^{-1} to the amide group (similar to sites S_4 and S_2 respectively). However, a direct comparison with our results is not possible because the distances between the water molecular oxygen atom and the carbonyl oxygen and amide nitrogen appear to have been kept fixed at 2.8Å. The two studies using molecular dynamics simulation techniques were aimed at looking for differences in water structure with and without the urea molecule being present. They both conclude that there is little difference to the water structure when urea is added even though the urea . . . water and water . . . water models are very different in those two calculations. However, the radial distribution functions for the water molecule oxygen atoms around the various functional groups of urea look considerably different. This is consistent with there being differences in the potential energy surfaces around urea in the two simulations.

Thus, differences in potential energy functions have lead to differences in water structure around urea in a computer simulation of aqueous solutions. In order to understand the origin of these differences, it is necessary to perform many simpler (energy minimization) calculations with one water and one solute molecule. Such calculations have lead to an understanding of the idiosyncrasies of various models used herein. Moreover correlations between the energy and geometry of the hydration sites and differences in the parameterization of these potential energy functions have been found.

Acknowledgements

JMG and FV would like to thank the SERC for travel grant number GR/D/14372 and the Scientific Council at the French Embassy (London) for monies respectively.

References and Footnotes

1. P. Rossky and M. Karplus, *J. Am. Chem. Soc. 101,* 1913 (1979).
   2. M. Mezei, P. Mehrotra and D. Beveridge, *J. Biomolecular Structure and Dynamics 2,* 1 (1984).
3. J.M. Goodfellow, J.L. Finney and P. Barnes, *Proc. Roy. Soc. B214,* 213 (1982).
4. V. Madison, D.S. Osguthorpe, P. Dauber and A.T. Hagler, *Biopolymers 22,* 27 (1983).
5. A.T. Hagler and J. Moult, *Nature 272,* 222 (1978).
6. M.Mezei, D.L. Beveridge, H.M. Berman, J.M. Goodfellow, J.L. Finney and S. Neidle, *J. Biomol. Structure and Dynamics 1,* 287 (1983).
7. F. Vovelle, M. Genest and M. Ptak, in *Intermolecular Forces,* Ed. B. Pullman, D. Reidel Publishing Co., pp. 299-315 (1981).
8. J.L. Finney, J.E. Quinn and J.O. Baum, in *Water Science Reviews,* Ed. F. Franks, Cambridge University Press, Vol. 1 (1985).
9. G.E. Schulz and R.H. Schirmer in *Principles of Protein Structure,* Springer Verlag, New York, p.29 (1979).
10. A. Caron, *Acta Cryst. B25,* 404 (1969).
11. J. Brandts and L. Hunt, *J. Am. Chem. Soc. 89,* 4826 (1967).
12. D.B. Westlaufer, S.K. Malik, L. Stoller and R.L. Coffin, *J. Am. Chem. Soc. 86,* 508 (1964).
13. M. Roseman and W.P. Jencks, *J. Am. Chem. Soc. 97,* 631 (1975).
14. H. Tanaka, H. Touhara, K. Nakanishi and Watanabe, *J. Chem. Phys. 80,* 5170 (1984).
15. R. Kuharski and P.J. Rossky, *J. Am. Chem. Soc. 106,* 5786 (1984).
16. W.L. Jorgensen, *J. Chem. Phys. 77,* 4156 (1982).
17. F.H. Stillinger and A. Rahman, *J. Chem. Phys. 60,* 1545 (1974).
18. F. Vovelle and M. Ptak, *Int. J. Pept. Protein Res. 13,* 435 (1979).
19. B.J. Gellatly, J.E. Quinn, P. Barnes and J.L. Finney, *Mol. Phys. 59,* 949 (1983).
20. J.M. Goodfellow, *Proc. Natl. Acad. Sci. USA 79,* 4977 (1982).
21. F.A. Momany, L.M. Carruthers, R.F. McGuire and H.A. Scheraga, *J. Phys. Chem. 78,* 1595 (1974).
22. S. Lifson, A. Hagler and P. Dauber, *J. Am. Chem. Soc. 101,* 5111 (1979).
23. S.J. Weiner, P.A. Kollman, D.A. Case, U. Chandra Singh, C. Ghio, S. Profeta Jr. and P. Weiner, *J. Am. Chem. Soc. 106,* 765 (1984).
24. Y. Orita and A. Pullman, *Theoret. Chim. Acta 45,* 257 (1977).

25. M.J.D. Powell, *Computer Journal 7,* 155 (1964).
26. J.L. Finney, J.M. Goodfellow, F. Vovelle and P.L. Howell, *J. Biomolecular Structure and Dynamics,* (1985), in press.
27. D.L. Beveridge, M. Mezei, P.K. Mehrotra, F. Marchese, G. Ravi-Shanker and S. Swaminanthan, Molecular Based Study and Prediction of Fluid Properties, J.M. Haile and G.A. Mansoori (Eds) *Adv. Chem. Am. Chem. Soc.,* New York, pp. 297-351 (1983).
28. U. Chandra Singh and P.E. Kollman, *J. of Comp. Chem. 5,* 125 (1984).
29. K. Jug, *Theoret. Chim. Acta 31,* 63 (1973).
30. J. Turner, J.L. Finney, J.P. Bouquière, G. Neilson, S. Cummings and J. Bouillot in *Colston Society Symposium on Water and Aqueous Solutions* (Eds. J. Enderby and G. Neilson), held in Bristol, April (1985).

Biomolecular Stereodynamics III, Proceedings of the Fourth Conversation in the Discipline Biomolecular Stereodynamics, State University of New York, Albany, NY, June 04-09, 1985, Eds., Ramaswamy H. Sarma & Mukti H. Sarma, ISBN 0-940030-14-4, Adenine Press, ©Adenine Press 1986.

A Sol-Sol Transition as the First Step in the Gelation of Agarose Sols

P.L. San Biagio,[*†] **J. Newman,**[*†‡] **F. Madonia**[*†] **and M.U. Palma**[*†]

*Istituto per le Applicazioni Interdisciplinari della Fisica, C.N.R.
†Istituto d Fisica, Universita' di Palermo Via Archirafi 36, I-90123 Palermo (Italy).

Abstract

Aqueous agarose sols were studied by static and dynamic light scattering techniques under isothermal conditions at which gelation, if it occurs at all, takes weeks to months. Experiments monitoring the light intensity at both forward scattering angles and at 90°, and experiments in the presence of 0.04 μm radii polystyrene latex spheres to monitor the microviscosity of the agarose sols, all indicate the presence of a structural transition in the sol state which is observed after a characteristic incubation time at constant temperature. This incubation time is strongly temperature dependent and it appears to diverge above a certain temperature. After the incubation time, the transition occurs in the continuous way expected for a nucleation-free spatial density decomposition (spinodal transition), and it causes no measurable changes in population/dispersity of polysaccharide species in the sol. Correlations between fluctuations of the scattered light and of microviscosity illustrate some aspects of the transition dynamics. The transition itself can be viewed as the first in a multi-step process which results in gelation under isothermal conditions.

Introduction

Agarose-water systems

The gelation of agarose sols is a multi-step process of great interest both from the point of view of science and from that of practical applications. Agarose is a widely-found, unbranched and electrically uncharged biostructural polysaccharide, having a molecular weight of the order of 100,000, and the structure of an alternating copolymer of 3-linked-D-galactopyranose and 4-linked 3,6-anhydro-α-Lgalactopyranose residues (1-5).

The gelation of aqueous agarose sols, its reversibility and the molecular and supramolecular structure of the polysaccharide as well as the structure of the surrounding water have been the object of many studies (6-18). These have shown that the

‡Permanent address: Phys. Dept., Union College, Schenectady, N.Y. 12308, USA.

underlying physics includes solvent-mediated solute-solute interaction, recognition between pairs of dimer units of different polymer strands, a transition of the polymer conformation from strands to double helices; recognition among double helical segments to form bundles; and a phase transition of percolative type, corresponding to the onset of percolation along polymer double helices. (The possibility of percolation is due to rare kinks exing in the repetitive polymer strand which allow the occurrence of branching when double helices are formed). The practical interests of hydrogels (of which agarose provides a distinctive example) cover an ample variety of fields, including gel filtration and chromatography, food technology, pharmacy, photography, ophtalmology, and emathology.

For practical reasons due to the long times involved, isothermal gelation of these systems has only been studied occasionally (9,10), and at temperatures (25°C and 40°C) sufficiently low to give rise to kinetics fast enough to be easily followed (of the order of 100 sec.). The main body of data available in the literature were obtained under conditions of scanned temperature (8,13-16). Only recently, modern computer-controlled instrumentation developed at our laboratories has allowed the start of extensive studies of very long kinetics of isothermal gelation (19,20). A quantitative characterization of order in gels so obtained has also been worked out (21,22), and used in the study of the effects of a stochastic dichotomous thermal noise on the sol-gel transition and on the final order so obtained (22).

Early kinetic studies of isothermal gelation at relatively low temperatures (9) provided indirect evidence for the operation of the mechanism of so-called spinodal decomposition, that is a spontaneous and non-nucleated clustering of polymers into regions of higher concentration, leaving corresponding regions of lower concentration (28,29). The technique then used, and the rapidity of the kinetics studied, however, did not allow tracing, in the succession of processes involved in gelation, the step where spinodal decomposition, if real, was in fact occurring. On the other hand, a study of the very early stages of the isothermal sol-gel transition in these systems was of great interest on its own. The reason for this is the fact that in a certain range of temperatures, isothermal gelation has been reported to begin with very remarkable time lags and to lead to a different final order (20,21).

This intrinsic interest was the original motivation of the present work, and we chose to use quasi-elastic light scattering for an experimental study of the early stages of the sol-gel transition. As we shall see in what follows, we have obtained direct as well as indirect evidence for the occurrence of a non-nucleated transition in the sol, preceding and perhaps favoring the interaction among polymer strands which in turn leads to the actual sol-gel transition. The question remains open concerning a more general role of non nucleated (spinodal) decomposition in the self-assembly of biostructures.

Spinodal transitions

It has long been known (e.g. in metallurgy and in inorganic glass making) that a stable single phase of a mixture of components can be made unstable with respect to its "decomposition" into a mixture of phases, that is a spatial modulation of its composition. The thermodynamics of this type of process goes back to J.W. Gibbs, and it was worked out in detail more than two decades ago by Cahn and coworkers for the simple case of a two-component system (23-25). We shall briefly recall it here. Fluctuations of concentration Δc contribute to the Gibbs free energy a term that, by expanding the free energy density of the homogeneous system in a power series and retaining the first nonvanishing term, can be written:

$$\Delta G = \left[\tfrac{1}{2}\, \partial^2 G / \partial c^2 \, (c - c_0)^2 + \mu \nabla^2 c \right] dc \quad (1)$$

Here $(\mu \nabla^2 c)$ is the additional free-energy density due to a gradient of composition. Let us consider the locus of $\partial^2 G / \partial c^2 = 0$, called the spinodal, and let us limit ourselves to consideration of a positive μ-value (a negative value would lead to another type of instability). Eq. 1 shows that, in this case, in the regions where $\partial^2 G / \partial c^2 > 0$ the system is stable (or metastable) against any (small) fluctuaction of density. Where instead is $\partial^2 G / \partial c^2 < 0$, we expect the system to be unstable against fluctuactions occurring on a scale large enough to make the gradient term in eq. 1 sufficiently small, and yet not so large as to meet a diffusion barrier. These fluctuactions will consequently grow. A "decomposition" of the originally homogeneous system will result, that is, a spatial modulation of composition. Actually, the Fourier components of concentration will be confined to a small range of wavelengths. As the growth proceeds (exponentially with time, at the start), this range will narrow further, and a peaked structure function will be observed. At a subsequent stage higher order terms will become effective, the exponential growth of the spatial modulation of concentration will gradually subside, and a coarsening of the structure (known as Ostwald ripening) will be observed (24,26,27,28). A particularly interesting and intriguing property of the two phases, covering also very important practical interests, is their connectivity which shows up as the onset of percolation at a certain stage of the phase transition (24,29,30,31).

The main distinguishing features of so-called non-nucleated (spinodal) decompositions are therefore: a pervasive, non-nucleated character of the phase transition; a resulting structure function observable in light or particle scattering experiments, and interesting connectivity properties of the two phases. It must nevertheless be remarked that a distinction between nucleate and non-nucleate phase transition is not necessarily always strictly meaningful (32,33). The early papers on spinodal phase transitions were followed by studies concerning polymer solutions (34-36), and more recently polymer gels, blends and melts (37-43). These latter cases, however, differ from that which interests us here, since at variance with what happens with biopolymers and water, two chemically different polymers are most often not compatible, even in melts (27). Finally, the origin of a network structure in a gel of rod-like polypeptides has also been traced to a phase separation of spinodal origin (44). At variance with the present case, however, the phase separation was not followed by mutual recognition and configurational transition of the biomolecules concerned.

It is appropriate to remark in closing that the structure factor observed in the angular dependence of the intensity of light scattered by biomolecular solutions, as reported by a number of Authors (45,46), probably originates from a spinodal decomposition. It would be of interest to follow the kinetics of such a decomposition.

Experimental

We used quasi-elastic light scattering (QELS) techniques for two purposes: i) to monitor a growth of larger species, if occurring, in the sol prior to gelation proper, and ii) to measure, at a semi-microscopic scale, changes of viscosity as probed by very small polystyrene latex spheres added to the specimens (47). The scattered intensity from these spheres was much greater than that from the agarose, as a consequence of the mass-dependence of the scattering efficiency.

Materials and preparation were as previously described (48). For photon scattering and correlation experiments we used either a Malvern Log-Lin K 7027 or a Malvern K 7023 (72 channel) correlator, with a Coherent 52 Argon Laser tuned at 514.5 nm and equipped with an internal etalon and intensity stabilization. Temperature was stabilized to within 0.1°C by a standard circulating bath, and measured by standard digital thermometers. Microviscosity was measured by QELS techniques as previously reported (47) by measuring the diffusion coefficient of small polystyrene latex spheres added in minute quantities in the specimens. Spectrometer and correlator provided simultaneously both the correlograms needed for the measurement of microviscosity, and the intensity of scattered light, averaged over the same interval of time used for the acquisition of the correlograms. Microviscosity and scattered intensity so obtained were further analyzed for cross-correlations by means of a standard correlation coefficient probability test (49).

QELS data on samples without added latex spheres were taken at various scattering angles, down to about 25°, recorded at four different sample times, and spliced together, allowing sufficient overlap between successive correlograms. The good quality of the splicing was assured by the agreement, within the noise level, of the long-time channels and the background level as independently calculated. At angles smaller than about 25-20° the spectrometer did not allow QELS measurements but the pattern of scatterd light was anyway observed on a white screen, after blocking the main laser beam with a small circular stop. Data analysis, by a non-linear least-square fit in terms of two exponentials plus a costant, or by the program CONTIN (50) kindly provided by Dr. Stephan Provencher, provided essentially equivalent results, and a similar equivalence was found also for data collected with logarithmic channel spacing (Malvern 7027 Log-Lin) and analyzed by exponential sampling techniques (51).

Measurements were taken by quenching samples at the desired temperature immediately after preparation, or after a short storage at 80°C under conditions preventing evaporation.

Results and Discussion

Preliminary QELS measurements were performed on sols at 60-70°C. At this temperature isothermal gelation does not occur even in a period of several months (20,21). Experiments were performed at two agarose concentrations, 0.5 and 0.75% (w/v) chosen near the overlap concentration threshold of the agarose coiled strands (48). Agarose strands in the coil form show a hydrodynamic radius of about 150 A, with a rather limited polydispersity (48). A much less numerous, and larger species having hydrodynamic radius of 1500-2000 A is also present (48).

Further experiments were performed in the 47-51°C range of temperatures where gelation was known to start with a conspicuous, concentration-dependent lag time and to proceed with a sigmoid-shaped time-dependence (20,21). At both 0.5 and 0.75% (w/v), correlograms taken within this lag time preceding gelation showed no measurable changes of either hydrodynamic radii or dispersity. Within this no-change period, however, a new phenomenon was observed and it occurred with a definite and repetable incubation time of its own, much shorter than that which precedes the start of gelation. We were led to observe the new phenomenon by a rather abrupt change occurring (following a characteristic incubation time at constant temperature) in the short time-scale fluctuations of the scattered light intensity. The change did not alter the average value of the scattered intensity, and the time of its occurrence depended upon temperature and concentration in the way shown in Fig. 1.

Simultaneously with the change in the fluctuations of scattered light intensity, an initially mobile but very quickly frozen-in pattern of scattered light appeared on the white screen. This pattern had an angular aperture of about ±2° and, when obtained with the unfocussed laser beam, it consisted of a fine and barely detectable pattern of random dots of the type shown in Plate 16-1 of Ref. 52. As the angular distance between dots depended on the thickness of the samples, precise structural information could not be readily extracted from the figure.

A third simultaneous occurrence was the equally abrupt onset of a strong correlation between the 100-sec. averaged fluctuations of scattered light intensity and microviscosity values measured by the diffusion coefficient of the polystyrene latex spheres (the time needed for each correlogram being also 100 sec. as explained in the text). A typical onset of correlations is shown in Fig. 2.

These observations find a simple interpretation in terms of a spinodal transition in our samples. We note that fluctuations in the intensity of scattered light, in the absence of dust particles, evidence the existence of similar fluctuations in polymer concentrations in the volume of scattering. On the other hand, fluctuations in the microviscosity measured by the polystyrene spheres evidence the existence of similar fluctuations in the polymer concentration probed by these spheres. The two spatial patterns, each with its time dependence, do not need to coincide in the early

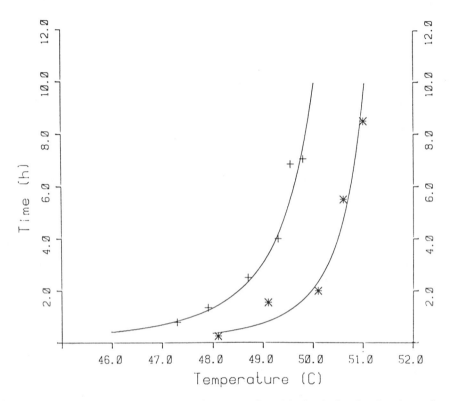

Figure 1. Dependence upon temperature and concentration of the incubation time for observation of the spinodal decomposition. Asterisks: 0.75% w/v. Crosses: 0.5% w/v. Note that under the same conditions the start of gelation proper takes days to weeks.

stages of a spinodal decomposition. This is a consequence of the initial spread of amplitudes and phases of the Fourier components of concentration. No correlation is then expected, nor is it in fact observed in the initial times of Fig. 2. The fact that after a certain time (that we have called before an "incubation time") a strong correlation is observed as shown in Fig. 2 is well consistent with the occurrence of a substantial coincidence of the two patterns (and of their time dependence as probed by the two different measurements), at the approaching of the Ostwald ripening. The coincidence of this length of time with that necessary for observing the appearance of the low-angle scattering pattern is also easily explained within this interpretation: prior to the start of the Ostwald ripening, the root mean square fluctuation of concentration is in fact expected to be small, so that no low-angle scattering pattern would be perceptible.

In conclusion, the observed phenomena find a clear interpretation in the occurrence of a spatial modulation of concentration due to a spinodal transition. Gelation proper occurs at a much later time. In view of its dependence upon polymer

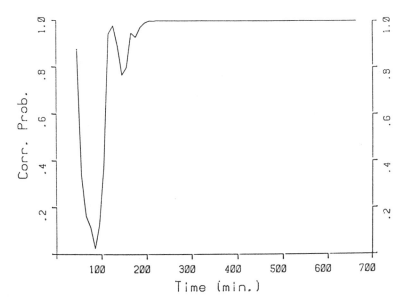

Figure 2. Agarose 0.5% w/v in water, at 48.9°C. Probability of correlation between microviscosity as probed by 0.04 μm polystyrene spheres, and fluctuations of scattered light as averaged over the same 100 sec. interval needed for acquisition of each QELS correlogram. The time at which the correlation probability becomes steadily 100% is the same as that of appearance of a low-angle scattering pattern.

concentration (15) it is difficult to imagine that the spatial pattern of gelation proper is not affected by the pattern of the spinodal decomposition that has occurred previously, and in fact the two are notably correlated (21). In other words, we think we have observed the occurrence of a spinodal transition which can be thought to influence and facilitate the biopolymer-biopolymer interaction leading to supra-molecular order.

Acknowledgements

This work has benefited from partial general support by CRRNSM and M.P.I. local fundings. We thank Prof. Beatrice Palma-Vittorelli for suggestions and criticism, Dr. Roy Pike for a discussion on the deconvolution of exponentially sampled correlograms, Prof. Stephen Provencher for the program CONTIN, Dr. Daniela Giacomazza for constant laboratory support, and Mr. Francesco Ficarra, Mr. Giovanni Duro, Mr. Antonio La Gattuta and Mr. Mario Lapis for skilled technical help.

Refrences and Footnotes

1. Rees, D.A., *Biochem. J. 126*, 257-273, (1972).
2. Dea, I.C.M., Mc. Kennon, A.A. and Rees, D.A., *J. Mol. Biol. 68*, 153-172,(1972).
3. Araki, C. and Arai, K., *Bull. Chem. Soc. Japan 40*, 1452 1456, (1956).

4. Rees, D.A., *Adv. Carbohydr. Res. 24,* 267-332, (1969).

5. Laurent, C., *Biochim. Biophys. Acta 136,* 199-205, (1967).

6. Obrink, B., *J. Chromatogr. 37,* 329-330, (1968).

7. Dea, I.C.M., McKinnon, A.A. and Rees, D.A., *J. Mol. Biol. 68,* 153-172, (1972).

8. Arnott, S., Fulmer, A., Scott, W.E., Dea, I.C.M., Moorhouse, R. and Rees, D.A., *J. Mol. Biol. 90,* 269-284, (1974).

9. Pines, E. and Prins, W., *Macromolecules 6,* 888-995, (1973). Feke, G.T. and Prins, W., Macromolecules 7, 527 530, (1974).

10. Wun, K.L., Feke, G.T. and Prins, W., *Faraday Disc. Chem. Soc. 57,* 146-155,(1974).

11. Amsterdam, A., Er-El, Z. and Schaltiel, S., *Arch. Biochem. and Biophys., 171,* 673-677, (1975).

12. Rees, D.A. and Welsh, E.J., *Angew. Chem. Int. Ed. Engl. 16,* 214-224, (1977).

13. Hayashi, A., Kinoshita, K. and Yasueda, S., *Polym. J. 12,* 447-453, (1980).

14. Indovina, P.L., Tettamanti, E., Giammarinaro-Micciancio, M.S., and Palma M.U., *J. Chem. Phys. 70,* 2841-2847, (1979).

15. Vento, G., Palma, M.U. and Indovina, P., *J. Chem. Phys. 70,* 2848-2853, (1979).

16. Leatherby, M.R., Young, D.A., *J. Chem. Soc., Faraday Trans. 1,* 77 1953-1966, (1981).

17. Key, P.Y. and Sellen, D.B., *J. Polym. Sci., Polym. Phys. Ed. 20,* 659-679,(1982).

18. Corongiu, G., Fornili, S.L., and Clementi, E., *Int'l J.Q. Chem., Q. Biol. Symp., 10,* 277-291, (1983).

19. Fornili, S.L. and Migliore M., *J. Phys. E: Sci. Instrum. 14,* 426-428, (1981).

20. Leone, M., Fornili, S.L. and Palma-Vittorelli, M.B., in *Water and Ions in Biological Systems, Eds.* Vasilescu V., Pullman B., Packer L. and Leahu L., Plenum Press, London, p. 677-684, (1985).

21. Leone, M., Sciortino, F., Migliore, M., Fornili, S.L. and Palma-Vittorelli, M.B.: submitted to *Biopolymers,* (1985).

22. Sciortino, S., Lapis, M., Carollo,C.M., Fornili, S.L., and Palma-Vittorelli, M.B.,: this volume pp. 287-297, (1986).

23. Cahn, J.W. and Hilliard, J.E., *J. Chem. Phys. 31,* 688-699, (1959).

24. Cahn, J.W. *Acta Met. 9,* 795-802, (1961). Cahn, J.W., *J. Chem. Phys. 42,* 93-99, (1965).

25. Cahn, J.W. and Charles, J., *Phys. Chem. Glass. 6,* 181-191, (1965).

26. Kawasaki, K. and Ohta, T., *Progr. Theor. Phys. 59,* 362 374, (1978).

27. de Gennes, P.G., *J. Chem. Phys. 72,* 4756-4763, (1980).

28. Pincus, P., *J. Chem. Phys. 75,* 1996-2000, (1981).

29. Heermann, D.W., *Z. f. Physik B (Condensed Matter) 55,* 309-315, (1984).

30. Hopper, R.W., *J. Non-Crystalline Solids 49,* 263-285, (1982).

31. Deutscher, G., Zallen, R., Adler, J.,: *Percolation Structures and Processes,* Adam Hilger, Bristol, U.K. and The Israel Physical Society, Jerusalem, Israel, (1983).

32. Langer, J.S., in *Fluctuations, Instabilities and Phase Transitions,* Ed., Riste, T., Plenum Press, New York, pp. 19-42, (1975).

33. Binder, K. *ibidem* pp. 53-86, (1975).

34. van Aartsen, J.J., *Eur. Polym. J. 6,* 919-924, (1970).

35. van Aartsen, J.J., Smolders, C.A., *Eur. Polym. J. 6,* 1105 1112, (1970).

36. Smolders, C.A., Van Aartsen, J.J. and Steenbergen, A., *Kolloid Z. u. Z. Polymere 243,* 14-20, (1970).

37. Tanaka, T., Swislow, Y., Ohmine, I., *Phys. Rev. Lett. 42,* 1556-1559, (1979).

38. Coniglio, A., Stanley, H.E., and Klein, W., *Phys. Rev. Lett.* 42, 518-522,(1979).

39. Goldburg, W.I. in *Scattering Techniques Applied to Supramolecular and Nonequilibrium Systems,* Eds., Chen, S., Chu, B., Nossal, R., Plenum Press, New York, pp. 383-409, (1981).

40. Reich, S., and Cohen, Y., *J. Polym. Sci., Polym. Phys. Ed., 19,* 1255-1267, (1981).

41. Sasaki, K., Hashimoto, T. *Macromolecules 17,* 2818-2825, (1984).

42. Snyder, H.L., Meakin, P., Reich, S. *Macromolecules 16,* 757-762, (1983).

43. Pozharskaya, G.I.,Kasapova, N.L., Skiripov, V.P., and Kol pakov, Yu.D., *J. Chem. Thermodyn. 16,* 267-272, (1984).

44. Tohyama, K., and Miller, W.G. *Nature 289,* 813-814, (1981).

45. Patkowski, A., Gulari, E., Chu, B. *J. Chem. Phys. 73,* 4178-4184, (1980).

46. Giordano, R., Mallamace, F., Micali, N., Wanderlingh, F., Baldini, G., Doglia, S. *Phys. Rev. A 28,* 3581-3587, (1983).

47. Madonia, F., San Biagio, P.L., Palma, M.U., Schiliro', G., Musumeci, S., and Russo, G. *Nature 302,* 412-415, (1983).
48. San Biagio, P.L., Madonia, F., Sciortino, F., Palma Vittorelli, M. B., and Palma, M.U., *J. Phys. (Paris) 45,* c7, 225-233, (1984).
49. Bevington, P.R. *Data Reduction and Error Analysis for the Physical Sciences,* McGraw-Hill Book Co., New York. pp. 123-127, (1969).
50. Provencher, S.W., *Comput. Phys. Comm. 27,* 213-227, (1982).
51. McWhirtier, J.G., Pike, E.R., J. Phys. A: *Math. Gen. 11,* 1729-1745, (1978).
52. Harburn, G., Taylor, C.A., Welberry, T.R., *An Atlas of Optical Transforms,* G. Bell and Sons, London (1975).

Biomolecular Stereodynamics III, Proceedings of the Fourth Conversation in the Discipline Biomolecular Stereodynamics, State University of New York, Albany, NY, June 04-09, 1985, Eds., Ramaswamy H. Sarma & Mukti H. Sarma, ISBN 0-940030-14-4, Adenine Press, ©Adenine Press 1986.

Self-assembly of Supramolecular Structures: Experiments on Noise Induced Order

F. Sciortino,* M. Lapis,† C.M. Carollo,* S.L. Fornili*† and M.B. Palma-Vittorelli*†

†Istituto Applicazioni Interdisciplinari Fisica, C.N.R.
*Istituto di Fisica, Universita' di Palermo Via Archirafi 36, I-90123
Palermo, Italy

Abstract

This is a preliminary report on an experimental study of the effects of noise on the self-ordering process of a biopolymeric aqueous system. The process occurs in vitro and it results in the formation of a gel. The system is significant, as the biopolymer is a polysaccharide (agarose) which is widely found in biostructures. In previous and more recent work at our laboratories, the ordering process in this system and the supramolecular order obtained under isothermal conditions had been studied in terms of an order parameter and of a space correlation function, quantitatively measured. These quantities were observed to depend on the isothermal gelation temperature in a way similar to that expected at a continuous phase transition.

In the present series of experiments, the ordering process was studied in the presence of a dichotomous markovian noise of temperature, with selectable correlation time and amplitude, making use of a purposely designed apparatus. Preliminary results are here presented and discussed in comparison with those obtained under isothermal conditions. In all cases we have observed noise induced order. The degree of order, obtained when the temperature fluctuated at random between two chosen values, and quantitatively measured by the order parameter and correlation length, was observed to be higher than that obtained under isothermal condition anywhere in the interval swept by the temperature noise. The effect is larger for larger noise amplitudes and shorter correlation times. These results suggest that the unavoidably noisy character of relevant parameters should not necessarily be regarded as a negative feature in the orderly self-assembly of biosystems.

Introduction

Self-assembly of order in biosystems occurs under conditions which are unavoidably noisy, as a consequence of fluctuations of temperature, pH, solute concentrations, etc. Interesting theoretical predictions are available on physical ordering processes under random fluctuations of at least one of the relevant parameters (1-8). Thus, asking what is the role of noise in the self-assembly of order in biosystems is not a trivial question.

Attempts to answer this question for processes involving biosynthetic and bio-chemical events would be unmanageable, for the time being. As a first significant step, however, we may consider simpler processes in vitro, of temperature-driven supramolecular ordering of biopolymers in water. An attractive example in this class is that of aqueous systems of a polysaccharide, agarose, which is widely found in nature as a main constituent of biostructures (9-11). Agarose is a widely spread unbranched and essentially uncharged biostructural polysaccharide, of molecular weight 100,000-120,000. Its primary structure (9-12) is that of an alternating copoly-mer of 3-linked β-D-galactopyranose and 4-linked 3,6-anhydro-α-L-galactopyranose residues. From X-ray data and computerized molecular model building, a model of double helices partially assembled in bundles has been derived for the gel structure (12). Branching of double helices occurs as a consequence of (infrequent) kinks in the repetitive polymer chain, and it is thought to originate the mechanical and percolative supramolecular structure of polymers in the gel (11-12). A Monte Carlo simulation of the agarose double helix and solvent water at 300°K has evidenced the occurrence of H-bond connectivity pathways around the helix (13). Gelation mechanisms and gel-sol reversion in agarose-water systems have been widely stud-ied and discussed (11-20). Recent photon-correlation studies have shown the occur-rence of a spinodal sol-sol transition (21-22), which distinctly precedes and likely promotes gelation (21-23).

A thorough study of gelation kinetics under isothermal conditions by turbidometric measurements has been recently performed (23-24). Making use of an ad hoc apparatus (25), based on techniques of programmable automatic instrument con-trol and data acquisition, gelation kinetics at different temperatures and agarose concentrations were followed for times up to several months (23). In the range of temperatures studied (35-52°C), gelation times, defined e.g. as times necessary to reach 50% of final turbidity, ranged from a few minutes in the lower temperature limit to months in the higher one. A behavior reminescent of a critical slowing down was observed when an upper temperature of about 52°C was approached (23). In that previous work, turbidity data were analyzed as outlined in the next section, and the mean square spatial fluctuation of agarose concentration $<\Delta c^2>$ was identified as a measure of the order parameter of the system. A correlation length was also determined. These two quantities, measured as functions of the tempera-ture of isothermal gelation kinetics, also showed critical-like behavior. Those previ-ous results, an example of which is given in Fig. 1, allowed characterizing quantitatively the order of aqueous agarose gels isothermally formed at different temperatures. Here they are used as a reference baseline, allowing a quantitative evaluation of the effects of a temperature noise on supramolecular ordering.

Basic ideas on the effects of noise

The theory of noise-induced order and noise-induced transitions (1-8) has been illustrated also in conferences of this series (6) and we are not going to review it in any depth. We shall here recall the very basic ideas, only to show that even in oversimplified cases a random fluctuation of a relevant parameter can bring about interesting results.

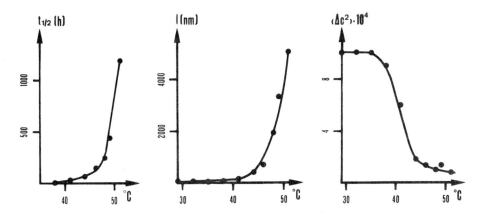

Figure 1. Results of turbidometric studies of the isothermal gelation process of an agarose water system (2% w/v agarose concentration). For all panels, the abscissa is the gelation temperature, in °C. *Left:* time (in hours) required to reach 50% of final turbidity. *Center and right:* correlation length and mean square amplitude of spatial fluctuation of agarose concentration in the gel, at the end of the gelation process. Note, for the three quantities, the critical-like behavior (redrawn from ref. 23).

Let us consider a macroscopic system out of equilibrium and take the classic view that its time evolution towards a steady state be described (3) by a deterministic equation. Let x and h be respectively a macroscopic variable describing the state of the system and a parameter, and let the equation describing the time evolution of the system be:

$$\dot{x} = f(x,h) \tag{1}$$

The simplest form of $f(x,h)$ allowing predictions of interesting effects of noise is:

$$f(x,h) = y(x) + hg(x) \tag{2}$$

where $g(x)$ is a non-linear function of x. Steady states are given by the solutions of

$$y(x) + hg(x) = 0 \tag{3}$$

Now let the parameter h be a randomly fluctuating quantity (3) of the form

$$h = h_t + \zeta_t \tag{4}$$

where ζ_t is a time dependent stochastic variable. In the case of white noise, the Fourier transform of ζ_t is flat

$$S(\nu) = \frac{\sigma^2}{2\pi} \tag{5}$$

and the correlation time is zero

$$< \acute\zeta_t \acute\zeta_{t+\tau} > = \delta(\tau) \tag{6}$$

The evolution of the system can be more easily studied by introducing the function

$$\frac{1}{\sigma} \int_0^t \zeta_{t'} dt' = W_t \tag{7}$$

which is a Wiener function describing a Brownian motion. Thus, Equation 4 becomes

$$h_t = h + \sigma \dot{W}_t \tag{8}$$

and Equation 1 becomes (4):

$$dx = \left[y(x) + hg(x)\right]dt + \sigma\, g(x)dW \tag{9}$$

This is a diffusion like equation, which contains a deterministic term (in square brackets) and a stochastic one. It can be studied by well known methods of statistical mechanics (1-6). The state of the system is described, rather than by the macroscopic variable x, by its probability distribution p(x). Stable/unstable states are given by the maxima/minima of p(x). A Fokker-Plank equation takes the place of the deterministic Equation 1 (4) and it is easily shown (3,4) that the extrema of p(x) are the solutions of

$$\left[y(x) + hg(x)\right] - \frac{\sigma^2}{2\pi} g'(x)\, g(x) = 0 \tag{10}$$

where g'(x) is the x-derivative of g(x). This takes the place of Equation 3. Here again, the term in square brackets describes the deterministic solution and the second term represents the effects of noise. Since g(x) is a non-linear function, solutions of Equation 10 can differ from those of Equation 3 in several ways, depending on the values of h and σ^2. Different possibilities range from simple shifts of the solutions, to the appearance of new solutions, or maxima of p(x) becoming minima for critical values of h and σ^2. We are in this latter case in the presence of critical behaviors, known as noise induced transitions (2-4).

Similar calculations have been performed for the case of a dichotomous markovian noise (2-3), i.e. for the case of a parameter jumping at random times from one to another of two values. Equations 4 and 6 become in this case:

$$h_t = h \pm \Delta h \qquad\qquad \acute\zeta_t = \pm \Delta h \tag{11}$$

$$< \acute\zeta_t \acute\zeta_{t+\tau} > = (\Delta h)^2 \exp\left(-\tau/\tau_c\right) \tag{12}$$

The latter defines the correlation time τ_c. The behavior of the extrema of p(x), namely of the stable and unstable states of the system is shown to depend in this case upon the mean value of the fluctuating parameter h, the noise amplitude Δh, and the correlation time τ_c (2,3,6). Interesting effects are expected when τ_c is short in comparison with the relevant times of the process considered. For critical values of these three variables, the system can be thrown from one state to another (noise induced transitions).

Calculations on the effects of noise can be worked out under simple assumptions only. Although experimentally "simple", the process that we have studied experimentally here would require complex evolution equations for its description. Further, its dependence upon the fluctuating parameter (temperature) is far from linear. The effects of noise are therefore not easily predictable, but this only adds to the interest of the actual experiments.

Materials

Materials and preparation procedure were identical to those used for the study of isothermal kinetics, and described in ref. 23. Agarose was Ultrapure Seakem HGT(P) from Marine Colloids Inc., lot n. 62933 (same as in ref. 23), with nominal sulphate content less than 0.15%. Water was Millipore Super Q filtered with 0.22 μm filters. Agarose powder was dissolved in H_2O, the solution was put in a stoppered tube, kept in boiling water for about 20 minutes with occasional gentle mixing and occasionally lifting the stopper to clear the solution of microbubbles. Then the solution was poured in previously heated spectroscopic cuvettes and great care was taken to prevent the solution from reaching at any time a temperature lower than 70°C. In the experiments here reported, agarose concentration was 2% w/v.

Turbidity measurements and analysis

Information on the gel structure was obtained from turbidity spectra, recorded in the range 350-800 nm. Turbidity, defined as $\tau = (1/d)\ln (I_{out}/I_{in})$ in the absence of absorption or reflection (here d is the optical path in the sample and I_{in}, I_{out} are the incident and transmitted light intensities), increases smoothly for decreasing wavelengths. As discussed in ref. 23, turbidity is due in our case to spatial inhomogeneities of agarose concentration. Now let c(r) be the deviation of agarose concentration from its mean value $\underline{r}$ and $<\Delta c^2>$ be the mean square fluctuation amplitude. The related correlation function is

$$g(r) = \frac{<\Delta c(r) \cdot \Delta c(0)>}{<\Delta c^2>} \qquad (13)$$

and let S(q) be the Fourier transform of the latter. Turbidity can then be expressed (18,23,26) as:

$$\tau = \kappa <\Delta c^2> q_0^4 \int_0^{2q_0} S(q)f(q)\ dq \tag{14}$$

Here q_0 is the wavevector of incident light, $q = 2q_0 \sin\theta/2$ is the scattering vector;

$$f(q)dq = (1 + \cos\theta)\ \sin\theta\ d\theta \tag{15}$$

a well known geometrical factor for scattered light (26), and

$$\kappa = (n_0^2/n^4)(dn/dc)^2$$

where n_0 and n are the refractive indices of the solute and of the solution. Equation 14 can be expressed as

$$\tau = \kappa <\Delta c^2>\ G(q_0) \tag{16}$$

By choosing an appropriate correlation function g(r), its Fourier transform S(q) and the $G(q_0)$ function can be calculated. In this way a calculated turbidity spectrum can be used for fitting the experimental one. An appropriate correlation function is for our case

$$g(r) = \exp(-r/l) \tag{17}$$

which provides good fittings of experimental data in the isothermal (23) as well as in the present case (27). The mean square concentration fluctuation $<\Delta c^2>$ and the correlation length were then determined by standard best-fitting procedures. Since $<\Delta c^2>$ essentially measures the order parameter (23,28,29), these two quantities well describe the degree of order of the system.

As a first rough approximation, a power law can be substituted for $G(q_0)$

$$\tau(q_0) \approx \kappa <\Delta c^2>\ q_0^s \tag{18}$$

The exponent s, when appropriate core is taken of the wave length dependence of refractive indices, is the slope of a log-log plot of turbidity vs wavelength (to be referred in the following as spectral slope). When the correlation function of Equation 17 is used, a one-to-one correspondence is found between the spectral slope and the correlation length, s being a monotonically decreasing function of l_0. Thus, the spectral slope, without any further elaboration of data, already contains the information on the correlation length (23).

Generation of dichotomous temperature noise

Thermoelectric Peltier modules (Midland Ross Co., Cambion Division, Cat. No. 801-2004-01-00-00) were used to obtain the stochastically variable dichotomous temperature. A schematic drawing of the special cells used is given in Fig. 2, left. A copper sample holder provided thermal contact with the Peltier modules, on both sides. Thermal contacts were improved throughout by the use of thin layers of Omegatherm paste. A thermostated circulating bath provided a reference temperature. The system was held together by a thermally insulating case. Current cycles of the type given in Fig. 2, right, were fed to the Peltier modules. Regulation of the

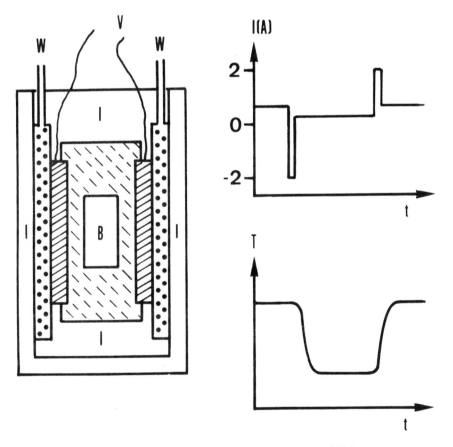

Figure 2. *Left:* spectroscopic cell for the experiments in the presence of dichotomous temperature noise. The copper cell (\\\ ), with a quartz window for the optical beam (B), is sandwiched between two Peltier modules (///). A circulating thermostated water bath (•••) provides a reference temperature. The whole system is contained within a thermally insulating case (I). *Upper right:* Current cycle in the Peltier modules. The time interval between pulses of opposite sign is determined by a random number generator (see Fig. 3). The steady values of the current are regulated to give the wanted higher and lower temperature values in the samples. The length of the (positive or negative) high-current pulses is regulated so that the temperature jump is as fast as possible. *Lower right:* Temperature cycle obtained typically in the sample.

current at the higher and lower levels and of the length of the high current pulses provided control of the higher and lower temperature in the sample and of the thermalization curve. Temperature cycles in the sample, typically as shown in Fig. 2, right, were so obtained. Up to twelve of these cells were simultaneously used, and the current in each pair of Peltier modules was independently controlled.

The block diagram for the control system is given in Fig. 3. The microcomputer Rockwell AIM65 no.2 acts as a random number generator. At random times, with a pre-selectable correlation time, current pulses adjusted as discussed above were fed to each pair of Peltier modules. Correlation times were chosen long in comparison with thermalization time and short in comparison with the characteristic times of kinetics.

The spectroscopic cells were mounted on a precision-machined carriage, shifting by means of a stepping motor controlled by another microcomputer (Fig. 3). Samples were so brought at pre-selected times across the beam of the spectrophotometer (a Jasco 320). This allowed acquisition of turbidity spectra, which were transferred to a PDP 11/34 computer for further analysis.

Results and Discussion

The effects of noise on the order obtained in the gel upon completion of the sol-gel transition are shown in Fig. 4. In this particular case three different amplitudes (± 3, ± 4, $\pm 5°C$) and two different correlation times (2 min. or 4 min.) were chosen for the dichotomous fluctuations of temperature. Parameters describing the order which is obtained under isothermal conditions (continuous line) and under the stochastic fluctuations of temperature (stars and circles) are presented. We recall that $<\Delta c^2>$ is essentially a measure of the order parameter (23,28,29) and that a higher degree of order is measured by a larger order parameter and by a shorter correlation length (30). Data in the figure show that a fluctuating, "noisy" temperature enhances the order obtained, and that this effect is larger for larger amplitudes and for shorter correlation times of the stochastic dichotomous noise. More specifically, the final order obtained when the temperature fluctuates in a certain interval, is higher than that obtained isothermally at any temperature within the same interval. This is consistently observed, except in the case when the smallest noise amplitude ($\pm 3°C$) goes with the largest correlation time (4 min). In this case the values of both l and $<\Delta c^2>$ fall very close to, yet not beyond the values corresponding to the highest order obtainable by isothermal gelation in the interval where temperature fluctuates.

Conclusions

We have presented a preliminary report on the effects of noise on a simple process of supramolecular ordering in a biopolymeric aqueous system. The process occurs in vitro and it results in the formation of a gel. The system is significant, as the biopolymer chosen is a polysaccharide (agarose) which is widely found in biostructures. Order in this system had been previously characterized in terms of

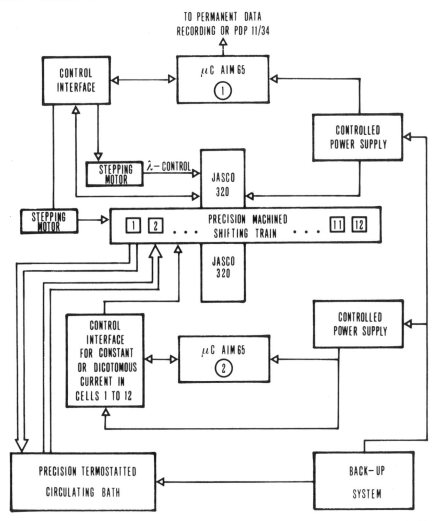

Figure 3. Block diagram of the experimental set up. The precision machined shifting train holds up to twelve spectroscopic cells (shown in Fig. 2). Each cell is so presented at pre-selected times across the beam of a Jasco 320 spectrophotometer. The train movement is driven by a stepping motor, controlled by the μC Rockwell AIM 65 labeled 1. The same μC controls, through another stepping motor, the wavelength sweep of the spectrophotometer and directly the acquisition of O.D. data. These are transferred in recorded form or directly to a PDP 11/34 computer. The μC labeled 2 acts as a random-number generator and it provides indipendently, to each pair of Peltier modules, a current cycle of the type shown in Fig. 2.

physically measurable parameters. In the present work we have measured and compared the final orders obtained under constant temperature and under noisy temperature. It was found that dichotomous stochastic fluctuations of temperature enhance the final order in a way which depends upon both amplitude and correlation time of the fluctuations. These results pose the question of a possible role of the inherently noisy character of relevant parameters such as temperature, pH, etc. in the self-assembly of biostructures. A more specific question, of notable physical interest (4) is whether the observed effects of noise induced order occur abruptly

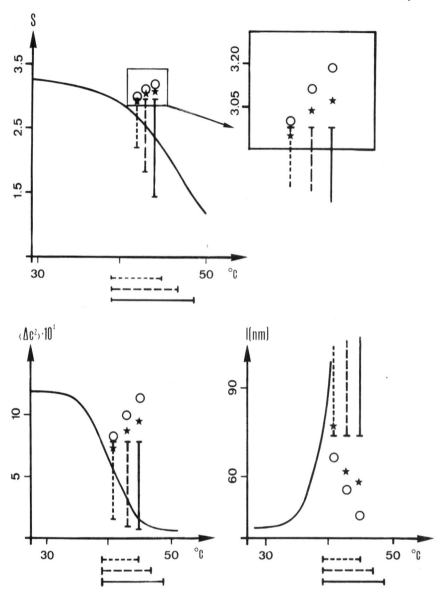

Figure 4. Comparison of parameters describing the supramolecular order obtained upon completion of a sol-gel transition, under isothermal conditions (continuous line) and under dicotomous stochastic fluctuations of temperature. In this typical case three different amplitudes (± 3, ± 4, $\pm 5^{\circ}C$) and two different correlation times (2 min. or 4 min.) were chosen for the dichotomous fluctuations of temperature. The actual intervals of variability of the dichotomous fluctuating temperature (39-45°C; 39-47°C; 39-49°C) are indicated by horizontal bars. The corresponding range of values obtained for each shown quantity in isothermal conditions, in the same interval of temperature, is indicated by vertical bars. Unterminated extremes of these bars indicate that the latter should actually extend beyond the space allowed in the figure. The system is 2% w/v agarose in Millipore Super-Q water. $<\Delta c^2>$ is the mean square amplitude of spatial fluctuation of polymer concentration and is essentially a measure of the order parameter; l (in nm.) is the related correlation length, defined by Equation 17 in the text; s is the spectral slope, defined by Equation 18 in the text. Data show that the final degree of order is unmistakably increased by the temperature noise, and that the effect is larger for shorter correlation times.

for definite values of noise amplitude and correlation time. Experiments are underway to test this point.

Acknowledgements

We thank Prof. M.U. Palma for suggestions and criticism, and Dr. M. Migliore for generous help with microcomputing, Mr. F. Ficarra for precision machining and mechanics, Mr. A. La Gattuta for skilled technical support with microelectronics and microcomputing, Ms. F. Cavoli for processing the text. This work has benefited from partial general support by CRRNSM and M.P.I., local fundings.

References and Footnotes

1. Arnold, L., Horsthemke, W. and Lefever, R., *Z. Phys. B 29*, 367 (1978).
2. Kitahara, K., Horsthemke, W. and Lefever, R., *Phys. Letters 70A*, 377 (1979).
3. Horsthemke, W. in *Stochastic non linear systems*, Arnold, L. and Lefever, R., Eds., Springer-Verlag, Heidelberg (1981) p. 116.
4. Lefever, R., *ibidem*, p. 127.
5. San Miguel, M. and Sancho, J.M., *ibidem*, p. 137.
6. Lefever, R., Horsthemke, W. and Michean, J.C., in *Structure and Dynamics: Nucleic Acids and Proteins*, Clementi, E. and Sarma, R. H., Eds., Adenine Press, New York (1983), p. 107.
7. Horsthemke, W. and Lefever, R., *Biophys. J. 35*, 415 (1980).
8. Horsthemke, W., Doering, C.R., Lefever, R. and Chi, A.S., *Phys, Rev. A 31*, 1123 (1985).
9. Araki, C. and Arai, K., *Bull. Chem. Soc. Japan 40*, 1452 (1967).
10. Rees, D.A., *Biochem. J. 126*, 257 (1972).
11. Dea, I.C.M., McKennon, A.A. and Rees, D.A., *J. Mol. Biol. 68*, 153 (1972).
12. Arnott., S., Fulmer, A., Scott. W.E., Dea, I.C.M., Moorhouse, R. and Rees, D.A., *J. Mol. Biol. 90*, 269 (1974).
13. Corongiu, G., Fornili, S.L. and Clementi, E., *Int'l J.Q. Chem. Q. Biol. Symp. 10*, 277 (1983).
14. Indovina, P.L., Tettamanti, E., Giammarinaro-Micciancio, M.S. and Palma, M.U., *J. Chem. Phys. 70*, 2841 (1979).
15. Vento, G., Palma, M.U. and Indovina, P., *J. Chem. Phys. 70*, 2848 (1979).
16. Pines, E. and Prins, W., *Macromolecules 6*, 888 (1973).
17. Feke, G.T. and Prins, W., *Macromolecule 7*, 527 (1974).
18. Wun, K.L. Feke, G.T. and Prins, W. *Faraday Disc. Chem. Soc. 57*, 146 (1974).
19. Letherby, M.R. and Young, D.A., *J. Chem. Soc. Faraday Trans. 1*, 77, 1953 (1981).
20. Key, P.Y. and Sellen, D.R., *J. Polym. Sci., Polym. Phys. Ed., 20*, 659 (1982).
21. San Biagio, P.L., Madonia, F., Sciortino, F., Palma-Vittorelli, M.B. and Palma, M.U., *J. de Phys. C7-45*, 225 (1984).
22. San Biagio, P.L., Madonia, F., Newman, J. and Palma, M.U., *Biopolymers*, submitted for publication.
23. Leone M., Sciortino, F., Migliore, M., Fornili, S.L. and Palma-Vittorelli, M.B., *Biopolymers*, submitted for publication.
24. Leone, M., Fornili, S.L. and Palma-Vittorelli, M.B., in *Water and Ions in Biological Systems*, Pullman, A., Vasilescu, V. and Packer, L. Eds., Plenum Publishing Corporation, London (1985) p. 677.
25. Fornili, S.L. and Migliore, M., *J. Phys. E: Sci. Instrum. 14*, 426 (1981).
26. Debye, P. and Bueche, A.M., *J. Appl. Phys, 20*, 518 (1949).
27. A smaller term of the same form, with a correlation length of about 10 nm. was also included in the analysis in ref. 23. We will neglect this small term throughout the present report, since it has no relevance in the present context.
28. Landau, L. D. and Lifshitz, E.M., *Statistical Physics*, Pergamon Press, London (1958).
29. Cahn, J.W. and Hilliard, J.E., *J. Chem. Phys. 28*, 258 (1958).
30. See, for instance, Stanley, H.E., *Introduction to Phase Transitions and Critical Phenomena*, Clarendon Press, Oxford (1971).

Biomolecular Stereodynamics III, Proceedings of the Fourth Conversation in the Discipline Biomolecular Stereodynamics, State University of New York, Albany, NY, June 04-09, 1985, Eds., Ramaswamy H. Sarma & Mukti H. Sarma, ISBN 0-940030-14-4, Adenine Press, ©Adenine Press 1986.

MOSES: A Computer Graphics Simulation Program in Real Time

Subhashini Srinivasan, Dennis McGroder, Masayuki Shibata, and Robert Rein

Unit of Theoretical Biology
Roswell Park Memorial Institute
Buffalo, New York 14263 USA

Abstract

A compact program for molecular model building was developed for extensive use with the Evans and Sutherland PS 300 local operations. In addition to the standard features such as stereoview, distance measurements, etc., the main features of this program are: 1) various types of representations with an extensive use of color options, such as various colors defining different atom types, 2) two or more segments (or molecules) can be manipulated simultaneously, 3) flexibility in the choice of several rotational bonds, 4) three different views can be displayed simultaneously, and 5) minimum use of host communication. This program provides valuable information in molecular model building, in conjunction with other tools such as distance geometry and molecular mechanics, as a part of MSM procedure as proposed (S. Srinivasan, G. Raghunathan, M. Shibata, and R. Rein, *Int. J. Quantum Chem, in press*). Due to the elimination of the extensive communication between the host and the graphics system, this program is applicable to various computer systems with standard asynchronous communication ports including micro-computers as a host system.

Introduction

Computer graphics has become an important tool in the study and visualization of the structure of molecules. The recently marketed Evans and Sutherland PS 300 graphics display system has made this field of endeavor more powerful. The three-dimensional architecture of the molecule, being the cause of their biological function, makes their manipulation and display in real time a gift to the molecular biologist and biochemist.

The Evans and Sutherland PS 300 graphics system is one of the most sophisticated systems available today due to the fact that most of the computational tasks can be performed locally without the aid of a host computer as in earlier systems. To our knowledge, only one Evans and Sutherland software package is available for academic use. The FRODO program, originally developed by Jones and modified by Quiocho for the PS 300 system, was designed for crystallographic studies in proteins. However, no program can be compact and at the same time general enough for all applications. We developed an application program and its supporting functions network to be able to manipulate two or more molecules at the same time with flexibility in the choice of representation. The choice of various representations makes modelling much easier as atoms are coded with colors so that their type can be recognized easily during manipulation. In this paper we will present the system configuration used, the flowchart of the host Fortran 77 program, and the PS 300 display tree structure and function network.

System Configuration

A) Structure of MOSES

MOSES consists of two separate parts, the host resident programs and the PS 300 local instructions as shown in Figure 1. The host resident programs are written in FORTRAN 77 using the HP 1000 minicomputer as a host computer which is connected to the Evans and Sutherland PS 300 by a standard asynchronous serial (RS 232-C) interface. The programs supply necessary information to create the image of molecules, retrieve the information which contains operations performed by the PS 300 system, and reconstruct the final structure in the host program. The PS 300 local instructions, written in the PS 300 command language, contain the display tree and the function network. The display tree defines the hierarchical data structure, linking molecules and the kind of operations which can be performed in connection with the function network. The function network enables us to manipulate molecules through various interactive devices such as control dials, data tablet and pen, function keys, and keyboard.

B) The Host Resident Program

Essential Features

The important features that are introduced in the Fortran 77 application program are the various representations such as ball and stick, colored surface, and color coded bonds. Apart from this, the important task for the host resident Fortran 77 application program is to help keep track of connectivity when creating the displayable objects on Evans and Sutherland PS 300 system. This connectivity is also used when creating various torsional fragments. This avoids appearance of superfluous bonds in case of a poor starting geometry which may contain short contacts. In cases where the connectivity data is not available, the program has options to create one by distance criteria.

C) PS 300 Data Structure

Essential Features

1) The local manipulations possible on the PS 300 include real time rotations, translations, and scaling transformations that are applied to the molecular

Present System Configuration

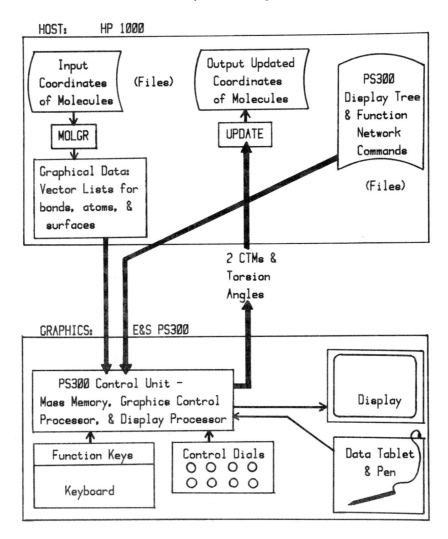

HOST: HP 1000

Input Coordinates of Molecules   (Files)

MOLGR

Graphical Data: Vector Lists for bonds, atoms, & surfaces

Output Updated Coordinates of Molecules

UPDATE

PS300 Display Tree & Function Network Commands

(Files)

2 CTMs & Torsion Angles

GRAPHICS: E&S PS300

PS300 Control Unit - Mass Memory, Graphics Control Processor, & Display Processor

Function Keys

Keyboard

Control Dials

Display

Data Tablet & Pen

images by the control dials. The display tree allows for two separate molecular 'objects' to be defined and manipulated. Both objects can be rotated and translated together with respect to a common 'world' coordinate system. Each object can be rotated separately with respect to its own object space coordinate system, and one of the two can be translated separately. Three orthogonal views can be switched on and off. Stereopairs can be viewed and stereo separation and angle can be adjusted.

2) The lack of interaction of host and Evans and Sutherland is compensated for by a distance monitoring function to compute the distance between two atomic

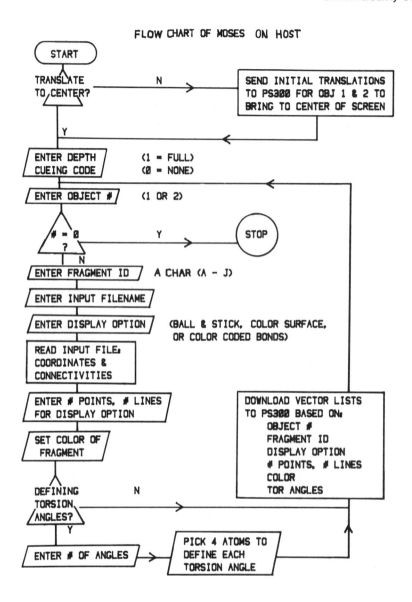

FLOW CHART OF MOSES ON HOST

centers using the local functions on the Evans and Sutherland PS 300. The two atoms are picked by the pen and data tablet; this function is always active so every two picks trigger the distance computation.

3) The final model is created in the host computer with the concatenation matrix (which is called the current transformation matrix (CTM)) retrieved for each manipulated object. These manipulated coordinates can then be used for multistep modelling (1).

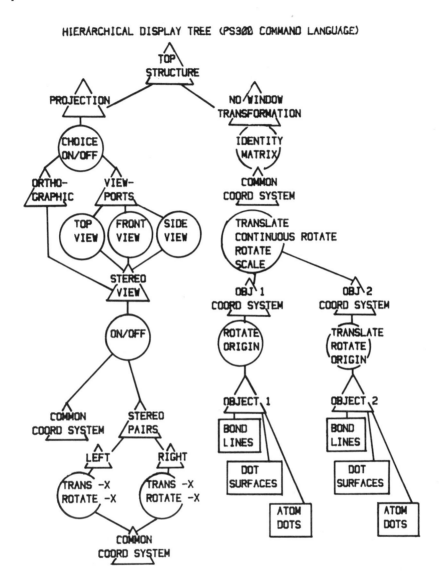

HIERARCHICAL DISPLAY TREE (PS300 COMMAND LANGUAGE)

4) A function for rotating torsion angles has been written for the function network. The host program defines the segments based on the choice of connectivity. The segments are colored and can be rotated around torsional bonds by using the control dials (maximum of eight torsional angles can be treated at one time which need not be in sequence). These rotations are accepted or rejected within the modelling session by retrieving the torsion angles from the PS 300.

User Advantages

1) The host program is small consisting of only 1,200 executable statements making it compact and readily transportable.

2) Does not necessitate the use of the PS 300 supported host routines making it usable by a non-VAX host.

3) Does not engage the host computer during the long hours of user manipulations.

User Disadvantages

1) No communication between the host to the PS 300 to make it interactive during manipulations.

Acknowledgement

This work was partially supported by a grant from NASA (NSG-7305) and pursuant to a contract with the National Foundation for Cancer Research. We acknowledge Professor F.A. Quiocho and J.W. Pflugrath of Rice University, Department of Bio-chemistry, P.O. Box 1892, Houston, Texas 77251 USA for making available the FRODO version for the HP 1000 and PS 300 configuration. We would like to thank Deborah Raye for her assistance in the preparation of this manuscript.

References and Footnotes

1. S. Srinivasan, G. Raghunathan, M. Shibata, and R. Rein, *120Int. J. Quantum Chem. QBS (in press)*.